LONDON

LONDON

The Selden Map and the Making of a Global City,
1549–1689

ROBERT K. BATCHELOR

THE UNIVERSITY OF CHICAGO PRESS
CHICAGO AND LONDON

ROBERT K. BATCHELOR is associate professor of history at Georgia Southern University.

The University of Chicago Press, Chicago 60637
The University of Chicago Press, Ltd., London
© 2014 by The University of Chicago
All rights reserved. Published 2014.
Printed in the United States of America

23 22 21 20 19 18 17 16 15 14   1 2 3 4 5

ISBN-13: 978-0-226-08065-9 (cloth)
ISBN-13: 978-0-226-08079-6 (e-book)
DOI: 10.7208/chicago/9780226080796.001.0001

Library of Congress Cataloging-in-Publication Data

Batchelor, Robert K., 1968– author.
    London : the Selden Map and the making of a global city, 1549–1689 / Robert K. Batchelor.
        pages cm
    Includes bibliographical references and index.
    ISBN 978-0-226-08065-9 (cloth : alkaline paper) — ISBN 978-0-226-08079-6 (e-book)
    1. London (England)—History—16th century. 2. London (England)—History—17th century. 3. Globalization—England—London. 4. Cartography—China—History.
    I. Title.
    DA681.B29 2014
    942.1206—dc23

2013028618

♾ This paper meets the requirements of ANSI/NISO Z39.48-1992 (Permanence of Paper).

CONTENTS

Translating Asia

THE VIEW FROM THE LIBRARY

In September 1687, the English king James II (known as James VII in Scotland) traveled to Oxford. The university threw a banquet for him in the Selden End of the Bodleian Library, which housed the substantial 1659 donation of manuscripts of the London legal theorist John Selden. The king had spent the summer trying to garner support in the west of England for his Declaration of Indulgence, rumored to be the brainchild of the Quaker William Penn, which would end the penal laws punishing Catholics and dissenting Protestants. But as he entered the room, other things appeared to be on his mind, and James showed a courtier the fabled Strait of Anian that offered a northern route to China on one of the library's two great globes. He asked the library's keeper Thomas Hyde, who was known to the court as a Persian and Arabic translator, about the recent visitor from Nanjing, the Catholic convert Shen Fuzong. James had commissioned a portrait of him to hang outside his bedchamber (figure 1). That same summer Shen, as part of group of French Jesuits sponsored by Louis XIV, had left London to help Hyde catalogue the library's substantial collection of Chinese books. Hyde acknowledged this concisely, out of deference and perhaps Anglican reticence in the face of a Catholic king, and then pointed to the collections of the High Church Anglican Archbishop Laud, which shared this part of the library with those of John Selden. Laud's and Selden's collections, assembled and previously housed in early seventeenth-century London, were two of the most diverse in the world, ranging from Mexica painted rolls to Chinese and Japanese printed books. But James asked Hyde only about the new printed Paris edition of the four books of Confucius, caring little about the longer history of collecting Asian manuscripts in London and Oxford.[1]

Fig. 1. Godfrey Kneller, "The Chinese Convert" (1687). Royal Collection Trust / © Her Majesty Queen Elizabeth II 2012

While far more cosmopolitan than a monarch like Henry VIII, James II still seemed to have little comprehension of the kinds of changes that had occurred in London over the space of a century and a half. But Laud's and Selden's collections remained as a kind of evidence of this shift—books, maps, and manuscripts from Asia that had been most recently studied by a young man from the Qing Empire.

When it opened in November 1602, Thomas Bodley's rebuilt library, the old Duke Humfrey's (1487) later with two new "ends" (Arts in 1622 and Selden in 1637), was a very different place than it had been in the time of Henry VIII. During the summer marked by the 1549 Prayer Book Rebellion, a royal visitation sent by the Lord Protector Somerset from London under the supervision of Oxford's new Chancellor Richard Cox (1547–52) discovered a trove of "popish" manuscripts there. A book burning from one of England's most important collections ensued outside, targeting illuminations and initials and possibly even red rubrics and "diabolical" mathematical diagrams. Under the Catholic Queen Mary, other manuscripts were apparently pilfered or sold, including the furniture in early 1556.[2] In some ways, events being driven by London were dissolving the library. The visitors in 1549 presented statutes drawn up by Somerset and the Privy Council, removing all previous practices of worship and declaring Edward VI supreme head of the Church. Thomas Cranmer's new prayer book arrived that month from the Fleet Street printing presses of Edward Whitchurch. Subduing resistance to these theological plans hatched in London also required outside talent brought to Oxford—notably the brilliant Florentine Protestant theologian Pietro Martire Vermigli, Edward's former tutor, as well as 1,500 German and Swiss mercenaries under the command of Lord William Grey, a veteran of wars in Scotland and France.[3] London in 1549 was already the gateway for translating European books, ideas, and indeed soldiers into the English countryside, but this did not necessarily result in a widening of horizons. "The Oxfordshire papists are at last reduced to order," wrote the Swiss medical student Johann Ulmer to Zurich from the university that year, "many of them having been apprehended, and some gibbeted, and their heads fastened to the walls." With their heads went Oxford's library, a symbol of an older and broader Catholic universal world with its sacred geographies and pilgrimages, although one that admittedly dated only to the visions of the last English Protector, Duke Humphrey (1422–29). There was no immediate sense that the printed prayer books and Continental theorists coming from London served as an adequate replacement for what had been lost.[4]

Bodley first saw the empty and derelict library as a student in the 1560s. Retiring from a diplomatic career to Oxford amidst the factional fighting between the houses of the Earl of Essex and the Cecils in the London of the 1590s, Bodley began collecting books for a new library in 1600. In 1608, while in London, Bodley asked Paul Pindar, the English consul to Aleppo and advocate for London's Levant Company, to search out "bookes in the Syriacke, Arabicke, Turkishe and Persian tongues or any other language of those Esterne nations, bycause I make no doubt but in processe of time, by the extraordinarie diligence of some one or other student they may be readily understood." In 1611 alone, he received twenty from Pindar as well as others from Sir Thomas Roe.[5] But even before acquiring the library's first Arabic manuscript, a Qur'an in 1604, Bodley had used a gift from the Lady Northumberland in 1603 to buy the library's first printed Chinese book, an edition of the classics containing parts of the Confucian Analects and Mencius. In 1606, with a £20 gift from Katherine Sandys, the wife of Sir Edwin, he obtained "Octo volumina lingua Chinensi" and two "Excusa in regno et lingua Chinensi."[6] With another shipment of Chinese books to the library in June 1607, he explained to Thomas James (Bodley's first librarian), "Of the China bookes, because I cannot give their titles, I have written on every volume the name of the giver."[7] Most of the forty-nine Chinese books collected by Bodley by the time he died in 1613 were medical texts, probably carried abroad by or for doctors associated with merchant networks from Fujian, and in general the Bodleian collection was a kind of snapshot of overseas Chinese reading habits—classics, medicine, novels, calendars—in the late sixteenth and early seventeenth centuries as gleaned through the activities of the London East India Company and its Dutch counterpart.

With these new kinds of books came new ideas about how to translate them. The losses at Duke Humfrey's in 1549 included substantial portions of the famous collection of the fourteenth-century bibliophile Richard de Bury, which ironically had been preserved there during the dissolution of Durham College a decade earlier. De Bury's *Philobiblon* (1345) was the classic explanation of the translation of ideas in medieval England. The Latin concept of *translatio imperii*, which compared the linear transfer of sovereignty between rulers and political entities to the rising and setting of the sun, was paralleled with the belief in *translatio studii*, the progressive transfer of books and knowledge from India to Babylon, to Egypt, Greece, and the Romans and then to the Arabs, Paris and finally Britain itself. In the 1590s, while Bodley made inquiries about restoring Duke Humfrey's, Thomas James had tried to salvage *translatio*. He published De Bury's manuscript, perhaps even using a copy preserved from the Durham library. But Thomas

James lived in a different world than De Bury. In 1627, James would donate the library's first Javanese and Old Sundanese texts, written on palm leaves. These had come through roundabout channels, his own family as well as networks established by the London East India Company through their factory on Java.[8] In donations like this, it became clear that *translatio* did not merely follow the sun, but now occurred through complex and often unseen chains of relations engaging with the myriad trading cities in Southeast Asia and the Indian Ocean and in books to which a bibliophile like De Bury had no access. Their very existence suggested that new books were still being actively produced on the other side of the world and that translation occurred in asymmetric and asynchronous ways, nothing like the linear and centralized conception of *translatio*. This change in understanding translation at Oxford reflected even broader changes in nearby London, which between 1549 and 1689 both leapt its medieval walls and overcame Protestant fears in order to engage the world on a vast new scale.

THE GLOBAL CITY

Now seen by scholars and indeed self-identified as a paradigmatic example of a "global city," estimates place up to twenty-one million people currently in the London metropolitan area and rings of suburbs.[9] It is one of the world's great financial capitals, with residents from every nation on the planet. But in 1549, the majority of London merchants understood foreign exchange through the financial markets of Antwerp, just across the Channel, or through the communities of foreign merchants like the Hanseatic League who had settled in the city itself. The questions faced by the city rarely went beyond Europe. Londoners took their understandings of translation, language, and the cosmology of the broader world from humanist and Protestant writers in the German-speaking cities of the Danube, Rhineland, and Low Countries or those of the Italian peninsula, authors like Peter Apian, Sebastian Münster, or Giovanni Battista Ramusio.

If one had in the 1540s looked for a potential global city in Europe, the most likely candidate would have been London's principal trading partner, Antwerp, which had emerged as the center of the German and Spanish silver exchange as well as the Portuguese spice and Spanish sugar trades in the early part of the sixteenth century. London's most wealthy guild, the wool-trading Merchant Adventurers, was technically located there, and Sir Thomas More had used the city's printing presses for the first edition of his *Utopia* in 1516. By 1540 it was the cultural, political, and economic center of Emperor Charles V's Thirteen Provinces, a truly imperial city that linked the Holy Roman Empire with Spain. But Antwerp outsourced its networks and became dependent rather than autonomous, "a world capital created by outside agency," as the historian Ferdinand Braudel put it.[10] This in turn encouraged the creation of concepts and techniques for translating the world that were abstracted from the city itself. The famous maps and cosmographic work of Gemma Frisius and Gerardus Mercator at Louvain and later the atlas maker Abraham Ortelius universalized the city's relationships to such a degree that they are often seen as products of Europe or even, in the case of Mercator's globe, pure science.[11]

By contrast, when London did eventually itself reorganize global space and time, the local signification—Greenwich after the Royal Observatory founded in 1675 and the Nautical Almanac produced there from 1767—proved remarkably durable as a sign of London's centrality. The new observatory was a telling indicator of about how much changed between 1549 and 1687 in London. In 1682, for example, the city hosted celebrated embassies from Morocco, Russia, and Banten, William Penn left to begin negotiations with Lenape villages for land, and the East India Company sent arms from its warehouses on the Thames to support an alliance with the Zheng family of merchant pirates on Taiwan in their last desperate attempt to resist the Manchu Qing Dynasty. Building on the sixteenth-century compilations of Richard Eden and Richard Hakluyt, London booksellers now filled their shelves with increasing numbers of accounts by travelers, missionaries, and merchants. The Royal Society (est. 1660) aimed to produce a globally relevant science. Outside the city, both Oxford and Cambridge now had literally world-class libraries. Industrious potters strove to compete with Chinese porcelains and Dutch delftware imitations of them. The first coffee houses and china shops had appeared. The variety of supplies necessary to send ships across the globe from Deptford and elsewhere kept thousands employed.

Why did this happen? The most obvious and the traditional answer, sometimes called the "simple model," is that London grew. It grew through immigration starting in the late fifteenth, and early sixteenth centuries. People came from England, Wales, Scotland and Ireland, from the Low Countries, German cities, France, Iberia, and Italy; on trading ships arrived smaller numbers of Chinese and Indian sailors, West and North Africans, creoles from the Americas. Historians have long recognized the uniqueness of London in both England and Europe in the early modern period. As most English towns still struggled to recover from the Black Plague in the early sixteenth century, London boomed. In 1550, London was a city of about 75,000 people, seven times the size of the next largest town in England yet still below pre-plague levels from the 1340s. A city in flux, it grew to 200,000 in 1600, 400,000 in 1650 and 575,000 in 1700. Despite a high mortality rate, it still grew faster than any other city in Europe.[12] By 1800 London had reached a population of one million, equivalent to Beijing or Edo (Tokyo), previously the world's largest cities. Some have suggested that the experience of living in a city like London was one of modern mobility, and arguably the English experienced both the movement and fractured nature of globalization in the urban setting of London even if they never had the experience of traveling and brokering deals elsewhere.[13]

Richard de Bury's *Philobiblon* was written at the end of a golden age, three years before London's medieval population plummeted by approximately two-thirds. It is hard to underestimate how traumatic the Black Death was for both London and England as a whole. For almost two centuries, travelers from Europe would describe the shocking emptiness of England, the population of which shrank from 5 or 6 million to about two-thirds of that number and was still only 2.8 million in 1547. Plague recurred at least sixteen times between 1348 and the last Great Plague of 1665. Thanks to immigration, London recovered more quickly and regained its pre-plague population by 1550, but outside of London, the empty spaces filled with sheep. In 1400, although it was the most prominent center of exchange in the English kingdom, London transmitted only about 30% of its raw wool and 50% of its cloth exports.[14] Over the course of the fifteenth century, London began to dominate cloth exports and foreign trade more generally. The vast majority of that trade, however, only went the short distance to Antwerp. Crisis and stagnation in cloth exports from mid-sixteenth century as well as troubles in Antwerp forced London to look beyond regional trading. But neither necessity driven by trauma nor opportunity generated by wool and the rise of the gentry implied a shift towards global relations.

Historians often contend that the enclosing gentry created English culture and through their "plantations" the Atlantic World, a country phenomenon merely enabled by city merchants. Among this gentry, who clearly played important roles from Ireland to Virginia, a national consciousness supposedly first emerged through print, including revised understandings of history, law, and politics and a concept of "empire" that meant dynastic independence.[15] Figures like John Hawkins and other sixteenth-century privateers in Africa and the Americas used the widening gate of provincial cities like Bristol and Plymouth and the rise of state interests in mercantilism to build an Atlantic World. The formation of a North Sea economy because of the Navigation Acts of the 1650s and 1660s secured "national control of England's entrepôt trade" making London a "multifunctional metropolis."[16]

Robert Brenner and Chris Isett have made this argument at a global level, suggesting that England diverged from East Asia "not as a consequence of any
advantage possessed by its domestic economy, but rather as a result of its
unique form of mercantile state and merchant companies that made possible its access to the land, raw materials, and above all slave labor, of its
American colonies."[17] This institutional explanation of English exceptionalism tied to the Atlantic economy goes back to those who often reluctantly
bought into the British state system in the eighteenth century—the Scottish (Smith and Hume), the Protestant Ascendancy in Ireland, the American
Colonies (from Canada to Guiana). Such groups often still cannot decide
what their special relationship is to the Whig narrative about the industrious and pastoral "domestic economy" of England let alone to the Tory
one of Church and King as shepherding institutions. The "Atlantic World"
thesis has nevertheless produced a rich scholarship that builds upon English agrarian and domestic exceptionalism to describe a world of complex
encounters.[18] Such narratives appear as concentric circles, little England,
greater Britain and finally the Commonwealth or Empire. They ultimately
imply a model for projecting rather than negotiating values, that gentry
plantation ethic so important to the formation of Munster and Virginia.

But as significant as the rise of the gentry was to the history of England,
so too was the shift in values resulting from London's early relations with
Asian trading cities between 1549 and 1687, especially for London itself.
Because this shift was connected with global trade and amassing bullion,
at times it simply appeared as an endless engine for making money and for
building ships, creating, as Cicero quipped, the "sinews of war."[19] From the
1550s, London began to import and re-export increasingly large amounts of
goods from the Americas and Asia. This created a favorable balance of trade
with the Continent, even when trade with Asia seemed to be dangerously
leaking bullion. The growth of this trade and its associated financial structures by the eighteenth century set the stage for the formal empire of the
nineteenth century and the Commonwealth of the twentieth, not to mention the continuing growth of London itself, that vast concentration and
networking of money and people.[20]

Writing in 1687 to the East India Company's factors in Bombay on the
eve of the Glorious Revolution, the governor of the East India Company,
Sir Josiah Child, claimed that pepper was the essential ingredient for both
Dutch and English prosperity and military power.

> If the present misunderstandings between the two Nations [the English
> and the Dutch] should ferment to an open war, it would be thought by

the Vulgar but a war for Pepper which they think to be a slight thing, because each family spends but a little of it. But at the Bottom it will prove a war for the Dominion of the British as well as the Indian seas. Because if ever they [the Dutch] come to be sole Masters of that Commodity, as they are already of nutmegs, mace, cloves, and cinnamon, the sole Profit of that one Commodity, Pepper being of generall use, will be more to them than all the rest and in probability sufficient to defray ye constant charge of a great Navy in Europe.[21]

Pepper, the most basic of everyday luxuries, was not simply profitable, it established patterns of exchange, as ballast it enabled the medium of shipping, and as a historical standard of value it defined a series of networked relationships among producers, middlemen, shippers, vendors, and a range of consumers. In this sense it was different from the plantation economies of sugar and tobacco in the Americas that were at the core of the "triangular" trade with Africa. Malabar, Java, and Sumatra all expanded pepper production significantly in the late seventeenth century, reorganizing and expanding rural commodity agriculture in relation to urban ports. If anything, the go-betweens of both the Native American and the African slave trades caused greater instability and war in the hinterlands. For Child, the action of connecting expanding centers of mass production of pepper with a broad consumer base in Europe, the Americas, and Asia itself had financed and helped develop the technical expertise for naval power in Europe, involving longer and more complex journeys and exchange patterns. London developed not just as a market town, pulling in goods parasitically from the surrounding countryside as Braudel might have thought, but a place that enabled and indeed fostered global commerce between Asia, the Atlantic World, and Europe. These interactions put London on a more solid footing than contemporary rivals like Lisbon, Seville, Paris, and Amsterdam, engaging in remarkably productive and durable ways with the cultural and economic dynamics of the Indian Ocean and maritime East Asia in the sixteenth and seventeenth centuries. These connections created a sense of autonomy, England's and especially London's political and economic autonomy from entities on the Continent—from the Hanseatic League and Antwerp to the Pope and Holy Roman Empire—even as frequent warfare in the Americas created problematic dependencies and by the eighteenth century extremely expensive wars. And through this engagement with Asia also came the recognition that English, like Latin, French, Arabic, Persian, Malay, or Chinese, could also be a language used beyond colonial space for global exchanges.[22]

THE QUESTION OF TRANSLATION

Child's arguments provide a useful reminder that it was not simply the local recognition of the free-born Englishman but the translatability of "British" institutions—the supposed universality of the nation-state, of law, of rights, of science—that earned them the title of modern. Whether it has been the sheer size and financial power of London, the importance of the Hobbesian, Lockean, and Newtonian languages of politics and science, or the success of the English language more generally as a factor of these historical developments, debates over sixteenth- and seventeenth-century British history have taken on a far more universal character than one might otherwise expect. This simple fact has served to hide the theoretical problem of translation, defined broadly as how meaning gets both handed down over time ("passing the parcel" as Alan Bennett has called it) and handed across apparent political, economic, linguistic, and cultural boundaries.

The historian's perspective can make it difficult to address translation because the very process of moving between languages seems to undermine the realm of direct experience. Translation calls into question the lens of national experience as constituted through a single language that remains the standard methodological frame of most historical writing. If experience takes place through language, then anything translated from another language always appears to have taken place outside of experience or at best in an alternative cultural and linguistic register, an incommensurable realm of experience. The repetitive appeal to a bounded ethno-linguistic tradition of English sources tied to the nation can also defend history against a series of challenges raised by philosophers from Friedrich Nietzsche, Benedetto Croce, and Michael Oakeshott down to Hayden White, who have expressed profound skepticism about the possibility of the historian as a translator of the "foreign country" of the past.[23] In response to this criticism, R. G. Collingwood argued that history was indeed an integral part of experience, but for him it was always a kind of technical "reenactment" rather than a virtual re-experiencing or even a witnessed performance. The unlived, unexperienced past is a kind of artificial already-there, a parcel or indeed a book waiting to be opened and translated. This is not unlike Walter Benjamin's idea of translatability being inherent in the form of a work. Certain texts were made for translation and by implication certain histories and places have in their experience translatability. London, as an emporial city in the early modern period, was one such place, but so were many of the trading cities of the Indian Ocean and East Asia. And "London" is a different kind of translatable space than "England," more friendly to certain kinds of reenactments of languages other than English that the homogeneous national space of language "England" arguably is not.[24]

Reenacting some of this past dynamism of language and translation requires a methodology that can reopen the relationship between history and translation, a methodology that would allow for work between at least two and often three or more languages and actors. It requires sidestepping the history and discourses of exoticism and images of the other that focus rigorously on English or more broadly European sources, to which there are several other excellent guides.[25] In examining this process, what appears to remain untranslated in the archive can be of equal importance in understanding the formation of relationships as materials assumed to have translated because they were in English or in a more limited sense Latin.

The relationship between languages, and the bringing of words and concepts from one to another, was in fact the theme of one of the palm-leaf manuscripts donated to the Bodleian by Thomas James, the *Bujangga Manik*. The story itself probably dates back to the fifteenth or early sixteenth century and addresses the question of how proper translation can take place between sacred languages and among the emporial spaces of commerce, in particular bringing certain sacred aspects of the Javanese language into Old Sundanese in order to renew its vitality. The protagonist Bujangga Manik travels on a pilgrimage to find spiritual enlightenment through refining his understanding of Javanese instead of staying home and marrying a princess.[26] His first journey into Eastern Java, the rump of the old fourteenth-century Majapahit Empire and the location of the newer late fifteenth- and early sixteenth-century principality of Demak, dramatizes the opening of Sundanese to Javanese language, texts, and trade practices. After that journey, he can speak Javanese (*carek Jawa*), and he knows the sacred texts (*tangtu*) and law (*darma*) (lns. 327–31). It is a remarkably confident story about the return of Sanskrit traditions from eastern Java (such as those that survive today in Bali) into Sundanese-writing areas, places that were losing their ties to Sanskrit as the language and culture of west Java and south Sumatra changed with the arrival of Chinese and Indo-Persian merchants.

Sheldon Pollock and Ronit Ricci have recently developed the concept of the layered and contested "cosmopolis" of texts and translations in relation to Sanskrit and Arabic in the Indian Ocean and Southeast Asia for such open-ended storytelling, collecting and archiving across languages and cultures.[27] Pollock and Ricci's approach is paralleled by the work of economic historians like Kirti Chaudhuri and Sugata Bose with the notion of the Indian Ocean as an "interregional arena," which as a space of trade and migration using particular currencies and documents had its own unique

logics of exchange and translation. Similar ideas appear in the developing critique of notions of a Chinese tributary world in work by John Wills,
Leonard Blussé, and Victor Lieberman.[28] The maritime spaces of Asia, in
some ways replacing as well as supplementing the older routes of the Silk
Road, were connected through extended diaspora networks of families and
religious orders as well as through contractual partnerships or, as they were
called in medieval Italian cities, *commenda*. Such arrangements where one
partner stayed and one traveled were quite common in most urban parts
of Asia and the Mediterranean by the time of Marco Polo, including Mongol relations with Islamic traders (the *ortogh*) and maritime relations in
the Indian Ocean; but the practice remained on the margins of economic
and legal practices in London until the sixteenth century.[29] Not surprisingly, these partnerships were explicitly competitive and necessarily multilingual, again a sharp contrast from the highly ordered guild structures of
London and the highly institutionalized presence of Hanseatic and Italian
merchants up until the mid-sixteenth century that tended to limit as much
as possible who translated and where and how they did. Conversely from
the interrelated networks of Asian port cities connected through merchant
diasporas, secular and religious travel, local go-betweens, relations with bureaucratic states, and even relatively stateless hill people and sea nomads
(*orang laut*) came a great deal of dynamicism in terms both of economics
and of language itself.[30] But starting in the second half of the sixteenth century, Londoners began to find ways of tapping into and conceptualizing the
complicated spatialization of cities, merchant networks, languages, seals,
and coinage in the Indian Ocean and East Asia; Josiah Child even went so
far as to describe a "medium" of trade rather than a simple exchange system
which created the conditions for building creative relationships, new kinds
of transactions and translations.[31]

   If the archive is often seen as the signature of the state, the consolidation of national language, and the preferred site of historical work since
Ranke, then the practice of translating across languages connected with
very different sacred and cosmological traditions emerges out of such networks of port cities as well as the house societies that collected manuscripts
in them.[32] The locked case and the unread manuscript in a library like the
Bodleian imply a sense of danger associated with such collection and translation, an attitude not unique to London and Oxford (figure 2). The *Bujangga Manik* itself begins with a critique of the dangers of house societies
as prone to collecting the wrong kinds of objects, particularly those that
might produce religious errors and a too comfortable domesticity. In this
passage, a Chinese box becomes a symbol of the potential bride of Bujangga

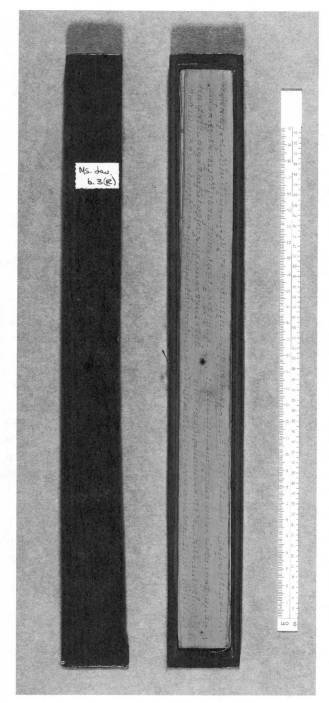

Fig. 2. *Bujangga Manik*, ca. late 15th century, acquired early seventeenth century and donated 1627, with lacquer case. © Bodleian Library, University of Oxford, 2009, MS Java b.1[R].

Manik becoming a dangerous temptation and drawing the monk away from
a life as a pilgrim.

> Teherna luguh di kasur,
> Ngagigirkeun ebun Cina
> Ebun Cina diparada
> Pamuat ti alas peutas

> Thus she was seated on a mattress,
> A Chinese box beside her,
> A gilt Chinese box,
> Imported from overseas[33]

Upon returning from his first voyage, the prince-pilgrim Bujangga Manik
rejects an offer of marriage with all the domestic trappings of the betel tray,
Chinese luxuries, fine traditional clothing, including a *wayang* figure belt
and the *keris* knife. The collected objects appear as falsely valued or translated commodities, and as a result the process of renewing language through
collecting and translating texts takes on even greater value. The house and
its objects are confusingly entangled in distant trading patterns and imbued
with complex localized, even personalized meanings. Bujangga Manik ultimately rejects the domesticated form of collection that connects the port
and the household when he sets off again with his own personal and much
narrower collection of sacred things that will guide him—a pilgrim staff, a
rattan whip, and two Javanese books, the "great book" (*apus ageung*) and
the *Siksaguru* ("teacher"). A certain kind of private subject thus emerges
from this kind of relationship between translation and collecting, one that
individually delimits the set of relevant texts and languages that are indeed translatable while nevertheless working to keep new editions in circulation and to keep translation and exchange vital. Even if they often could
not read the books like the *Bujangga Manik* that they collected, Londoners
nevertheless learned from the very process of obtaining such texts to move
between the processes of translating and collecting and to build new kinds
of selves and new kinds of institutions appropriate to the world of exchange
they encountered in maritime Asia.

THE SUBJECT OF THE BOOK

The nineteenth-century discipline of Orientalism, which advocated reading Asian books in terms of a distinct hermeneutics and thus in separate

reading rooms, along with the sheer growth in collections, made obsolete the seventeenth-century universalist approach in which Latin and Greek jostled in libraries with Arabic, Persian, Chinese, and Telugu. With archives in Chinese, Japanese, Sundanese, Javanese, Malay, Persian, Avestan, and Arabic as well as several European languages including Greek and Latin, the problem of describing London in terms of translation requires a multilingual approach that is no doubt beyond any one individual. But to open up that potential for dialogues, I have used the "contrapuntal readings" described by Edward Said in which divergent texts are played off of each other and the "striking connections" (datong 打通) of Qian Zhongshu's essays on European and Chinese literary and philosophical concepts in order to open the rhythms of inter- and intra-urban translation and collecting in the sixteenth and seventeenth centuries.[34]

In the course of my research, I also found that in tracing the provenance of the books, one could find hints about why they were collected, and at various stages, how they were translated and read.[35] In 2006, while in the Arts End of Duke Humfrey's tracking down entries from old library catalogs, I called up a bound volume that Elias Ashmole had assembled in the 1680s. It began unpromisingly enough with an odd manuscript fragment labeled "nothing."[36] Amidst other fragments of Chinese books on mathematics and medicine as well as paper samples and French emblems, there was a previously unrecognized copy of the 1677 edition of the calendar produced by the Zheng Ming loyalist regime on Taiwan (Yongli 31) with note on cover "A Chyna Almanack. Given me by Mr. Coley, 28 Sept. 1680. E. Ashmole" (figure 3). The little collection has since been broken up for preservation and access. Ashmole had received it from Henry Coley, the foremost publisher of almanacs in London during the Restoration, and it is clear that someone had known enough about its contents to give it to Coley, who likewise thought Ashmole could answer certain questions. The binding of these books together, however, seemed to indicate a loss of knowledge—what Coley knew to be a kind of almanac had become simply a paper curiosity.

The next summer I was in London retracing Thomas Hyde's personal library, a small research collection that he had kept separate from the Bodleian's larger collections that is now scattered throughout the British Library's holdings. It had ended up in the possession of Sir Hans Sloane by way of the Royal collection, and it too had to be retraced and recovered through old catalog entries and by moving between the British Library's departments of Rare Books, Western Manuscripts, and Asian Studies.[37] On Hyde's copy of the 1671 edition of the Yongli calendar there were extensive translation notes dating from 1687. These indicated that through Shen Fuzong's efforts,

Fig. 3. Calendar made in Zheng Jing's Taiwan, *Daming Yongli sanshiyi nian datongli* [大明永曆三十一年大統曆], reign year of 31 (AD 1677). Bodleian Library, University of Oxford, Sinica 88 (formerly part of MS Ashmole 1787), photo by Robert Batchelor.

Hyde understood such objects not just as "almanacks" but as collections of *tu* (圖), a word that encompasses maps, and other kinds of diagrams and tables that convey information and in the seventeenth century could even apply to boardgames. Here was a significant concept. Shen had even brought a printed boardgame with him, the *Sheng guan tu* (陞官圖, lit. "promotion official diagram"), which indeed seems to have been the only printed material that Shen brought out of China and which Hyde both translated and printed in a book on Eastern games. Hyde and Shen translated *tu* with the Latin "tabula," and understanding *tu* allowed comparisons across what appear to European eyes as distinct genres.[38] Hyde's loose papers also contained strange notes made jointly with Shen about a Chinese compass and inscriptions on two different maps. These seemed to be part of the research on *tu*, but at the time it was entirely unclear to which maps these notes referred.[39]

Then in January 2008, I returned to Oxford for a conference and had the chance to follow up on the maps in the Bodleian. Hyde and Shen had catalogued a map in the collections of John Selden, so I suspected it related to the notes at the British Library. I was hoping that along with the Codex Mendoza, Selden might have also acquired from Samuel Purchas's estate the original of a Chinese sheet map obtained by the East India Company captain John Saris in Banten, which had been reengraved and printed in Purchas's *Pilgrimes* (1625). I asked David Helliwell if he knew where the map was, and he responded that he had seen it as part of his recataloging of the Sinica manuscripts. After locating it on the shelves, he said we were going to need a bigger room, and he found a table for us somewhere between two reading rooms big enough to unroll Selden's map of China. There on the Selden Map I noticed the fine traces of a network of merchant shipping routes across East Asia, the only diagram of its kind to survive from an early modern Chinese merchant organization. Three years later, in January 2011, after the map had been restored, we made another discovery. On the back was a draft of the routes, meaning that these had actually been drawn first. It was a diagram after all, one that revealed a forgotten form of Chinese mapmaking. See figure 4.

When examining the calendars and the annotations in other Chinese books, I had still thought of them within a narrative of conceptual and linguistic difference, what Benjamin Elman, an early mentor for my work, described as thinking about Chinese science "on [its] own terms."[40] Translation historically became a problem as the lingering medieval sense of universalism embodied in the *translatio studii* had given way by the seventeenth century to a mutual awareness of culturally distinct concepts like

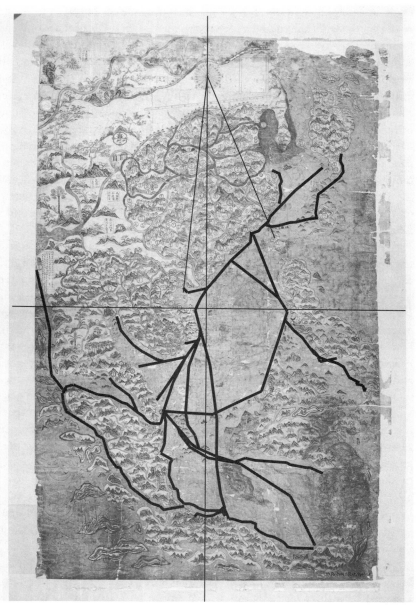

Fig. 4. The Selden Map with routes highlighted in black. © Bodleian Library,
University of Oxford, 2012, MS Selden Supra 105.

tu that could seem incommensurable. But Selden in his book *Mare Clausum* (1635) suggested that even across linguistic difference mutual recognition and more precise technologies of measurement could be used to make claims upon ocean spaces in terms of *dominium* or legal property.

For me, this alternative form of mapping that nevertheless translates became a kind of overarching theme, thus the title of this book, and I began to see that same theme in a range of not only Chinese maps brought to London but also maps made in London like those of Sebastian Cabot and Edward Wright. Selden's Chinese map had been the centerpiece of his collection in Whitefriars, and it had come with a compass. It seems to have been drafted around 1619, out of the ferment emerging from late Ming shifts in trade, the rise of Tokugawa Japan, and corresponding shifts in the global trade in silver or "silver cycle" that connected Spanish mines with the Ming economy. The Selden Map was not only a rare and probably the earliest surviving example of Chinese merchant cartography but also a kind of mutually recognized technical achievement, one that by 1651 was part of an active debate in London about whether it was legitimate to use legislation like the Navigation Act to close off the Atlantic. Thus the Selden Map's approach to space, branching trees of shipping routes that connect to both clearly sovereign spaces like the Ming and to distant trading spaces from the Persian Gulf to Spanish silver mines, gave in many ways a deeper insight into early modern "globalization" than the traditional all-encompassing European world map of Mercator. It was a world of densely networked information, suggested by the plethora of ports on the Selden Map, but that information was neither easily translated nor readily shared (figure 5).

Suddenly the Zheng calendars and all those Chinese medical books, philosophical classics, and novels that at first seemed unread in the archive now appeared to have been produced, collected, and read in and for multiple contexts. Each had layered meanings, to paraphrase Haun Saussy, like a series of translations of an uncertain original. This was hardly surprising in an era when commodities were transferred and revalued among many ports, languages, and peoples as in the spatial logics of Asian maritime trade described by Kirti Chaudhuri, but it meant that London was far more entangled in Asian patterns of exchange and translation than has been previously recognized.[41] Noticing the routes and ports on the map for the first time in 300 years revealed something that had been occluded about early modern London—how to understand its global character.

One can read this book as a story about how that global character emerged, the making of a global city, but it is also a kind of extended essay

Fig. 5. The Selden Map with ports and significant sites labeled. © Bodleian Library, University of Oxford, 2012, MS Selden Supra 105.
Note: Only provinces and locations outside the empire have been marked.
Fenye and prefectural seats or *fu* 府 are not labeled.

Features taken from the Encyclopedia Map:

A. Zongji city (總戛城 *Zongji cheng*)
B. Jade Gate Pass (玉門關 *Yumen guan*) and Shaanxi Administrative Region (陝西行都司 *Shaanxi xing du si*)
C. "River City" (河州 *Hezhou*)
D. Kunlun Mountain (崑崙山 *Kunlun shan*) and snow-capped mountains (一名雪山 *Yi ming xue shan*)
E. Horse Lake west of Yellow River, three thousand li (馬湖府西至黄河三千里 *Ma Hu fu xi zhi Huang He sanqian li*)
F. Li River northwest of Yellow River, one thousand five-hundred li (麗河西北至黄河一千五百里 *Li He xibei zhi Huang He yiqian wubai li*)
G. Yellow River Source (黄河水源 *Huang He shuiyuan*)*
H. Star Constellation Sea (星宿海 *Xingxiuhai*)
I. Shanxi Province (山西)
J. Beijing District (北京)
K. Liaodong Defense Region (遼東)
L. Shangdu (金阮上都 *Jinruan Shangdu*)
M. Yalu River (鴨綠江)
N. Shandong Province (山東)
O. Nanjing District (南京)
P. Henan Province (河南)
Q. Shaanxi Province (陝西)
R. Sichuan Province (四川)
S. Yunnan Province (雲南)
T. Guizhou Province (貴州)
U. Huguang Province (湖廣)
V. Jiangxi Province (江西)
W. Zhejiang Province (浙江)
X. Fujian Province (福建)
Y. Guangxi Province (廣西)
Z. Guangdong Province (廣東)
*Selden map adds the character "水" (water)

Features unique to the Selden Map:

1. Only Huangwali foreigners, Henanli foreigners, beyond this point. (黄哇黎番呵難黎番俱在此後 *Huangwali pan, Henanli pan, ju zaicihou*)
2. North rapids here (北886在此 *Bei ta zaici*)
3. Unreadable cartouche
4. Sado Island (佐渡州 *Zuodu zhou*)
5.? (總州 *Zong zhou*)
6. Mount Fuji (七鳥山, *Qidao shan*, lit. "Seven Island Mountain" referring to the Izu Islands)
7. Echigo Province (越 *Yue*)
8. Shanghao Province (上好州 *Shanghao zhou*)
9. Mt. Hou (母後 *Mu Hao*)
10. Edo (出王城 *Chu wang cheng*)
11. Ise Province (伊勢 *Yishi*)
12. Imperial Residence (所居地 *Suoju di*)
13. Kyoto (大王城 *Dawang cheng*)
14. Matsusaka Domain (松佰王 *Songbai wang*)
15. Kanto Region (?) (菅東 *Guandong*)
16. Korea (朝鮮王 *Chaoxian wang*)
17. Tsushima Island (水剌馬 *Shuishengma*)
18. Osaka (沙捷王 *Shajie wang*)
19.? (一插花大王 *Yichahua dawang*)
20.? (温子米王 *Wenzimi wang*)
21. Nagoya (居仔 *Juzi*)
22. Sasebo (闇色保王 *Shesebao wang*)
23. Kagoshima (空打剌馬 *Kongdashengma*)
24. Seto Inland Sea (萬島 *Wandao*, "Ten thousand islands")
25. Ike Island (衣戈 *Yige*)
26. Hirado (魚鱗島 *Yulin dao*, "Fish scale island")
27. Hyogo (兵庫 *Bingku*)
28. Goto Islands (五島 *Wu dao*, "Five Islands")
29. Nagasaki (笕仔沙机 *Longzishaji*)
30. Kochi Castle (?) (亞里馬王 *Yalima wang*)
31. Satsuma (殺子馬 *Shazima*)
32. Kagoshima Bay (殺身灣子 *Shashen wanzi*)
33. Koshiki Islands (?) (天堂 *Tiantang*, "Paradise Island")
34. Qiantang River (钱塘江)
35. Yakushima passage "flowing eastward, very tight". (野故門水流東甚緊 *Yegu men shuiliu dong shen jin*)
36. Ryukyu (琉球國 *Liuqiu guo*)
37. Beigang, Taiwan (北港)
38. Jiali Forest, Taiwan (加里林 *Jiali lin*)
39. "This passage, flowing eastward, very tight" (此門流水東甚緊 *Ci men, liushui dong shen jin*)
40. Eastern Sea (東海 *Donghai*)
41. Penghu Islands / Pescadores (澎湖 *Penghu*)
42. Pratas Reef (南澳氣 *Nanao qi*)
43. Aparri (大港 *Dagang*)

44. Malolokit (射昆美 *Shekunmei*)
45. Pagudput (月投門 *Yuetao men*)
46. Burgos (香港 *Xianggang*)
47. Laoag (南旺 *Nanwang*)
48. Vigan (台牛坑 *Tainiukeng*)
49. Lingayen (玻琉 *Daimao*)
50. Bataan (扶鼎安 *Fudingan*) and Capones or Turban Reef (頭巾礁 *Toujin jiao*)
51. Manila (呂宋王城 *Lusong wangcheng*, "Luzon Royal City")
52. Ten Thousand Shell Gate (甲万門 *Jiawan men*)
53. "Spanish go back and forth from this harbor to Luzon [Manila]" (化人番在此港往來呂宋 *Huaren fan zaici gang wanglai Lusong*)
54. Oton (福堂 *Futang*)
55. Cebu (束務 *Shuwu*)
56. Maguindanao (馬軍礁老 *Majunjiaolao*)
57. Sulawesi or Celebes (蘇橼 *Sulu*)
58. Maluku Is./Ternate (萬老高 *Wanlaogao*); Spanish Place (化人住 *Huaren zhu*); Dutch Place (紅毛住 *Hongmao zhu*)
59. Calicut (古里國 *Guli guo*)
60. Sailing directions to Aden (阿丹国 *Adan guo*), Zufar/Salalah (法兒国 *Far guo*) and Ormuz (急魯謨斯 *Hulumosi*).
61. Hanthawaddy/Toungoo (放沙 *Fangsha*)
62. Siam (逞羅國 *Xianluo guo*)
63. Laos (樓里 *Laoli*)
64. Tonkin (東京 *Dongjing*)
65. Thanh Hoa (清花 *Qinghua*)
66. Nghe An (?) (新安 *Xinan*)
67. Dong Hoi (布哎 *Buyao*)
68. Hue (順化 *Shunhua*)
69. Quang Nam (廣南 *Guangnan*)
70. Da Nang Peninsula (在峴港 *Zai Xiangang*) and Hoi An (會安 *Hui'an*)
71. "The Paracel islands (10,000 isles) resembling the shape of a sail." (萬里長沙似船帆樣, *Wanli changsha si chuanfan yang*)
72. Hoang Se (lit. 'Yellow Sands', Vietnamese name for the Paracels) (嶼紅色 *Yu Hongse*, Chinese "Red Islands")
73. The Spratleys (?) or another part of the Paracels (萬里石塘 *Wanli shitang*)
74. Qui Nhon (新州 *Xin zhou*)
75. Champa (占城 *Zhancheng*)
76. Phan Rang (羅鸞頭 *Louwantou*)
77. Diamao (Con Dao) Island (玳瑁洲 *Daimao zhou*); West Orienting Island (西童 *Xi dong*); East Orienting Island (東童 *Dong dong*)
78. Mekong Island (毛蟹洲 *Maoxie dong*)
79. Cambodia (柬埔寨 *Jianpuzhai*)
80. Nakhon (六坤 *Liukun*)
81. Phattalung (無頭郎 *Foutoulang*)
82. Patani (大泥 *Dani*)
83. Kedah (吉礁 *Jijiao*)
84. Sumatra that is Aceh (蘇文達即亞齊 *Suwenda ji Yaqi*, an older name for 蘇門答 *Sumenda*)
85. Malacca (麻六甲 *Maliujia*)
86. Pahang (彭坊 *Pengfang*)
87. Johor (烏丁礁林 *Wuding jiaolin*)
88. West Snake Dragon (Anambas Islands) (西蛇龍 *Xi shelong*)
89. East Snake Dragon (Natuna Islands) (東蛇龍 *Dong shelong*)
90. Kun Lun Island (崑崙 *Kunlun*)
91. Brunei (汶萊 *Wenlai*)
92. Banda (援丹 *Yuandan*)
93. Ambon (唵汶 *Anwen*)
94. Makassar (傍伽剌 *Bangjiashi*)
95. Banjarmasin (馬神 *Machen*) [erased]
96. "Strait entrance is here" (峽門在此 *Xia men zaici*)
97. Palembang (旧港 *Jiugang*)
98. Jambi (占卑 *Zhanbei*)
99. Indragiri (丁机宜 *Dingjiyi*)
100. Padang that is Boli (?) (茅陣即貓離 *Maozhen ji maoli*)
101. Pariaman (巴里野雁 *Baliyeman*)
102. Banten/Sunda (順塔 *Shunta*)
103. Jakarta/Kelapa (咬留吧 *Yaoliuba*)
104. Indramayu (?) (吧那 *Bana*)
105. Cirebon (緒營 *Zhuman*)
106. Gresik (饒洞 *Raodong*)
107. Bali (磨屋 *Moli*)
108. Bima (里嗎 *Lima*)
109. Timor (池汶 *Chiwen*)

on how my own encounter with the Selden Map changed my views about British, Chinese, and indeed world history. To demonstrate the power of translation, I have structured the narrative around five traditional turning points in the history of London. The first involves the simultaneous political, religious, and economic crisis of 1549–53. This had some of its most profound effects on London, when under Edward VI a staggering variety of changes were experimented with out of recognition of the weakness and incompleteness of the Henrican Reformation and of traditional models of corporate governance. The second concerns the development in the late Elizabethan period of ideas and languages about the English and British nation as a coherent space ("empire") that could support the growing global ambitions of London's trade and navigation strategies. The third relates to the efforts in the Inns of Court and London intellectual circles to define an autonomous legal and mercantile framework under James I and then Charles I that could achieve results for the city similar to the independent Elizabethan nation on a global and especially oceanic scale, the collapse of this legal order in the English Civil War and its reconstitution under the first Navigation Act and in the first Anglo-Dutch War. The fourth focus is on the period of the Restoration and the efforts to use the image of the king to create a coherent sense of state authority to buttress the newly reformed exchange institutions and financial practices of the city. Finally, the fifth period leads up to the Glorious Revolution, when London's relation to both French and Dutch global strategies as well as to dramatic shifts in Asia resulting from the expansion of the Qing and Mughal empires demanded radical changes in both urban and national governance. In each of these five periods, traditional accounts have tried to subsume London's interests into a local and national framework. The arguments in this book will suggest the reverse, how changes taking place in Asia and on a global scale were translated through London into apparently national developments.

In brief then the development of classic signifiers of modernity—the corporation, the nation, the rule of law, the state, and political revolution—are all considered in the book's five chapters in relation to changes in Asia. In each a claim is made about how the process of translating affected these developments in London. Chapter 1 begins with the publication of London's first world map in 1549, the founding of London's first joint-stock enterprise, the Cathay Corporation, between 1549 and 1553, and the publication of London's first global travel anthologies by Richard Eden (1553, 1555). In a context of profound religious and dynastic shifts and uncertainty as well as difficulty with the Antwerp trade and state debt, the corporation founded by Sebastian Cabot, the former *piloto mayor* and cosmographer of

Spain, offered a secular strategy for linking London to Asia that relied on a valuation process outside of Christian Europe. Rather than a developing solidity of state and religious institutions driving London's development, as in the classic thesis of Geoffrey Elton that informed Douglas North's Nobel Prize–winning work in economics, or even the older notion of a gentry community of landed trusts put forward by Frederic Maitland, this institution of the global corporation was strikingly new and dynamic in London.[42] It had substantial roots in the longstanding forms of partnerships and merchant ventures of the Asian maritime world and the Mediterranean, but a new conception of "Cathay" also carried with it a remarkably sophisticated understanding of the beginning of the global silver cycle in which Asian and especially Ming demand for silver as a standard of commercial value was met by the production of Spanish, German, and Japanese mines. As a result, unlike more temporary partnerships or less flexible guilds, the emerging joint-stock corporation also depended on a basic conception of and data about global networks and trading cycles centered on the as yet unknown Ming Empire.

While the proliferation of the new institution of the joint-stock company provided a basis for London's increasing fiscal and commercial autonomy from both Antwerp and the Spain of Philip II in the early years of Elizabeth's reign, it did not offer a clear sense of autonomy for the nation or the "imperial crown of England" of the intertwined sovereignties of England, Wales, Ireland, and the ever problematic Scotland. Chapter 2 argues that the circumnavigations of Francis Drake and especially Thomas Cavendish provided the data on the silver cycle and the reality of the size and coherence of the Ming Empire necessary for constituting conceptions of such sovereign entities in a period when Catholicism, Spanish invasion, Irish rebellion, and even merger with the rebellious Dutch all threatened to tear apart the English "empire." The translation of Ming population data from a map captured by Cavendish as well as the interviews about the use of the compass with the two Japanese sailors he brought to England suggested definitively the size of "China," its distinction from "Cathay," and the fact that its sovereign independence was defined both internally and externally through the collection of population data in the census and the technical achievements of the compass. The active awareness of a historical process of sovereign boundary definition going on in East Asia between China and Japan set the stage for the revaluation in London of cartographic, population, and navigational data as a tool rather than mere representation of power. Assembly of data by the corporation in the absence of a state bureaucracy capable of doing so became a key element during the 1590s for the founding of the

East India Company, and for early theorists of scientific and mathematical knowledge like Edward Wright.

Chapter 3 examines early seventeenth-century London, where the rise of legal institutions like the Inns of Court worked to rationalize and devolve authority from the monarch in order to define legal processes on a contractual basis. Most of this chapter focuses on the important figure of John Selden, who has recently been understood to advocate a kind of Hebraic Republicanism in this period, but in fact was much more broadly concerned over issues of how laws and technologies translate and define legal ownership or dominion. The seas for him were historical sites of negotiations and technical encounters—"closed" through a dense mesh of historical contract-making rather than open and uncharted as his Dutch counterpart Hugo Grotius understood them. Selden's collection of books and manuscripts from around the world as well as the map of Chinese shipping routes and the geomantic compass he obtained suggest how broader investigations into translating languages, laws, histories, and technologies in this period by Thomas Bodley, Samuel Purchas, William Laud, and many others gave Londoners a sense of alternatives to asserting dominion imperially and according to Dutch natural law strategies. At the same time, such investigation of legal and contractual agreements about trade and conceptualizing commerce more generally in East Asia helped shape commercial legislation like the first Navigation Act (1651), which made historical rather than imperial claims on Atlantic trade.

Chapter 4 examines the emergence of an image of absolute state authority during the Restoration under Charles II, which served to hide the complex foundations and settlements of authority that had been developed in London over the previous decades. This image was produced both through a series of new negotiations over two Navigation Acts and through important court cases that defined both a free trade in bullion and monopolies on imported East India goods and slave trading with the new American colonies. Clear failures of this image, however, began to appear early in Charles II's reign in the form of messy court cases and failed royal projects overseas. Both the East India Company and the burgeoning printing and engraving industry in London stepped in to fill these gaps in sovereign authority even more aggressively than Charles II and his court, further increasing the importance of London and its global networks. Particularly important was the work of John Ogilby, who translated a series of Dutch atlases and travel accounts about the new Qing Empire and composed a set of roadmaps inspired by those itineraries. The Qing could be used to demonstrate continuity after civil war because of the ways that it employed the tools of creating an image of absolutism, including road and postal systems. At the same time, developing relations with Banten on Java and the Ming-loyalist merchant-pirate dynasty of the Zheng family on Taiwan suggested models for constructing state authority on a commercial and emporial basis. Towards the end of Charles II's reign, the failure of East India Company projects in Banten (1682) and Taiwan (1683) as well as revolution in Bombay (1683–84) and military collapse in Tangiers (1684) in turn reflected the superficiality of this image of absolutism. Despite efforts by Charles II and James II to build a slave-based empire in the Caribbean and the North American colonies from the Carolinas to New York, failures in Asia indicated that London's situation had much more in common with commercial city-states like those at Banten or small states connected with shipping networks like Siam or the Ming loyalists on Taiwan than with the vast Qing or Mughal Empires or the assertive French and Dutch.

Chapter 5 concludes by analyzing the Asian roots of the Glorious and Newtonian Revolutions in politics and science, what many have seen as "the birth of the modern." The collapse of East India Company projects in the early 1680s suggested the need for new political forms. Qing success against both the Zhengs on Taiwan and the Russians in the Amur River basin as well as technical and philosophical aspects of the Chinese language became subjects of intense critical scrutiny. At the same time, experiments in hybrid forms of sovereignty in Siam and various Indian ports and attempts by the East India Company to wage war in the Indian Ocean from London suggested the need for radically new understandings of translation. Thomas Hyde and Shen Fuzong in particular suggested that Indo-Persian traditions as transmitted in part through the Farsi community at Surat deserved greater attention than the degenerate Qing, where the meritocracy of the exam system and bureaucracy had become a kind of game. Into this debate came both Isaac Newton's new "system of the world" and the new understandings of sovereignty in the Glorious Revolution resulting in a fiscal-military state centered on London. Rather than the central and increasingly territorial authority of the Qing Emperor, so admired by the French Jesuits, the longer lessons of the devolution of authority in the commercial networks, port cities, and domestic collections in maritime Asia during the Ming period as well as the more recent examples of Bombay and Siam suggested the need for new kinds of mixed authority and mathematical understandings of multiple centers of political and economic gravity. Rejecting absolutism was not just a rejection of France, Catholicism, or even the newly perceived despotism and elitism of the Qing. London's connection to networks of Asian trade had proven more essential than the burgeoning slave colonies of the Atlantic world—a longer tradition of translation and commerce as it had been defined since the 1550s rather than absolutism as it had emerged during the Restoration. The way for Newton's mathematical "system of the world" and the Parliamentary choosing of the Dutch king William of Orange had thus been prepared.

The principal argument in this study is that cultural encounter does not have to be merely relativizing or productive of a kind of negative distance, what is often lumped under the term "Orientalism." More often than not it involves a kind of dynamic entanglement. London between the 1550s and the 1680s is a good example of a fertile era of translation brought about by an equally dynamic period in Asian commerce and indeed the global silver cycle that moved American silver into the East Asian and Indian Ocean economies. Translation in London was a sign of vitality rather than decline, a key to innovation rather than a fetish for ancient and pre-Babel languages. Translation was also a historical process, and unlike broader ethical stances like "cosmopolitanism" it emerged through particular relationships and in some cases, such as the Native American and African slave trades, a kind of exclusivity. In general, efforts towards translation were far greater in relation to Asia than in relation to Africa or the Americas. However, the basic argument of the book in terms of translation is about its multipolar, decentralized, and in some ways sequential occurrence across particular historical contexts. In this way, translation prepared the ground for broader processes of Enlightenment and the political, economic, and scientific "revolutions" of the late seventeenth century. It is not clear that as a whole early modern Londoners were particularly cosmopolitan in their outlook, but in some cases proactively and often by necessity Londoners translated, and in many ways did so successfully, to make their city work in dynamic and innovative ways.

CHAPTER ONE

The Global Corporation

1553: THE JOINT-STOCK COMPANY

In 1549, Sebastian Cabot, the former *piloto mayor* and cosmographer of Spain and future founder of the Cathay Company, partnered with Clement Adams to publish London's first printed world map. Cabot knew part of the coast of South America and his father had explored in North America, but the rest of his information came from Spanish and Portuguese maps or was largely speculative, especially in East Asia. He prominently displayed the Great Khan of Cathay seated on a throne as well as the fabled island of Ciapagu (Japan) in the blank spaces between Baja California and the south China coastline (figures 6 and 7). The map suggested a kind of continuity, as Sebastian returned to England to fill out the legacy of his father. John Cabot, the memory of whom may have been dim some fifty years on, had reached the rather explicitly named "new founde land" or on Sebastian's map "*prima tierra vista*" in several voyages from Bristol under a royal patent that lasted from 1496 to 1500. This time Londoners would support the new enterprise of his son Sebastian in 1553, as investors in a privately organized joint-stock rather than royally patented voyage.[1] It was a profoundly important moment for the city, in which citizens from old medieval guilds, who normally invested in shipping across the Channel to Antwerp, came together to fund an explicitly global project. It was also a sign that the informational world of London had begun to change.

In the 1550s, "Cathay" hovered on the verge of fiction, a term not ready to disappear from cosmographies and maps but increasingly crowded out of the picture by more tangible places.[2] The accounts of Marco Polo and Sir John Mandeville as well as papal legates to the Mongols popularized the idea of the existence of the distant realm of Cathay during the thirteenth

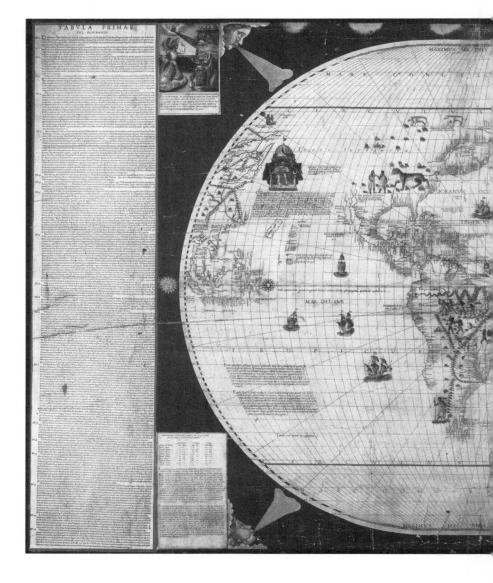

Fig. 6. Sebastian Cabot, *Mappa Mundi* (Antwerp ? ca. 1544–8).
BnF, cartes et plans, rès ge. AA 582.

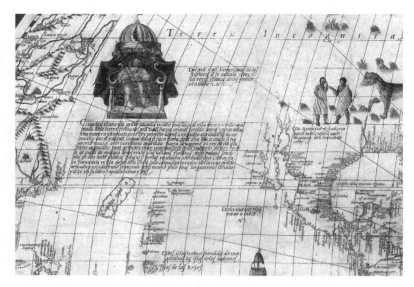

Fig. 7. Detail, Cabot, *Mappa Mundi*, upper left corner, with Cipangu, the upper Marianas (the "Ladrones"), and Baja California. BnF, cartes et plans, rès ge. AA 582.

and fourteenth centuries, describing chains of cities across the landed routes of the Silk Road. This was still true at the turn of the sixteenth century in London; Mandeville in particular became quite popular with the new printing industry, with at least three editions of the *Travels* appearing in London by the major printers of the day between 1496 and 1503.[3] But the term "Cathay" did not come from Europe—the Mongol *Secret History* (13th–14th century) made reference to Khitad for the Liao Dynasty (907–1125, Qidan Guo 契丹國), and this usage had passed into Russian (Китай or Kitay) while Persian (ختن Khitan) as well as Chagatay, from which Uzbek and Uyghur emerged in the fifteenth century, retained earlier usages and applied them to the Jurchen Jin, 1115–1234 and later peoples.[4] This also gave it a kind of traction through translation, as servants of the Cathay and later Russia Company would hear such usages in Russia and Persia and transmit the stories back to London, keeping hopes alive well into the later years of Elizabeth's reign for a northern passage to the cities of East Asia. In fact, Cabot through his map, his company, and the connections it made with Central Asia should be understood to have revived a term that his father had largely taken out of English geographical discussions with the "New Found Land" in what became the Americas.

It is thus wrong to think of Sebastian's company as merely looking backwards towards medieval cosmography, the way that Valerie Flint has suggested Columbus and John Cabot understood the world.[5] The Cathay Company's initial privateness despite a subsequent royal charter from Mary and her husband the future Philip II required, because of its multiple shareholders, a greater degree of confidence in and knowledge about the enterprise coming from a broad base of merchants of the city. The detailed map, which helped generate this confidence, also required networks gathering and translating contemporary information—from the Spanish Casa de la Contratación in Seville and the Portuguese Casa da Índia in Lisbon as well as French chartmakers in Dieppe and German cosmographers—rather than a simple revaluation of medieval texts. Finally, behind the investment in the map and the investment in the company was a much clearer sense perhaps not of the destination, for which Cathay was almost a placeholder, but certainly of the routes and processes of exchange themselves—the cities of the Indian Ocean and Southeast Asia visited by the Portuguese and the silver exchanged at Antwerp that offered tangible evidence of the significance of Asian maritime commerce. One of the most compelling aspects of Cabot's map would have been the density of names along the coasts, data gathered from Cabot's own archives as well as the world map at the Casa de la Contratación and the French chartmakers at Dieppe. Valuations in London were based on emerging global cycles of trade in the 1540s, and these could only be understood as translations formed like the chains or ropes on ships, composites of threads and links from the uncertain port cities of East Asia. Cathay was thus a speculative element that brought together people and information in what natural scientists call a "polythetic" manner—precisely a non-Aristotelian class or concept in its diverse composition out of various threads of exchange and history.[6]

Cabot's ability to compile global maps and to revive a term like "Cathay" suggests his ability as a translator par excellence. At a personal level, he retained ties to two monarchs and claims to rights in four cities, multiple subjectivities and allegiances. The founding of the Cathay Company seems to have been an effort to capitalize on the expertise that Cabot brought as the Spanish *piloto mayor* from Seville while at the same time resolving some of the difficulties created by Cabot's uncertain status as a retainer of the Holy Roman Emperor Charles V and a Venetian citizen. Invited to England by the Privy Council in 1547 during the transitional year of Henry VIII's death, Cabot first traveled to Bristol in 1548, where he had spent time as a youth with his family. Still in Spanish employ, Cabot sought and obtained a series of documents and funds from the Privy Council that would establish his presence in England, including a previous pension as "grand Pilot" (dated January 1548 and beginning September 1548) and travel warrants (from October 1547 and September 1549), moving to London by

1549. At the same time, in an attempt to keep Cabot in Spanish service, Charles V gave a flurry of last-minute orders in October and November of 1548 for payments and privileges directed to Cabot after Cabot came to visit him in Brussels that summer.[7] The emperor then began to demand Cabot's return in January 1550, especially after he printed the 1549 world map, which copied many of the secret details of the one used by the Casa de la Contratación. Protector Somerset claimed that Cabot had been discharged from the emperor's service and that he was now an English subject although free to return to Spain. Cabot had new letters patent from the English archives drawn up about his father's discoveries to replace ones that had been issued in the 1490s.[8] Meanwhile Cabot tried from September 1551 to claim Venetian citizenship with the help of the travel writer and secretary to the Senate, Giovanni Baptista Ramusio, so he could claim rights outside of his service to England and Spain and potentially share in the profits of his father John's discoveries. Sebastian continued until Charles V abdicated in 1555 to communicate to the emperor's representatives that he wished to remain in Spanish service and had important information to convey.[9] At that point it no longer mattered so much, for during the final two years of Cabot's life Philip, the future king of Spain, was also the king of England.

Cabot's Cathay Company, later under royal charter from Mary called "the Merchant Adventurers for the discovery of regions, dominions, islands and places unknown" and finally after its initial voyages shortened to the Muscovy or Russia Company, was the first private long-distance trading organization not based on dynastic, religious, or familial principles.[10] When on May 21, 1553, the optimistically and Italianately named ships, *Bona Speranza, Bona Confidentia,* and *Edward Bonaventure*—hope, confidence, and good fortune—set sail for Cathay, Clement Adams wrote that having sailors dress in blue and climb on the rigging of the ships had been for the benefit of the fifteen-year-old convalescent king Edward VI at Greenwich.[11] The expedition, however, had been organized without him. The weakness of Church and King as signs, especially compared with the activities of Henry VII and Henry VIII in this regard, was a key element to the founding of the joint-stock. The ships were a pageant for London, elaborate and expensive civic displays of unity like the ones that Henry VIII began attending during Midsummer's watch in Cheapside in 1510 and then banned after 1539.[12] The Lord Mayor Sir John Gresham had restarted such pageants in 1548, the year before Cabot arrived, but the ships were an even more dramatic demonstration of the ability of the merchants of London to surpass the achievements of the monarch. They launched from the Deptford royal dock complex, constructed in 1513 by Henry VIII. But designed for long-distance commerce, rather than fighting along the shallow coasts, they must have made the war-ships of the Royal Navy seem obsolete. The English navy's "narrow seas" navigational capabilities using soundings and "coaster" (sight of coastline) methods kept it south of the Baltic and north of the Bay of Biscay even with the expansion of shipbuilding in the 1540s.[13]

Over the previous century, some of the ground had been laid in London for such a change. Large-scale shipbuilding, printing, and other new indus-tries all made the enterprise possible. Increased collection of data about the city by ward officers and the building of more urban infrastructure, whether roads and water or almshouses and hospitals, had slowly encouraged new kinds of civic institutions and cooperation across guilds. This longer history helps explain the flurry of new civic institutions that appeared after the dis-solution of the monasteries and the death of Henry VIII, notably Bridewell prison (1553) and the four hospitals—Christ's for orphans (1552–53), Beth-lem for the insane (1547), and St. Bartholomew's (1547) and St. Thomas's (1551) for the sick (figure 8).[14] But even though the population and civic and financial resources of London may have reached a kind of tipping point by 1549, western ports like Bristol and Plymouth seemed in many ways more dynamic. London was still dominated by both foreign merchants into the 1550s and 1560s—the powerful Hanseatic League and Italian merchants—and the Merchant Adventurers trading almost exclusively with Spanish Antwerp.

But the Cathay Company was a group that cut across the various guild and government institutions of the city. Over two hundred investors, called sometimes a "company" and at others a "mysterie" (mastery, another name for a guild), organized in fact as a *societas* or contractual partnership. They used an equally weighted joint-stock, all without any corporate charter like the Merchant Adventurers, who traded with Antwerp.[15] Despite his recent arrival from Spain by way of Bristol, Cabot garnered the support of the Lord Mayor Sir George Barne and London's sheriff William Garrard, both mem-bers of the Haberdasher's guild. The group also had a number of aristocratic and courtly investors (the Captain General Hugh Willoughby was a knight), but three-quarters of the investment came from London merchants, espe-cially those from livery companies not directly involved in the wool cloth trade, the principal staple of the city. In a period of fluctuating prices and demand for English cloth at Antwerp and when Spanish wealth through silver and Portuguese through spices had begun to dramatically shift ex-change practices on the Continent, after "much speech and conference to-gether" "certain grave citizens of London, and men careful for the good of their country, began to think with themselves, how this mischief might be

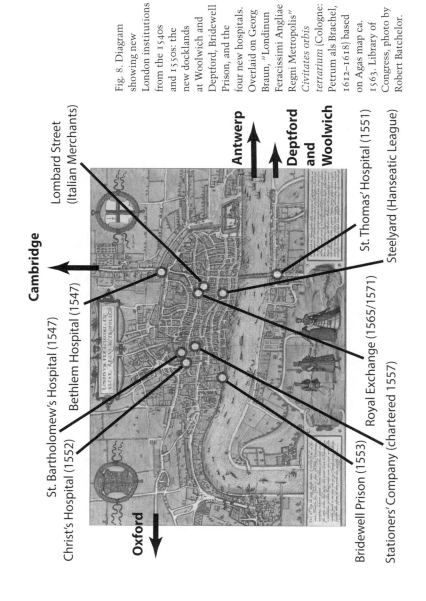

Cambridge

St. Bartholomew's Hospital (1547)

Christ's Hospital (1552)

Bethlem Hospital (1547)

Lombard Street
(Italian Merchants)

Antwerp

Deptford
and
Woolwich

St. Thomas' Hospital (1551)

Steelyard (Hanseatic League)

Royal Exchange (1565/1571)

Bridewell Prison (1553)

Stationers' Company (chartered 1557)

Oxford

Fig. 8. Diagram showing new London institutions from the 1540s and 1550s: the new docklands at Woolwich and Deptford, Bridewell Prison, and the four new hospitals. Overlaid on Georg Braun, "Londinun Feracissimi Angliae Regni Metropolis" *Civitates orbis terrarium* (Cologne: Petrum als Brachel, 1612–1618) based on Agas map ca. 1563. Library of Congress, photo by Robert Batchelor.

remedied."[16] This new "thinking with themselves," which fundamentally involved a new kind of thinking with others, marked a key shift in the history of the city.

Edward and the Privy Council could have given the Cathay Company a charter, as they did the Bristol Society of Merchant Venturers the year before in 1552; but they did not do so, probably reflecting the political and even religious uncertainty of the period. Unlike the patented expeditions of Sebastian's father John to the "Newfoundland," which reportedly planted both English and Venetian flags in a kind of double claim, the Cathay Company could initially make no claims to territorial conquest or missionary efforts.[17] Edward did write a letter of introduction, carried by the expedition leaders Hugh Willoughby and Richard Chancellor in Latin, Greek, and "divers other languages," but it was addressed ecumenically to "all kings, princes, rulers, judges, and governours of the earth, and all other having any excellent diginitie on the same, in all places under the universal heaven." It described in rather leveling terms the proper function of merchants as moving "good and profitable things" to places they would be "commodious," creating mutual dependencies in which "every one seek to gratifie all."[18] This notion of mutual gratification was an urban ethic, echoed over the next century and a half, one derived from the contractual basis of the "society" itself as well as the need to make trading agreements with foreign princes in the absence of significant direct support from the monarchy.

The "Ordinances for the direction of the intended voyage for Cathay," dated May 9, 1553, less than two weeks before the launch of the ships, was the most remarkable document to come out of this process.[19] It suggests that Cabot and his investors understood that a voyage to Cathay required a new practice of the person and the subject, one that he referred to as "worldly." Sailors needed to regulate their behavior "not only for duty and conscience's sake towards God . . . but also for prudent and worldly policy, and public weal, so to endeavour yourselves as that you may satisfy the expectation of them, who at their great costs, charges and expenses, have so furnished you in good sort."[20] Those writing the ordinances assumed that rather than encountering a single sovereign entity or commercial possibility, any voyage would involve multiple exchanges. Long-distance trade for a society of merchants depended upon the building up of exchange relationships in a serial or horizontal manner along routes and through the careful avoidance of possible controversy and conflict.

The Cathay Company voyage would not be an expedition like that of Columbus or even Sebastian's father John, which extended religious and

dynastic authority, and the investors explicitly set out to limit such dangerous claims in relation to trade. The charter deliberately avoided London's
sectarian problems that had emerged in the aftermath of the Henrican Reformation. Cabot gave strict orders, "Not to disclose to any nation the state
of our religion, but to pass over it in silence, without any declaration of it,
seeming to bear with such laws and rites, as the place hath, where you shall
arrive." "Our"—given Cabot's own ties to Catholic Spain and Venice as
well as the uncertainty and contemporary conflict over religious issues in
England—was as fictional as "Cathay." Empty religious institutions stood
as monuments in London to the possibility of confiscation of property held
collectively on "superstitious" grounds. For much of the 1540s and into
the 1550s, with uncertainties after Edward died in July of 1553 surrounding the succession of Lady Jane Grey and then Mary to the English throne,
laws confirming the property ownership of both trading guilds and London corporations left the continuation of ownership potentially a matter of
royal prerogative. The mortmain rights of any group of people as a *societas*
remained limited, apparent, and contingent. Even the legislation in 1547 at
the beginning of Edward's reign, which confirmed the exemption of companies and guilds regarding seizures, implied the possibility of change by
its articulation of such exemptions.[21] Because of its particular goals and
translation practices, the Cathay Company needed to demonstrate a degree
of unprecedented independence from religion and religious debates. This
was compounded by the particular historical moment of its first voyage—
notably May 1553, at the time of the conflicted succession among Jane,
Mary, and Elizabeth. It was also in direct contrast with the more activist
and contentious religious policies of Spain and Portugal. The Jesuit missionary Francis Xavier had just died on an island off the coast of China in 1552
after supposedly converting more people in his travels from India to Japan
than anyone since Paul.

First exchanges as a moment of radical uncertainty and translation in
relation to both sovereignty and religion thus became for Cabot particularly
important

> For as much as our people, and ships may appear unto them strange and
> wondrous, and theirs also to ours: it is to be considered how they may be
> used, learning much of their natures and dispositions, by some one such
> person, as you may first either allure, or to take to be brought aboard
> your ships, and there to learn as you may, without violence or force, and
> no woman to be tempted, or entreated to incontinency, or dishonesty.

Translation between collectivities of people without agreed-upon universals, "theirs" and "ours," had to take place through the individual *persona*. Any person, regardless of their religious predilections or citizenship, had to be prepared to serve in this role, which would involve a double estrangement in which each side appeared "strange and wondrous." The commodity itself thus became the principal site for translating desires, and Cabot explicitly recommended getting the person encountered drunk in order to "know the secrets of his heart" and with the hope that they will "allure others to draw nigh to show the commodities." At the same time, and through the metaphorical institution of fraternity ("brotherly love") and manners, the company's servants would hide any indication of their own desires or opinions. "Disdain, laughing, contempt" about foreign customs must be tempered by "prudent circumspection, with all gentleness, and courtesy."

Should this strategy be understood in any way differently from the Italian courtly ethics of Machiavelli or Castiglione, the secular dissembling and opportunism of the Prince or the polite conventions of the courtier, ambassador, and diplomat, that Stephen Greenblatt has called Renaissance "self-fashioning"?[22] The *persona* was after all a Roman trope, literally derived from Etruscan theater masks but usually referring to the voice used by satirists like Juvenal as a kind of disguise. But Cabot's own *persona* as governor of the Cathay Company stemmed from his ability to pull together a wide variety of contradictory sources of authority and knowledge into a corporate persona that had the potential not only to disguise religious motivations but also to translate the unfamiliar. In this sense, the *persona* of the "Ordinances" was based on the idea of incoherent identity within a group and between groups, implying the fundamental necessity of translation not only in Asia but also among London sailors and merchants themselves.

The most significant Mediterranean legacy came not from the courtly life but from trading practices. Sebastian's father's own early work in the Eastern Mediterranean for both Genoa and Venice involved cross-cultural *commenda* deals with merchants in the Burji Mamluk cities of Alexandria, Beirut, and Mecca. According to the Milanese ambassador to London, John Cabot had learned that the Arabs did not know where spices grow, but that "caravans come to their homes with this merchandise from distant countries, and these caravans say they are brought from other remote regions." Things passed in a chain "from hand to hand" so that the last people to receive them are those in "the North toward the west." The Bristol voyages of the 1490s were intended to "establish in London a greater storehouse of spices than there is in Alexandria."[23] The customs registers of Henry VII, unfortunately fragmentary for London and Bristol, are nevertheless filled with the records of commenda-like partnerships among denizens and aliens. Bristol had in particular become since 1480 a center for exploration as well as collecting and translating books and manuscripts, drawing on Portuguese speculations about the "Island of the Seven Cities" from the 1450s and the Icelandic fishing voyages sponsored by denizens. But the legacy of John Cabot's voyages died out in the latter years of Henry VII's reign, and Portuguese and French sailors exploited the cod fisheries of the Outer Banks and the potential for fur trade far more than Bristol.[24]

Meanwhile, in most of England during the early sixteenth century, outside of the special needs of places like Bristol for North Sea fishing, the guild-like fellowship used its power to either shut down or keep out entirely the *commenda* and its derivatives. This was especially true of London's Fellowship of Merchant Adventurers of England (1407, chartered 1505), which derived from the earlier efforts to monopolize the wool trade beginning with the Merchants of the Staple (est. ca. 1240–1260s).[25] Technically headquartered in Antwerp, where cloth prices were set and currency exchanges negotiated, it effectively outsourced its translation needs through Antwerp's networks and practices. By the 1540s, more than 80% of all goods that passed through English customs went through London, and most goods going through London went to and came from Antwerp on Merchant Adventurer ships.[26]

So Cabot's company was a truly distinct intervention—a translation of practices from Italy and ultimately the Mediterranean trade with Asia that had few precedents in the world of London merchants. At the same time it used the quasi-fictive Cathay to open up a vast space of potential translation and trade far beyond the ambitions of most London merchants and the court. At the core of the birth of the joint-stock company was the threefold problem of how to translate between mercantile and religious factions within the city, how to translate between citizens who were guild members and those like Cabot who came from other mercantile cities in Europe, and then finally how to translate from that regional European context into a broader set of relations with the commercial cities of Asia. The joint-stock with its abstraction into shares was an ideal form for such translations, making investors into equal participants and separating the company as much as possible from the sectarian and dynastic concerns of church and king. But it was Cathay that grounded the possibility of value distinct from church and king and the claiming of lands with crosses and flags, a value that in fact resided in the chain of exchanges itself.

The actual 1553 journey that set off to reach Cathay by the northeast was a story of accidents and disasters. Yet the fiction of the company and its search for Cathay held. Only the *Edward Bonaventure* with Richard Chancellor as captain reached Russia, anchoring at the Northern Dvina River in the White Sea. The crews of the other ships, including Sir Hugh Willoughby, froze to death just before the Murman coast on the Kola Peninsula. Chancellor received an invitation to proceed to Moscow, where he met with Ivan IV. The Romanovs were in the process (1552–56) of conquering the khanates of Kazan and Astrakhan in the south, shattered remnants of the long-dissolved Mongol Empire, which would give Muscovy access to both the whole of the Volga and the Caspian Sea. Chancellor described Moscow as larger albeit poorer than London, dominating vaster territories, and he made sure to inquire about possible connections to Cathay.[27] Richard Eden, the first significant travel writer in English during the age of print and the great propagandist of the voyages, interviewed Chancellor after his return in the summer of 1554. Chancellor explained that the "Duke of Muscovy" (or "Emperoure of Russia") had said "there could not any navigation bee imagined so commodius and profitable to all Christendome as this might bee yet by this way the voyage woulde be found open to India to come to the rych contrey of CATHAY which was discoursed now two hundredth yeares sense by Marcus Paulus." The czar supposedly added that a northeast passage would be much shorter than either the Portuguese or Spanish route.[28] Ivan had thus confirmed the older manuscript and print stories about Cathay, giving it a new Russian reality and a new possible set of networked relations.

The reported words of Ivan convinced the Catholic Queen Mary and her new Spanish husband Philip, the future king of Spain who had just arrived in England with a large entourage in July 1554, to give the first charter to a London long-distance trading company in February 1555. John Lok, who left London for Guinea in October 1554 and returned with gold, pepper, elephant's teeth, and five slaves that spring, received nothing of the sort, the Spanish having no interest in the English supplying slaves to the Caribbean. Unlike the "Ordinances," the royal charter gave the company traditional monarchical and Catholic authority "to subdue and occupy all cities, etc. of infidelity newly found by them as the vassals and subjects of the king and queen and to get the dominion and jurisdiction of the same to the crown forever."[29] Even if Philip had no real powers as an English king and the charter assumed that English sovereignty would like Antwerp operate as a kind of tributary realm to the great Spanish Empire, it is still surprising that he and Mary supported this project. The goal seems to have been to convert Cabot's company back into a traditional entity. Chancellor did not survive the return from a second voyage to Russia, but the czar's envoy

Ossip Nepeja sent back on that voyage did make it to London. In 1557, the same year Cabot died, Mary and Philip (technically as of January 1556 also Philip II of Spain and Duke of Burgundy but still residing in London after his father's abdication) sent a formal letter to Ivan to confirm commercial relations and request guaranteed safe passage to Persia.[30] This encouraged the new and shorter name Muscovy or Russia Company. Yet despite Philip's and Mary's attempts to co-opt the company, the joint-stock as an institution aimed at translating commerce transitioned smoothly into the reign of Elizabeth, and its charter subsequently served as the basic model for English long-distance trading companies generally.[31] Even if it did not reach Cathay, the company had proven adept at building a strikingly complex set of translations between very different Christian sovereignties—Russia, England, and Spain and Burgundy. It had opened the potential for translation through Russia of agreements with the Safavids in Persia and the Shaybanids in Bukhara, where Anthony Jenkinson would go in the early years of Elizabeth's reign (figure 9).

REDEFINING THE TRANSLATOR

Mid-sixteenth-century London was rife with highly localized and weighty religious debates over translation, and Cabot with his links to Spain and Venice had to tread carefully. In sixteenth-century Europe, most translation had at its basis a religious rather than an economic motive—an expectation of fulfillment, the fulfillment of the word of God. Cabot's image of the annunciation in the upper left corner of the map he printed for Charles V before coming to England retained this sense of expectation. Likewise, the letter carried by the first Cathay Company voyage from the pious boy king Edward VI, with its suggestion that "every one seek to gratifie all," was as Christian and humanist as Thomas More's *Utopia*. Translation mediated in order to magnify the true word, as in the case of Mary's annunciation and "magnificat" (Luke 1:16, "my soul doth magnify the Lord") depicted prominently on Cabot's map. The good interpreter could, as in Wyclif's translation of the oft-cited Daniel 5:16, "interprete derk thingis, and vnbynde boundun thingis," opening up and enlightening the unseen world.[32] Yet if translation held great hopes, it also posed equally great dangers. In the 1540s and 1550s in London as well as Oxford and Cambridge, proper translation of language, institutions, and authority like the Prayer Book became a central issue. It is remarkable in such a contested environment that exploratory projects by a Spanish-Venetian defector got any traction whatsoever.

By 1553, when Cabot wrote the Ordinances, translation in relation

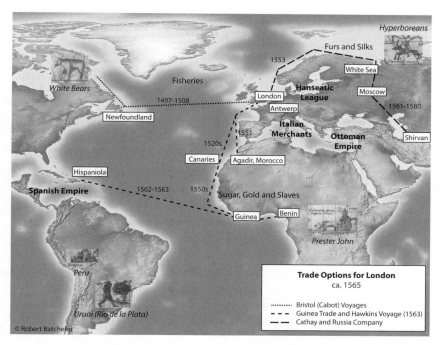

Fig. 9. North Atlantic/Western Asia map showing trading options and routes for London ca. 1550–1565 with insets from the Cabot map.

to religious issues had divided London. Most prominent was a moderate humanist position associated with reforms at Cambridge between 1535 and 1546. This group was open to translation of Protestant ideas from Germany but wanted to limit how much this should shape the institutions of the church or kingdom. Cambridge molded a generation of people who would be important in Cabot's enterprise—including the travel compiler Richard Eden.[33] Sympathetic to the polyglot scholarly culture of Antwerp as well as the English gentry's ties to the cross-Channel cloth economy, this group generally weathered the shifting dynastic and religious storms of the mid-sixteenth century quite well, dominating the increasingly important Privy Council and the Royal Household. And such ideas were spreading. Richard Eden's translations of the Portuguese humanist Damião de Góis (1502–74) on Ethiopia and the Lapps in *Decades of the New World* (1555), suggested broader debates about the definition of Christianity. Likewise the first English translation of More's *Utopia* (1551), which the livery companies supported for the "publyque weal," offered a vision of global networks of "people, cities and towns, wherein is contynuall entercourse and occupyinge of marchandyse and chaffare, not onelye amonge themselfes and wyth theyre borderers, but also wyth marchauntes of farre contreys bothe by lande and water."[34]

But almost immediately from the 1540s a strong conservative position developed from within the church and university arrayed against these moderate humanists. A strain of English "Catholic" orthodoxy exemplified by Stephen Gardiner, later responsible for rendering Philip toothless as king of England, paradoxically advocated royal supremacy over the church and the limitation of translation in deference to more local tradition. His was the old dense Catholicism of place in London, where within the city walls in the 1520s there had been over one hundred parish churches, a cathedral, and thirty-nine religious houses, a highly local world of neighborhood loyalties that valued such local ties over humanist and mercantile worldliness.[35] Even Thomas More's late writings, notably the *Dialogue of Comfort against Tribulation*, written in the Tower in 1534 but not published until 1557, could be used to buttress Gardnier's position against more urban humanist texts like *Utopia* that played off of Antwerp's cosmopolitanism. Amidst a series of disparaging references to the "worldly," More had warned against "scattering our minds abroad about so many trifling things," and advised "withdraw[ing] our thought from the respect and regard of all worldly fantasies, and so gather our faith together into a little narrow room."[36] For Gardiner the "little room" was the king-in-parliament, which had important differences from Somerset's learned humanist Parliament or later Cardinal Reginald Pole's learned Church. Like the late repentant work of More, Gardiner tried to avoid the problem of translation and "worldly fantasies."[37] The conservatives remained highly suspicious of any translation—whether Protestant books coming across the Channel from Germany or the work of Catholic scholars from the Mediterranean.

Countering both the humanists and the conservatives, radical Protestants emerged, especially among the preachers of London, who saw translation not as interpretation but in the evangelical mode of those who spoke in tongues, expounding or declaring the truth. They rejected both Catholic and Protestant humanist gestures towards a broader and more tolerant European intellectual culture. Supportive of translating the Bible itself into English, they placed little emphasis on the practice of translation or interpretation itself, which they associated with "priestcraft" and the mediation of the Church. To them, both the humanists and the conservative Catholics like Gardiner seemed "wily" with language.[38] They supported iconoclasm and altar stripping. Book burnings like that at Duke Humfrey's in 1549 seemed to be an effort to placate them. They attacked "Eastern" predilections whether Catholic or "pagan" towards idols and mediation, fearing the assimilative consequences of idolatrous kings, notably the loss of dynasty and becoming intermingled in Assyrian cities, but like Gardnier their main concern was autonomy.[39] William Tyndale's English Bible, first printed in Cologne and Worms (1525–26), finished by John Roger using the translations of Myles Coverdale from the mid-1530s in Antwerp, and only published in England as the Great Bible (1539, translation revised six times by 1541), appealed in part because it replaced the Latinate figure of the interpreter with the direct speech of the "expounder," who "hath the holy sprete of the holy goddes within him." The direct speech of true prophecy, the expounding and declaring of the word, transcended translation, and thus did not need to worry about the scholarly limits set on translation by both the humanists and the localists.

Cabot's group came from a fourth faction in the city, those willing to keep their religious views to themselves in order to escape such local translation problems. For them, inflation, currency devaluation, and troubles in the cloth trade with Antwerp pushed translation in even more radical directions away from Europe and the standard range of Christian humanist languages (including Arabic). Through Cabot they were ready to make deals with Cathayans and more directly the Russian and Safavid empires. Somewhat paradoxically, the more openly universalist and engaged Catholicism of Mary, itself directly linked with Philip II's Spain as well as papal reformers like Cardinal Pole, proved as fruitful for the survival of this fourth faction as did Elizabeth's early ambiguity about more radical strains of Protestantism and relations with Catholic Spain. The solution for translating had to be double—to keep one's skeptical and local religious views on translation and interpretation to oneself and at the same time to recognize that in a world of widely differential languages and desires informed by complex traditions of sacred languages, translation was necessary at the level of exchange. But if neither the individual prophetic self nor the institutions of religious authority could be trusted to carry out such translations, who or what could?

Cabot and his backers thus came up with the idea of a *societas* or company that could regulate witnessing and confession. To give it value, he had to connect that organization with the external exchange and translation values of an unknown but not Utopian destination—Cathay. Oliver Williamson has argued that the key to the contract at the basis of the firm is "continuing interests in the identity of one another" rather than the rationality of its own internal legal or contractual structures. Traditionally, this meant a relationship between the subject and the king, as John Cabot at least partially framed his relation with Henry VII or more commonly the practices of citizenship through guild membership. The same could be said of the relationship between the pilgrim like Sir John Mandeville and the church. In the case of Cabot's enterprise, and indeed later instantiations like the Levant or East India companies, the firm is defined by the need to maintain translation between the *societas* and Cathay.[40]

The personas articulated by the textual traditions of Marco Polo and Sir John Mandeville, who used the Great Khan of Cathay as the model and focus of Asian translation, remained important myths in defining how to engage in exchange relations. But these were articulated by Christian pilgrims operating under the authority of the Church. In the "Ordinances," Cabot made the religious figure of the *persona ficta*, one who interprets and translates authority in the name of the church, return to its Latin linguistic roots as *interpres*, an agent mediating between prices and desires.

The figure of the mediator, who as the go-between still figured prominently in Asian trade, had in many cases disappeared in Christian Europe, often due to explicit policies like the expulsion of Jews from London between 1290 and 1656. The *persona ficta*, an institutional mediating body, had been the medieval Church's solution to the interrelated problems of translation across sovereign spaces, institutional autonomy, and duration over time. Introduced in 1243 by Sinibald Fieschi and developed in bulls between 1245 and 1247 once he became Pope Innocent IV, it shored up the Church's authority in relation to challenges by the Holy Roman Emperor, Byzantium, Muslims, the Mongols, and Church prelates themselves, the "five wounds" of the 1245 Council of Lyon. The creation of the *persona ficta* thus coincided with the earliest stories of Cathay and the arrival of the Mongols, who invaded Poland and Hungary in 1241. Innocent sent the Franciscans John of Plano Carpini and Benedict the Pole as envoys to the court of Kublai Khan in 1245, about fifteen years before the Polos supposedly set off from Venice. Much of Plano Caprini's text was taken up with the practical problem of translating Innocent IV's Latin letters into Mongolian by way of Ukrainian interpreters, and the letters explicitly referred to them as substitutes for the actual presence of the Pope as *legate missi*. According to Innocent IV's formulation, the church was a *universitas* and as such a *persona* in Boethius's sense of the word. It could create a fictional person, an artificial or juristic "person" out of its authority, one that could both translate and possess things under its proper name.[41] Any creation of such a *persona* relied upon the authority of the Church. This is technically why John Cabot had planted both a cross and a flag as a way of reenacting *imperium*, the sacred and translatable authority of the church in tandem with the limited sovereignty of the king. One problem created by the Henrican Reformation was the disappearance of such a mediator, and the founding of the Jesuits and later the sending of Reginald Pole as a papal legate during Mary's reign had been an effort to address this broader problem in the Catholic Church itself.

The legal fiction of the fictive person had inspired a whole genre of medieval literature in which semifictional characters like Marco Polo or especially Sir John Mandeville set off as personas with the authority to stage negotiations with non-Christian princes without endangering the souls of either the storyteller or the reader.[42] Polo did this with the framing device of having the whole story narrated from a Genoese prison by Rustichello of Pisa, so that even if it was true, the story as a prison tale was also clearly fictive. The Mandeville story too had a framing device, set in the Abbey of St. Albans, about twenty-five miles northwest of London, where the supposed knight errant Mandeville sat writing out his memoirs. The journey towards Cathay in Mandeville was ultimately a romance about relations among cities, in particular the *urbs*, or the place of the city as a point of exchange, rather than the Christian community of the *civitas* as a point of gathering.[43] Like Marco Polo's account as well as John Cabot's report from Mecca, it contained within it the idea of a chain of cities and merchants that successively translated Asian goods into Europe without any overarching authority. According to the earliest English manuscript of Mandeville, the southern tributary of Cathay, Manci or Manzi, had two thousand great cities, including Latorym, a city "much more [populous] than Parys"; Cassay (sometimes Kinsay or Quinsay and usually taken for Hangzhou), "ye moste [populous] cite of al the world"; and Chibence, with its sixty stone bridges and twenty-mile city wall. Next to Manci lay Cathay, the emporial center of trade of the whole region, where "cometh merchauntis eche yere to feche sipicery and other merchaundiz more comunly than to other cuntreez."[44] Goods and, more precisely, values that made Venice and Genoa rich in their own right originated in Manzi and Cathay. Urban networks in Asia thus represented an alternative and powerful value system that challenged the universal claims of Christian *civitas*. As a piece of late fifteenth- or early sixteenth-century English popular verse summarizing both Marco Polo and Sir John Mandeville suggested, "the Grete Caan, Emperour of Tartaria . . . [l]eveth not on the Crystyn laye . . . [he ne holdis the Cr]istyn fay, [for ellis] the warld he might concord."[45]

Cabot, through the medium of the company, tried to reverse the medieval understanding of the world and realize Cathay as a space outside of the system of inheritances both spiritual and temporal (the papal donation

of 1493 to Spain and Portugal and the subsequent Treaty of Tordesillas).
London merchants who bought shares received no promise of return in
property in the sense of landed trust (John Cabot gave islands to his partners) or dynastic privilege secured by the monarch. They instead received
a stake in an exchange promising the reduction of transaction costs and
the best price possible, a stake in pure *interpretes* between two universal
and distinct systems of exchange—the Asian and the European. The Ordinances thus inverted More's Utopian critique of the seemingly unbreakable
relation between the enclosing gentry of the English countryside and the
Antwerp trade in cloth of London merchants. The other world was not an
impossible utopian place that had no property, as some among the gentry
would later try to claim in relation to Ireland or Virginia. Instead it was a
world of radical exchange, in which anyone willing to buy a share in a joint-
stock company could circumvent both the local control of corporate guilds
like the Merchant Adventurers and the global claims of the Iberian powers.

All that remained was to prove that such a separate value system existed
and was not itself fictive. Inspired by Cabot's company, the first significant
piece of English cosmographical literature appeared in print in London in
June 1553, just weeks before the death of Edward VI, in an effort to prove
that distinct systems of trade and exchange existed in Asia. Rather than a
synthetic cosmography or collection of travels, Richard Eden's *A treatyse
of the newe India with other new founde landes and islandes* redacted and
translated a section of the second Latin edition (1552) of the Basel scholar
Sebastian Münster's "boke of universall cosmographie" as Eden's subtitle
had it. Münster's popular book, originally in German in 1544 and then in
Latin in 1550, had arrived in London bookshops most likely in the Latin
edition in 1550 or 1551.[46] The heavily-illustrated *Cosmographia* had first
expanded in German editions (Basel: 1544, 1545, 1546, 1548, and subsequently), but the 1550 Latin translation, followed by editions in 1552, 1554,
and 1559, was the key to the work's European popularity. Upwards of 50,000
German copies and 10,000 Latin ones circulated during the sixteenth century along with five editions in French. Wanting his book to reach all of
the German cities, Münster took care to remove anything that could be
perceived as a criticism of either Spain or Catholic Universalism, charges
that the Portuguese humanist Damião de Góis had nevertheless made at
Louvain before the first edition even appeared.[47]

Eden's printer was the clearly partisan Antwerp Protestant Stephen
Mierdman, who received a special license after moving to London in 1550
to produce Protestant tracts and to "employ printers, English and foreign."[48]
But Eden did not want the whole picture as envisioned from the imperial

German cities or even German Protestant humanism nor the kinds of debates that it inspired. His translation was more targeted, largely coming from Münster's fifth book *De Terris Asiae Majoris*. Münster, framing his ideas with Strabo and to a certain extent Polo, had tried to erase such translations by alluding to the classical and universalist idea of the *periegesis* or the progression or sequential pilgrimage through the world albeit without a clear imprimatur from the church. Eden mostly skipped Münster's extensive use of Strabo, beginning instead with a section about the relationship between "Sinarum regione" (the Chinese region) and western India, depicted as two distinct trading and tributary systems.[49] He was in fact taking apart cosmography.

Eden's selection laid out a circular route connecting the emporium cities of the Indian Ocean: Aden, Hormuz (Ormus), Cambay (Khambhat in Gujarat), Narsinga (Vijayanagara), Canonor (Kannur), Calicut (Kozhikode), Zaylon (Sri Lanka), Tarnasseri (Tanintharyi on the Malay Peninsula), Bangella (Bengal), Pego (Burma), Malacca, Taprobane (Sumatra), Bornei (Borneo), Giava along with Java the greater and lesser, Madagascar, and Zanzibar (figure 10). Arab, Persian, and Gujarati merchants from the ninth century had pioneered this circular navigation of the Indian Ocean, using point-to-point compass bearings and following the currents that shot across from Sri Lanka to the Malay Peninsula (described in terms of days traveled) and then the South Equatorial Current from the Straits of Sunda between Sumatra and Java across to northern Madagascar and Zanzibar. Beginning with Vasco da Gama, Portuguese sailors had used pilots from the Indian Ocean and then the South China and Java Seas to teach them the traditional routes. According to Eden, "Turks, Syrians, Arabians, Persians, Ethiopians, Indians," along with the transregional Portuguese and Cathayans, all used the Indian Ocean system, but Cathay and "great Cham of Tartaria" with its tributary Manzi had its own system of trade in eastern Asia, which he translated as "Superior" or "High India," a division undermining the Portuguese claim to all trade in the East Indies.

These two Indias within the east—High and Low—interacted through commerce but remained separate as economic systems. In particular, paper money ("four square with ye Kinges ymage printed theron") made Cathay distinct in terms of economic practices, and the Muslim-Portuguese routes of the Indian Ocean did not directly interact with the Cathayan network of tributaries. The two Indias also allowed Eden to think in terms of challenging Ottoman domination of the Mediterranean and Red Sea, not through an imperial alliance with Portugal and Spain let alone adopting their universalist concepts, but by making connections with Persia and Cathay as cultural and

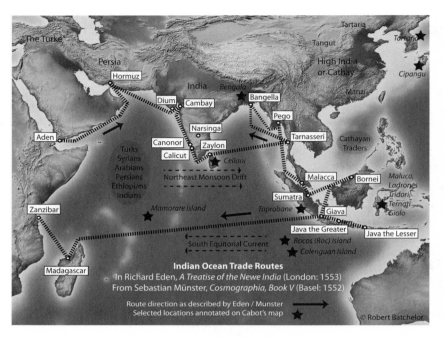

Fig. 10. Map of Indian Ocean with trade route described by Richard Eden and locations annotated on Cabot map marked.

political entities that dominated distinctly regional ocean trading systems. Politically pragmatic, in 1555, with the future Philip II soon to be crowned king of England, Eden could then argue that "there can nothynge be imagined more effectuall for the confusion of the Turke" than an alliance with "the Great Cham of Cathay and the Sophie of Persia."[50] Arguably, the same could be said of Christian humanism, for this was a formula for escaping debates about translation rooted largely in Hebrew, Greek, Latin, and Arabic.

THE COSMOGRAPHIC BREAK

Cabot himself was extremely well informed about the contemporary European debates and achievements in cosmography. He was among a handful of people in Europe privy to current data about both Spanish efforts in the Americas and those of the Portuguese along the southern coast of Ming China. His map shows that he knew for example about Chincon, the island of Xiamen (Amoy) in Fujian province where the Portuguese arrived to trade in 1541. This was one of a number of clusters of named and unnamed islands on the map where the Portuguese traded illegally and collaborated with Chinese and Japanese fishermen, traders, and pirates in the 1540s, after the bans and imprisonments of the official embassy under Tomé Pires in the 1520s.[51] But Cabot's maps of ca. 1544 and then 1549 in London did not merely translate Spanish and Portuguese geographic knowledge; they also revealed the limits and margins of the Iberian enterprises.

The only surviving copy of Cabot's printed map was composed sometime after 1544 and most likely printed in either Antwerp or Nuremberg by 1548 with an accompanying Latin pamphlet. Clement Adams either reengraved this map or made a copy using a modified set of the old plates in London in 1549, but no copies of the full London map survive, only a printed transcription of the new set of inscriptions.[52] Cabot also sent two other maps in manuscript, one of which may have been the original 1544 map, to Whitehall and Seville respectively. The now lost Whitehall map, which included a portrait of Cabot, ended up in the Privy Gallery, and it was seen in the late sixteenth and early seventeenth century by Humphrey Gilbert, Richard Willes, and Samuel Purchas among many others.[53] Most Londoners knew the printed rather than the manuscript version of the map best. Well into the 1580s, Cabot's map remained a standard visual element in the homes of London merchants and investors, giving a sense of historical direction to the efforts of the growing metropolis.[54] Through Cabot, London merchants had their own printed world map before they had a printed map of England or the city of London itself.

The classic and still largely accepted account of the development of Tudor geographic understanding by E. G. R. Taylor contends that lacking institutions and courtly support, English geography and indeed cosmography in the early and mid-sixteenth century should be seen as a series of derivative imitations of more coherent Continental theories and practices.[55] London had no central institutions like the *casas* of the Spanish or Portuguese, and it had to rely upon quite divergent practices of translation in order to bring in the work of navigators and cosmographers in Basel, Dieppe, Seville, Lisbon, Venice, and other European cities. Cabot's map demonstrates how much more cognitively complex the demands made by the joint-stock company in terms of the translation of navigational and cosmological technique were for Londoners than those required by earlier trade guilds dedicated to particular routes or the confines of the narrow seas. Absent significant numbers of English pilots, all such knowledge had to be translated, something that had been a target of satire by cosmopolitan Londoners as early as More's *Utopia*.[56] But when Cabot left a large blank space between North

America and Cathay, he did what contemporary makers of printed maps in Europe did their best to conceal: the bounded character of European knowledge and most importantly sovereignty.

Cabot's map was explicitly designed as a technical rather than simply representational object, full of annotations and explanations that could be used for building a research program. It also included instructions for using Cabot's unique system of latitude and climates, which had been rejected as a result of political infighting at the Casa de la Contratación.[57] The profuse catalog of geographic labels on the coasts followed the lead of Mercator's and Frisius's globe, commissioned by Charles V in 1541 and first brought to Cambridge in 1547 by the young John Dee. Cabot's map made this kind of knowledge and technical work much more widely accessible in London. Cabot had previously planned to publish such a map in Seville, making a contract in 1541 with the printers Lazaro Noremberger and Gabriel Miçel but then most likely failed to get the approval of Charles V. Perhaps the 1544 map, which includes the crest of Charles V, was a second attempt to gain favor in Antwerp. Mercator himself was imprisoned as a heretic in 1544 for several months, suggesting the potential dangers involved with such work. Cabot gave striking detail for coastal regions like the Caribbean, South and Central America, Africa, the Indian Ocean, and Southeast Asia, areas well-known to the Spanish and Portuguese, with many areas including the Pacific coast of Spain's American colonies including far more names than Frisius either had access to or felt comfortable showing. It is therefore not surprising that he published the first edition without indicating a location and the second in the safety of provincial and Protestant London.[58]

At court, Cabot's map competed most directly with a map hanging in Whitehall by the Florentine mapmaker Girolamo Verrazano, presented in 1527 to Henry VIII while his brother Giovanni was exploring the coast of North America on an expedition for Francis I. Despite Henry's marital and religious difficulties, Charles V emerged as England's clear ally in the wars with Scotland and France from 1543. This artifact of a possible Anglo-French alliance slowly had become obsolete. The Verrazano chart was particularly vague about East Asia, loosely placing islands off the coast of Vietnam and Guangdong Province to define the South China Sea, clearly far less than both the Portuguese and the Spanish after Magellan knew even in the 1520s. The French wanted to claim their own passage to Cathay, through the supposed "Sea of Verrazano," the term for the water inside North Carolina's Outer Banks that Giovanni saw in 1524. That sea, which later inspired Walter Raleigh, also featured also in more widely distributed world maps of Münster at Basel as the locus of hopes for a Huguenot-Calvinist-Anglican or even Anglo-French alliance to challenge the Iberian dominance of Asian trade. Against such speculation tied to particular dynastic and religious alliances, Cabot's map suggested the need for further organized research tied to the uncertain symbolic icons of pre-Protestant texts.

The Portuguese in particular had been busy voyaging and working since their arrival in Malacca (Melaka) in 1511 with Javanese and Malay mapmakers as well as Guangzhou and Fujian merchants to create a relatively detailed picture of the locations of Southeast Asian trading emporia. Like the Selden Map discussed in chapter 3, these Southeast Asian maps showed routes (ji 計 meaning "plot" or "vector") defined by length of time sailed (geng 更 or watch), measured by compass directions and solar movements. These gave precise directions for getting from one port to another, for avoiding reefs and catching currents and winds.[59] In 1511, during the invasion of Malacca, the Viceroy of India Afonso de Albuquerque captured a large chart made by a Javanese mapmaker. The chart, influenced by Islamic and Chinese cartography, supposedly showed everything from Portugal and Brazil to Taiwan and Ryukyus including, as the account goes, "the navigation of the Chinese and Gores [Ryukyu Islanders], how their lines and routes are followed by the ships, and the hinterland, and how the kingdoms border on each other."[60] This map was lost in a shipwreck in 1512, but Albuquerque had Francisco Rodrigues translate the eastern portion from Javanese into Portuguese, writing to King Manuel, that the king could thus "really see where the Chinese and Gores come from, and the course your ships must take to the Clove Islands, where the gold mines are, and the islands of Java and Banda, of nutmeg and mace, the land of the King of Siam, and also the end of the navigation in the land of the Chinese, the direction it takes and how they do not navigate farther."[61] It is only with the discovery of the Selden Map that it has become clear precisely how these features were translated onto Portuguese portolans and rutters and ultimately the printed descriptions of European cosmographers.

An example of this is the complex reef and island region in the South China Sea of the Paracels (萬里長沙, Wanli Changsha) and what some have identified as a name for the Spratleys (萬里石塘, Wanli Shitang). In fact, the distinctions predate such modern groupings of the islands and show a complexity of understanding based on multiple routes that most European maps avoided until the nineteenth century. The Selden Map also seems to include the Vietnamese name for the Paracels, Hoang Se (lit. "Yellow Sands") transliterated into Chinese as Yu Hongse (嶼紅色, lit. "Red Islands"). The groupings appear in simplified form on Cabot's map as they were described and depicted in Chinese maritime literature predating the arrival of Europeans

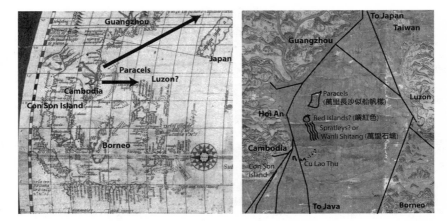

Fig. 11. Left: Cabot map, detail of the South China Sea and Japan. BnF, cartes et plans,
rès ge. AA 582. Right: Selden Map, detail of South China Sea. © Bodleian Library,
University of Oxford, 2012, MS Selden Supra 105.

and on the Selden Map dating from ca. 1620, as a sail (*yang* 帆) in the case
of the Paracels or the tail end of a long snake (*she* 蛇) for the broader set
of islands (figure 11).[62] The "snake" had been described by the fourteenth-
century Chinese traveler Wang Dayuan, who claimed to have gone as far as
Zheng He would sixty years later. Wang wrote that "Wanli Shitang origi-
nates from Chaozhou [in Guangdong Province]. It is tortuous like a long
snake lying in the sea. Its veins (脉) can all be examined. One such vein
goes to Java (爪哇), one to Boni (渤泥) and Gulidimen (古里地闷) [Borneo],
and one to the west side of the sea toward Kunlun (昆仑)."[63] The four lines
on the Cabot map get merged past "Pulunga" (Cu Lao Thu?) into the routes
going west from "Polucandor" or Con Son Island route, but they retain the
clear sail-like square of the Paracels and the long tail of the snake.

The islands also offer a directional hint to the broader navigation of East
Asia, as they did more explicitly in relation to the directional routes on the
Selden Map. The label on Cabot's map wrongly indicates a passage through
the Paracels to the Lequois or Ryukyus ("canal parasos lequios"), which
Magellan's sailors had described as a meeting place for sailors from China
and the Malukus. This interestingly parallels the division between the
Wanli Changsha and the Wanli Shitang on the Selden Map. Another chain of
islands at the top of the Paracels suggests that following a parallel course to
the upper reaches of these island reefs would lead to Ciapagu or Japan, itself
is placed directly under the realm of the Great Khan. For Cabot, Ciapagu
thus became the hinge between on the one hand the older overland Silk

Road stories about Cathay and the route systems of Southeast Asia that had been mapped by the Portuguese through Javanese and Malay translations. It offered a way of bypassing completely what Cabot labeled as "la china," where before gaining access to Macao in 1556 the Portuguese had failed to establish legal trade or missions and were being executed and imprisoned even as Cabot was finishing his map.

Cabot like Eden thus engaged in selective translation—taking apart the universalist cosmographies and maps being produced in German cities or the Spanish and Portuguese centers of navigation. He also carried over icons like the snake of the Paracels that could prove important in shifting debates in relation to those same cosmographies. Far from being purely abstract and mathematical, navigation for Cabot still linked and translated between iconic geographic structures that had been passed down and translated through multiple traditions—English translations of Spanish translations of Portuguese translations of Javanese translations of Chinese iconography. The technical method of maintaining icons like the sail and snake across languages as navigational devices remained even when the metaphor had become obscure; the Cabot map and the Portuguese maps it relied upon no longer directly indicated a sail and snake. Despite all the cartographic trappings of the *mappa mundi* and the effort to relate the data about locations to a grid of longitude and latitude, the method had similarities to navigation practices in both the Indian Ocean and East Asia in which well-traveled routes were defined through chains of signifiers, compass directions, and distance in ships watches.

How did Cabot learn to put all of these layers together? He initially went to Spain in 1512 after the Bristol ventures of the 1490s and 1500s had ended, serving as a chartmaker in the service of Robert Willoughby under orders from Henry VIII to support Ferdinand II's war effort against France. Amerigo Vespucci had just died of malaria in Seville, and Cabot, through contacts at Brugos, managed to get a position with the Spanish as a naval captain (*captain de mar*) affiliated with the Casa de la Contratación. In 1519, Cabot became the Casa's Piloto Mayor y Examinadór de Pilotos, a month before the young king Charles I became emperor-elect Charles V and just as the debate over Iberian claims in the "South Sea" or Pacific was beginning to cause conflicts.[64] The Spanish Casa had grown on the model of the Portuguese, and both institutions consolidated knowledge into a central database that took the form of a map—the Portuguese *Padreo Real* or the Spanish *Padrón Real* after 1507.[65] To universalize this strategy, the Spanish changed the name to the *Padrón General* (generic, broad or general) after August 1527.[66]

Ideologically Cabot's map followed the basic pre-Magellan image of the world presented in the great Spanish navigational classic Martín Fernández de Enciso's *Suma de Geographia* (1519), published in Seville the year Cabot became the *piloto mayor* of Spain.[67] Much of Enciso's book was an extended and somewhat abstract meditation on spheres, an artifact of what Ricardo Padrón has called the "new culture of abstraction," which in 1519 before Magellan, Cortés, and Pizarro represented Spain's only possibility for defining a relation with East Asia.[68] But Magellan had brought a Malay interpreter back as a slave from the Portuguese siege of Malacca in 1511, and that interpreter also went in an effort to build Spanish knowledge of East Asian routes on the famous circumnavigation voyage of 1519. This engagement with Malay along with Portuguese experience is one reason why islands were consistently called "pulao," referencing a now lost Malay mapping tradition. From 1523, after the return of what remained of Magellan's crew, Cabot himself worked with the expedition's chartmaker Diogo Ribeiro. The Portuguese Ribeiro received an appointment to the new post of instrument-maker and cosmographer to the Casa, and the detailed depiction of south China on Cabot's map comes in part from Cabot's relation with Ribeiro and his family of chartmakers. Both participated in the Junta of Badajoz in April 1524 to determine whether Portugal could claim the Malukus, which the Spanish ultimately gave up in the Treaty of Zaragoza (1529).[69] Unlike Cabot, Ribeiro in his surviving 1529 maps that circulated in manuscript outside of Spain (surviving in Rome and Weimar) left much of the cartographic space beyond southern Sumatra and the Gulf of Siam blank. He did follow the lead of Enciso and split China and the Malukus from the rest of Asia so that they appeared on the western rather than the eastern edge of the map, as they would on Cabot's map as well. All of these gestures helped substantiate Spain's claims through Magellan.

Having no direct experience in Spanish exploration projects and seeing an opportunity to make his chart work tangible, Cabot handed over pilot training at the Casa to Alonso de Chaves (royal cosmographer after 1528) and mapping to Ribero and left "for the discovery of Tharsis, Ophir and Eastern Cathay" in April 1526. He brought with him two Englishmen recruited from Bristol—Henry Latimer and Roger Barlow. For a brief moment in 1527 and 1528, with Cabot still at sea and the line between Spanish and Portuguese claims still uncertain, Robert Thorne tried to get Henry VIII interested in Cabot's efforts, writing from Seville almost precisely at the time Henry had declared his love for Anne Boleyn.[70] But Cabot did not find an easy route to East Asia or a simple negotiation of sovereignty. He did experience radical linguistic diversity in the Rio de la Plata and then at Porto dos Escravos (São Vicente), a station settled by Portuguese for the Native American slave trade.[71] When Cabot's expedition ended disastrously in the Rio de la Plata, investors in Seville and Bristol were furious, including Barlow's friends Robert and Nicholas Thorne. Cabot found himself in a Spanish prison.

Subsequently released, Cabot tried to redeem himself through abstraction, by coming up with a new method for deriving longitude from onboard ships that could ground Spanish claims in measurement, explained in long textual annotations to his 1544 map. As early as the 1520s, Cabot claimed that he could navigate based on magnetic variation using a compass and the declination of the sun, similar to techniques published by João de Lisboa. His system worked on the false assumption that the shifting position of magnetic poles or magnetic declination for the compass had something to do with the sun's declination.[72] The map he later printed in London in 1549 was to be a tool for developing these ideas, using as he had since 1533 a grid or graticule system distinctly different from the diagonal compass rose and rhumb lines that covered Diego Ribeiro's 1529 portolan-style master map. But Cabot's gesture was not entirely a retreat from translation of older mapping systems. The Ptolemaic graticule could directly address the distinction between the known and the unknown as newly determined locations were relativized to the abstract grid of latitude and longitude, thus systematically organizing navigational data gleaned from a multitude of sources.[73] Competition from 1538 and litigation in 1545 at the Council of the Indies by Pedro de Medina in relation to Cabot led to the adoption of a merged form of the portolan and the graticule as the Casa's standard in mapmaking, decreasing the importance of the debates about longitude upon which Cabot had grounded his reinvention.[74]

At the same time as Cabot's efforts in Seville floundered, London merchants and certain well-placed humanists at court began to expresss interest in breaking away from dependency on French and Iberian cartographic knowledge and, especially after 1549, from trade with Antwerp itself. In the 1530s and 1540s, most longer-distance voyages involving English sailors were actually coming out of Plymouth, often in cooperative efforts with Huguenot pilots from Dieppe who made charts using captured Portuguese and Spanish examples. These merchants traded illegally in West Africa, Brazil, and the Caribbean with the support of Francis I.[75] The Guinea and Brazil voyages under William Hawkins of Plymouth from 1530 were directly influenced by the connection with Dieppe, while the *Barbara*, which left London for Brazil in 1540, had twelve Frenchmen as part of the crew including the Dieppe pilot Robert Nycoll.[76] These voyages were navigated through an extension of Mediterranean and North Sea techniques, which the French and Portuguese had taken up in the aftermath of John Cabot's voyages, coupled with a by now relatively good understanding of basic Atlantic current and wind patterns. But as political and military alliances shifted to Spain in 1543 and 1544, these joint projects with French navigators came to be seen as politically dangerous, and at the end of Henry VIII's life, the Privy Council started to look for alternatives to this navigational dependency. By 1546, John Dudley, then Admiral Lord Lisle, had brought the French cosmographer Nicholas de Nicolay and the Dieppe pilot Jean Ribault to London. Dudley also probably drove the effort to provide Cabot with pensions from the Privy Council in the late 1540s.

By 1550, with trouble in the Antwerp cloth markets, Cabot was gleaning information from a coterie of Spanish, Portuguese, French, and Muslim pilots tied to the Mediterranean and Atlantic world, all of whom had navigational skills and could help with the translation of new routes. Cabot certainly had experience and expertise as the head of Seville's Casa, which gave him an air of authority, but the actual practices he employed in England did not rely upon that.[77] In November 1550, Cabot helped organize a ship owned by Sir Anthony Aucher. Under the captain Roger Bodenham of Bristol, they sailed to Chios with the help of a pilot from Cádiz named Nobiezia and another from Greece along with a new generation of long-distance English sailors, including Richard Chancellor. Cabot's servant John Alday was to sail in 1551 on the *Lion of London* to Morocco, but he fell sick and was replaced by Thomas Windham and two Muslim pilots on two successful voyages. Two Portuguese pilots made the longer voyage to the coast of Guinea in 1551 and 1552 with Windham to trade for gold, although they failed to reach the Niger Delta to trade in pepper. By early 1551, Ribault and Cabot were supposedly locked up in the Tower of London to make plans for the Indies voyage.[78] If true, that particular synthesis of expertise from Seville and Dieppe under the aegis of the Privy Council did not materialize, nor did the more limited ambitions of the Dieppe and Portsmouth privateers inspire anything as novel as the joint-stock of the Cathay Company.

Regardless of secret efforts in the Tower and trial voyages, which did have longer-term effects on London's trade in West Africa and the Mediterranean, Cabot had already made a different and more enduring public synthesis of translated knowledge by printing his map and demonstrating London's need for new kinds of information, new kinds of translation, and indeed new institutions of exchange. One of the basic points of Edwin Hutchins' book *Cognition in the Wild* about the cognitive aspects of navigation at sea is that it requires the superimposition and "loose coupling" of a wide variety of types of knowledge, gathered through the medium of the chart. For Hutchins this defines its "computational power." The computational power of cognitive artifacts like a map thus stems not from the correctness of the overall methodology (note Cabot's mistaken longitude theory) or even the specific data therein, but from the potential of the chart as a schema of "patterns of interconnectivity" to coordinate among a series of what Hutchins calls "constraint-satisfaction networks."[79] Cabot's greatest asset appears to have been his ability to keep what were claimed as closed knowledge systems—cosmography and particular houses or schools of navigation—open to translations from other practitioners, and at the same time to recognize the limitations of making universal and imperial claims. Rather than simply appealing to London humanists, Cabot's map must have caught the eye of both skeptical localist Catholics and more radical and prophetically oriented Protestants, not to mention actual merchants and sailors. Cabot was in this sense a translator's translator.

ASIAN DEMANDS: THE EMERGING SILVER CYCLE

All of this ultimately leaves the question of why Cabot's map and the "Ordinances" succeeded in gathering investment in London and set up a more durable and novel set of translation practices there, distinct from those developed by the Iberian Casas, Dieppe, or the cities of the Rhineland. Cabot's mapmaker Clement Adams gave two reasons that "certain grave citizens of London" changed their minds about taking a risk on the profitability of long-distance trade instead of simply retaining their cozy relationship with Antwerp. The first was the falling price of cloth in relation to inflation: "[O]ur merchants perceived the commodities and wares of England to be in small request with the countries and people about us, and that those merchandises were now neglected, and the price thereof abated." As prices fell and war debts rose, the Privy Council under Edward VI lacked sufficient access to silver or gold. It continued to debase the currency as Henry VIII had done late in his reign, causing inflation on both imported and domestic goods, although not on cloth exports. This made the Cathay Company as well as the concurrent Guinea voyages efforts to moderate the price of imports and to raise the price of exports by generating demand in the face of challenges emerging after 1549 with the Antwerp trade. Local difficulties gave new incentives for difficult translations.

But Adams offered a second and more important reason for investment in the company: the clear profitability of Iberian long distance trade. "Seeing that the wealth of the Spaniards and Portuguese, by the discovery and search of new trades and countries was marvelously increased, supposing the same to be a course and means for them also to obtain the like," he wrote, "they thereupon resolved upon a new and strange navigation."[80] The payoff of access to both commodity silver (Spain) and Asian trade (Portugal) had only in the 1540s become clear not merely to courts with territorial ambitions but also to private merchants, who had no motives for territorial expansion. London clearly lacked good information about markets beyond Antwerp, especially global markets like those accessed by the Portuguese in Asia and the Spanish in the Americas. As Kenneth Arrow has argued, "In conditions of disequilibrium, a premium is paid for the acquisition of information from sources other than prices and quantities."[81] Cabot's "Ordinances," the map, and the Cathay Company offered ways for Londoners to address such disequilibrium without having either to fall back on unrealistic or at best fragile sovereign claims to *imperium* or to rely upon confused religious cosmographic claims to translating a world picture.

The traditional economic interpretation by historians of the shift in trade in this period concerns Adams's first claim about local disequilibrium due to currency debasement and inflation. Increasingly from the 1540s, the crown and the Privy Council had used London's cloth trade with Antwerp as a way of raising money needed for warfare against Scotland and France. From 1542 to 1551, Henry VIII and then the Privy Council under Edward debased English silver coinage. The first experiment had occurred as early as 1526, but momentum grew especially from 1544 with the reorganization of the Tower mint. The debasement from 1542 had made it more difficult to get silver in exchange for cloth at Antwerp. This in turn encouraged development of the trade in luxury imports in exchange for kerseys, the cheapest finished cloth available in England, which replaced "short" and "long" cloths in this period. From 1549, the Privy Council tried to game the relationship between London and Antwerp through new devaluations of silver coinage, arbitrage, and interest rate negotiations. This had the immediate effect of driving cloth exports to new highs in the late 1540s, but returns diminished over time.[82] At the same time, again from 1544, the English crown began to accumulate large foreign debts in exchange for silver and munitions purchased on the Antwerp market. From 1549 to 1551 the terms of exchange with Antwerp were debased even more dramatically, with Edward Seymour, Duke of Somerset, using progressively less silver in English coinage to pay for England's war with Scotland. Antwerp bullion sellers were willing to offer loans at interest rates from 2% to 5%, better than those given to the emperor himself, almost irresistibly bringing the English crown deeper into the system of debts.[83] A massive revaluation in 1551 wreaked havoc with political and commercial relations. English cloth merchants recovered, even in the face of overall stagnation in cloth exports over the course of the sixteenth century, but primarily because they took on the old Hanseatic Steelyard business after its privileges were revoked in 1552.[84]

In the middle of this "crisis," two basic solutions appeared. The first was put forward by a Cambridge humanist, Thomas Smith, Henry VIII's appointee as the Regius Professor of Civil Law and the tutor of both Eden and Cecil. Under Edward VI, Smith became clerk of the Privy Council (March 1547) and then one of the two secretaries of state (April 1548). One of his first tasks in the summer of 1548 was to restore English trading privileges at Antwerp by meeting with Charles V at Brussels, around the same time Cabot started negotiating his own position in London. As Cabot prepared his map for publication, Smith wrote in 1549 his *Discourse of the Commonweal* (published 1581), a dialogue between a knight, a learned doctor, a merchant, a husbandman, and a capper (representing artisans).[85] The complaints began with the usual suspects of decay and enclosure, but the merchant shifted the discussion to a "general dearth of all things," especially imports, which supplied most of England's manufactured needs, noting that (pace Thomas More) enclosure could not be the cause. The knight and the husbandman then had a debate about inflation—rents against food prices—but the learned doctor blamed the rising price of imports on the fact that English coin was not "universally currant" or "commodyous." At this point in the dialogue, the lesson sunk in for the knight and the husbandman. Neither could exercise significant control over price. For Smith the solution was a conservative one: reverse the policies of Somerset and literally restore the old "commonwealth" by reverting to the "old rate and goodnes" of the coin.[86] This, however, did not happen, and Edward VI, no doubt as the mouthpiece of the Privy Council, threatened to call in the liberties of London's mayor and aldermen in June 1552 if they could not control the "unreasonable prices of things."[87]

Thomas Gresham, who would later be instrumental in founding London's Bourse (1565, known as the Royal Exchange from 1571), put forward the practical alternative to Smith's conservatism—break away from Continental debt relations and create a money market in London to raise assets. The repeal of usury laws in 1546 had encouraged the development of London moneylending. From 1552 under the Privy Council, during the reign of Mary and into the early years of Elizabeth, Gresham as the royal agent in Antwerp played complicated arbitrage games to break the court out of debt relations. The crown would borrow money from the Merchant Adventurers trading at Antwerp in exchange for the promise of English money. The

Adventurers then sent their wool to Antwerp, where they paid the crown's representative (Gresham from 1553) the silver or gold they received in exchange for their sale. He then used the money to pay off crown debts in Antwerp and to float more loans based on the fluctuating exchange rate. Out of this process silver and some gold would be sent back to Englandand coined at the Tower mint, and the crown would pay merchants from this debased currency. This system remained effective in reducing crown debt but highly fragile. If merchants sold too much cloth, the exchange rate would rise in spite of debasements and English silver coin would flow to Antwerp. It also meant that the wool trade became less profitable as the crown through arbitrage indirectly skimmed from it. On April 16, 1553, just a month before Cabot's voyage, Gresham wrote to Protector Dudley that the exchange rate had begun to move upward because English merchants were selling too much, threatening paradoxically to drain actual gold and silver out of England because of higher exchange rates. Gresham's initial recommendation was to tighten the apprenticeship rules for Merchant Adventurers so they would act more like a cartel and less like individuals; in other words, limit the number of London merchants involved in the Antwerp trade.[88]

The Cathay Company and the joint-stock more generally were the inversion of this emerging strategy of reducing the number of merchants engaged in the export trade. Open broadly to merchant investors as shareholders, it would bring spices and Asian luxury goods directly to London in order to make it less dependent on cloth exports to Antwerp and reducing the pressure on the Antwerp exchange rate. London should be more like an Asian emporium engaging in a wide variety of exchange and interpretive practices instead of remaining dependent on a larger empire for the value of its currency. Of investors in the Cathay/Muscovy Company in 1553, several came from backgrounds in raising loans and arbitrage in Antwerp, including virtually everyone (successes and failures) who worked to raise crown loans in the late 1540s and early 1550s—Thomas Gresham, his uncle Sir John Gresham, John Dymocke, Sir William Dansell, and Christopher Dauntsey. Ideally Cathay would allow the new company to bring in goods for reexport to Europe in exchange for woolens sold around the globe, giving London the power to leverage the relationships between silver, spices, and the broader European economy and allowing the city to define its own interest rate and currency policies despite having no direct connection to German or American silver or West African gold. As much as the Antwerp Bourse was the model for what slowly emerged as Gresham's and later Elizabeth's Royal Exchange (1571), the Cathay Company was the institution in London that first generated a new premium on the collection of information on a global scale as well as the use of a variety of reexportable luxury commodities for mediating questions of global economic disequilibrium.

There was in fact an even broader reason for this shift. Simply put, the regional price system set at Antwerp was no longer an adequate source of information, either for London or in more global terms. In part this can be traced to this period's sheer growth in regional markets due to population in a number of places around the world.[89] Additionally, starting in the 1540s, the massive influx of silver first from German mines and then from the New World into Europe created vast amounts of liquidity in terms of commodity metals in relation to credit markets accustomed to relatively fixed bi- and tri-metallic ratios (gold and silver in relation to more common currencies from copper and tin to cowries). A famous old monetarist argument dating to the 1930s suggested silver supply as a cause of inflation in this period.[90] The silver boom and new mining techniques in Germany (esp. 1516–25, peaking between 1536 and 1540), as well as in Japan from the 1520s, followed by significant imports of Spanish silver in late 1530s and especially the discovery of new mines in Mexico and Peru in the mid-1540s, all began to feed Ming demand for silver.[91] From 1519 onward, the Spanish crown, although theoretically controlling all of this silver, fell into significant debt to the Fugger banking syndicate, transferring most of its precious metals arriving in Seville to Bruges and Antwerp to pay off loans. For this reason, Antwerp surpassed Bruges as a money market in the 1520s, dramatically so after the expansion of the Antwerp Bourse in 1531.[92] By 1553, it was well known among London merchants that almost all transactions at Antwerp took place in Spanish silver *reales* from American mines.[93]

The missing part of this account is, as a number of historians have shown over the past two decades, the dynamic demand generated in East Asia for silver. The lively struggles and cooperative efforts in the South China and Java Seas among Chinese, Portuguese, Malay, Javanese, Moro, Ryukyu, and Japanese pirates and merchant diasporas also opened up new opportunities for London to break free of dependence on Antwerp. After the failure of early Ming experiments with paper in the early fifteenth century, its copper cash had become an export commodity because of the strength of its merchant networks. Lacking enough of this copper cash, counterfeiters made more, overvaluing official bronze coins in relation to the private debased coinage. As a result silver became the standard of value and the basic medium of savings in the empire even though most coinage and everyday exchange employed counterfeit cash made of a mix of copper, tin, and iron.[94] Demand for silver eventually encouraged the Japanese to begin mining in the 1520s and would subsequently attract the Portuguese back to China in the 1540s. Also by 1540, Fujianese merchants from Quanzhou and Zhang-zhou had opened up a trade for silver with Japan, and the Ming voiced frequent worries of their collaboration with Japanese *wokou* ("dwarf bandits" 倭寇) or pirates.[95] Then in 1543, Portuguese ships arrived in Japan with the help of informants from the South China Sea and also began to participate in this trade. By coincidence this occurred just as the Spanish found out about deposits of silver at Potosí (1545) and Zacatecas (1546). From the late 1540s to the 1640s the famous silver cycle developed that literally pulled silver from Japan, Europe, and the Americas towards, China where it could be disposed of at a 50% premium.[96]

As silver increasingly became a global standard from the 1540s, the former Portuguese system of global arbitrage between merchant networks underwent profound transformations. Portuguese fortified *feitoria* (Cochin in 1503, Diu in 1509, Goa in 1510, Melaka in 1511, Ormuz in 1515), which were in many ways hubs for storing bullion, depended heavily on networks of Chinese, Malay, Gujarati, and other intermediaries. They increasingly drew profits from disparities between European, South Asian, and East Asian monetary systems.[97] But the Portuguese royal monopoly on trade needed to raise capital to send ships and maintain forts, and it did so by giving banking syndicates a share of the pepper profits. In 1499, the Portuguese factory moved from Bruges to Antwerp, and from 1501, Manuel had agents in Antwerp exchanging spices for silver, copper, and naval stores. By 1508, he had contracted with the Affaitati and Gualterotti syndicates of Antwerp to purchase spices (pepper, nutmeg, mace, ginger, cloves) in exchange for loans to buy bullion and supplies through the Casa da Índia's Feitoria de Flandres. But in 1543, the year after Henry VIII began to debase the English currency, João III could no longer regularly meet the interest payments on his loans. In 1549, the Portuguese actually closed their factory in Antwerp and began to sell large quantities of spices in Lisbon to stop such hemorrhaging. The problem of payments would not be solved until 1554, when João renegotiated his loans, broke off relations with the Affaitati syndicate, and pulled the spice trade out of Antwerp entirely.[98] In general, older regional emporiums like Antwerp, tied to particular imperial strategies, saw their authority decline in favor of a much more fragmented and increasingly privatized or devolved global trade that could reduce transaction costs for moving silver by shifting among different networks. Cabot's Cathay Company, even though it was not explicitly designed to move silver, was an early example of such devolution that occurred in a number of regions, including the Fujianese-Japanese-Portuguese collaboration in East Asia, in response to the demands of an emerging global silver circuit. Its unique explicitly fictive dimension as a company stemmed precisely from the fact that those involved did not understand what was going on at a global level, but they needed to develop translating mechanisms that potentially could.

Most cities with worldly pretentions up to the sixteenth century were either like Antwerp, an emporium at the mercy of surrounding empires, or the capitals of empires like Beijing or Istanbul that were delimited by them. London after 1553 began to develop differently. The joint-stock company and mapping strategies that emerged from the efforts of Cabot's group of Londoners were mechanisms for remaining open to and translating a wide range of directly and indirectly related global processes. Rather than simply supplying a range of commodities either from England or from various and uncertain Asian locations, the company and map verified the chain of emporial cities as a source of value because of the pull of demand for silver. The mixture of political-religious crisis at home and the emergence of the silver cycle on a global scale undermined the Iberian drive towards *imperium* and reopened the question of the network of routes between Asian emporial cities. The pull of demand from Ming China along with a host of other port cities in maritime Asia explains why against all odds Cabot's group succeeded in gaining recognition not only from the Protestant Privy Council led by Dudley in 1553 but also from the Catholic and Spanish sovereigns Mary and Philip in 1555. Asian networks and exchange practices allowed London's ambitions to become global, and this was one reason that Selden would later value the map that came into his possession. Cabot's corporation and map domesticated translation processes that would make the economies, knowledge, and languages of Asia an integral part of the future success of the city.

CHAPTER TWO

National Autonomy

Lucius Florus in the very end of his *Historie de Gestis Romanorum* re-
cordeth as a wonderfull miracle that the Seres (which I take to be the
people of Cathay, or China) sent ambassadors to Rome, to intreate friendship, as moved with the fame of the Romane Empire. And have not we
as good cause to admire, that the Kings of the Moluccaes, and Java major,
have desired the favour of her majestie, and the commerce & traffike of
her people? Is it not as strange that the borne naturalles of Japan, and the
Philippinaes are here to be seene, agreeing with our climate, speaking our
language, and informing us of the state of their Easterne habitiations?
—Richard Hakluyt to Sir Francis Walsingham, London, November 17,
1589, "Epistle Dedicatorie," *Principall Navigations, Voiages and Discoveries of the English nation, made by Sea or over Land, to the most
remote and farthest distant Quarters of the earth at any time within
the compass of these 1500 yeeres* (London: George Bishop and Ralph
Newberie, 1589)

You people going and coming all exist in my dreams. [汝往來者皆在吾
夢中耳]
—Attributed to Luo Hongxian (羅洪先, 1504–64) by Huang Zongxi
(黃宗羲, 1610–95), *Mingru xue'an* [明儒學案] (1691), 18:1a

1588: READING A CHINESE MAP IN LONDON

London's population increased from around 70,000 in 1550 to 200,000 in
1600, a bustling urban area with an increasingly complex and diverse
economy. Yet, the city still felt ungrounded and fragile, teeming with mi-
grants and refugees from the countryside and the wars in the Low Countries,

rumbling with religious tensions, fearful of Spanish invasion. And then, in the momentous year of 1588, Thomas Cavendish returned to London from his circumnavigation of the globe with a Chinese map of the Ming Empire, a Chinese compass, two Japanese sailors, and three young boys from the Philippines, all captured from a Spanish ship off the coast of Mexico. This was the first of three important maps from China—the second printed by Samuel Purchas in 1625 and the third being John Selden's map—that would shape ideas about East Asia and global trade in London. The map brought back by Cavendish was tangible if still indirect evidence not only of English success in navigation and discovery, the rumors that had swirled around Francis Drake after his return, but also of what was driving the global success of the Spanish and Portuguese: those networks of Asian trading cities, notably the populous Ming Empire, which produced and exported manufactured goods from textiles to porcelains and pulled in silver as a result. The encounter with East Asian urbanism had already taken place for the Portuguese from the 1510s and the Spanish from the 1560s; it was perhaps more jarring in England, at war with the now joint monarchy of its former king Philip II. By the 1570s, it was also more or less known in Europe that the early Portuguese efforts in China, those of Fernão Peres d'Andrade and Tomé Pires in the 1510s, had been rebuffed.[1] In London, there was a strong incentive similar to that emerging in the rebellious United Provinces to see beyond Iberian reports and to understand the political geography of the world in new ways. The map, compass, and sailors from the trading world of the Pacific presented such an opportunity.

Understanding the world in the late sixteenth century meant comprehending vast shifts in populations, sovereignties, and cosmologies, the kinds witnessed not only by Hakluyt from his vantage in London but also by the man whose map he saw, Luo Hongxian. Luo was a Ming cartographer and philosopher who created the oldest surviving printed atlas in China, the *Guang yutu* (廣輿圖 "Extended World Chart"), which was first printed in Jiangxi around 1555 (figure 12). Like Münster, whose cosmographic project emerged out of attempts to map relationships between German cities of the Rhineland, Luo created his atlas as the economic and political changes pioneered by the Ming in the fifteenth century began to have effects on a global scale.[2] Although his map came out of a tradition that dated at least to that of Zhu Siben's Yuan-era (Mongol) map *Yutu* (輿圖 ca. 1320), Luo used more recent Ming census figures and tributary reports to compile data in ways not dissimilar to the Antwerp atlas-maker Abraham Ortelius. But while European cosmographers focused on organizing place names, Luo with more extensive numerical data compared numbers of households, administrative

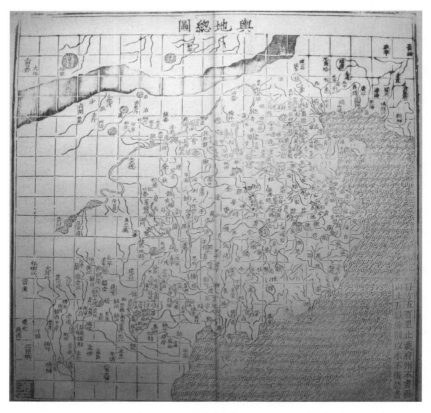

Fig. 12. Luo Hongxian, "Yudi zongtu" [輿地總圖] in *Guang yutu* [廣輿圖 "Extended World
Chart"] (ca. 1555). Library of Congress, Asian B 182.07 L78, photo by Robert Batchelor.

categories, and the size of armies. The word *guang* (廣) in the title *Guang
yutu*, which means something like "propagate" or simply "extensive," suggested the desire to extend the cartographic image of authority outward,
and he included maps of Japan and the Ryukyus.[3] Even though Luo was dis-
graced at court in factional fighting and forced into retirement, new editions
(notably 1579) and numerous borrowings followed his first printing, giving
the book the kind of afterlife that those of Münster and Ortelius enjoyed.
The census data tied to cartographic depictions of networks of cities would
be used by generations of late Ming merchants in these various popularized
and reworked forms as a kind of advertisement for the commercial power
of China.

     One such Chinese map derived from Luo's atlas arrived in London in
late 1588, just as England and its excommunicated monarch Elizabeth had

narrowly escaped Spanish invasion, and Hakluyt for his new book decided to publish a translation of the data. The military figures alone were staggering and precise, even if somewhere between Luo's and Hakluyt's editions errors crept in, so 454,728 horsemen and 7,459,057 footmen for a total standing army strength of 7,923,785 [sic; the correct total is 7,913,785].[4] Equally precise numbers for households and soldiers gave an indication not merely of the strength of Ming commerce but a sense of the logic behind the fiscal-military state supported by data and revenue gathered through census and taxation. With the map also came two Japanese pilots, who understood the use of the Chinese maritime compass and who revealed its technical aspects to the navigational theorists of the London of the 1590s. These very particular translations helped foster the realization of Cathay in the 1580s and 1590s, a model of the rationalized and data-driven state and its support by mathematically savvy navigators that could ground speculative efforts, including the founding of the East India Company itself in 1600. It was a period in which the territorial and dynastic integrity of "Britain" was emerging, through bloody conflict in Ireland and with the ruthless killing of Mary Queen of Scots, an emergence that paralleled the equally violent consolidation of the United Provinces and distant Japan. Both Ming China and Philip II's Spain played important roles in all of these "national" struggles tied to the three most rapidly urbanizing areas on the planet. For London, 1588 was not simply a resistance of external dominance, but a critical year in a series of exchanges that laid the groundwork for imagining a world of as independent states or nations interacting outside of a comprehensive religious cosmology.

By the 1580s and 1590s, London as a city was torn by forces more powerful than mere population and commercial growth. Within the city itself, the number of joint-stocks expanded rapidly, offering a new kind of common identity distinct from the citizenship provided by the guild. Many guild members no longer lived in close proximity to other guild members, and successful Londoners often had connections to more than one guild. Richard Hakluyt himself was a Skinner, a guild that had strong connections with the Russia Company, but his education at Oxford was paid for in part by the Clothworkers. London printers increasingly made the provincial gentry imagine the city as the center of intellectual and economic life, and the young especially came to live there.[5]

Outside the old guild and political structures of the city, an even greater dynamicism could be seen in the suburbs. To the west, great houses along the Strand proliferated—most famously William Cecil's Burghley House (1560), Robert Dudley's Leicester House (1575, inherited by Robert Devereux in

1588 to become Essex House), and Walter Raleigh's Durham House (1583).
These became centers of patronage, "big science" projects, and political intrigue.[6] To the east, poor rural immigrants and radical Protestant dissenters
denied housing in the city began to fill Spitalfields, notorious for counterfeiting everything from coin to cochineal, while along the northern bank
of the Thames, sailors and those who serviced the booming shipbuilding
and maritime supply industries filled Wapping, Ratcliff, and Limehouse.
Increasingly across the East End, French, Dutch, and other sailors, weavers,
and artisans could be found pushing the linguistic, religious, and technical limits of London in new directions. While Raleigh had Thomas Harriot
working to translate the Algonquin of Manteo and Wanchese in Durham
House, an even more complex collection of languages and religions could
increasingly be found near the docks in the East End; Elizabeth I felt it necessary to send the lord mayor an open letter of complaint about "blacka-
moors" in 1596.[7] On the opposite side of the Thames in Southwark and
Bankside as well as in Shoreditch to the east of the city, the entertainment
industries of the sixteenth century grew up, notably the famous theaters
the Rose, Swan, and Globe out of which came much of what is now associated with "English" culture. John Norden's 1593 "Map of the City of
London," itself based on the more substantial Agas map of ca. 1563, is almost rigorously nostalgic in denying this new suburban growth and trying
to frame a London within the walls by surrounding it with the stable crests
of the "twelve companies," creating an image of a stable and oligarchic elite
choosing the lord mayor (figure 13). Suburban and global forces slowly dissolved this old London; perhaps more importantly, they supplanted it.

The voyages of Drake and Cavendish held great symbolic value not only
because of the technical achievement of replicating Magellan but also because as war dragged on with the Spanish and various Elizabethan colonization schemes consistently failed, even in nearby Ireland, they suggested a
tangible way of grounding the national autonomy of England and realizing
its imagined global status. This was not a desire for empire, except perhaps
in the limited sense of unifying the British Isles, and the concept of empire
gives only dim insight into the shifts occurring in this period.[8] Economically, the voyages presented an alternative to the array of new Elizabethan
companies aimed at regulating trade from the North Sea and Baltic down
to the Mediterranean—Merchant Adventurers at Hamburg (1567–78), Spanish (1577), Eastland or Baltic (1579), Levant (1581), Barbary (1585) companies that competed in a crowded field with a potential for trade imbalances
as commodities like Baltic grain were increasingly necessary to supply the
food needs of London. Jodocus Hondius's (Joost de Hondt) celebratory map

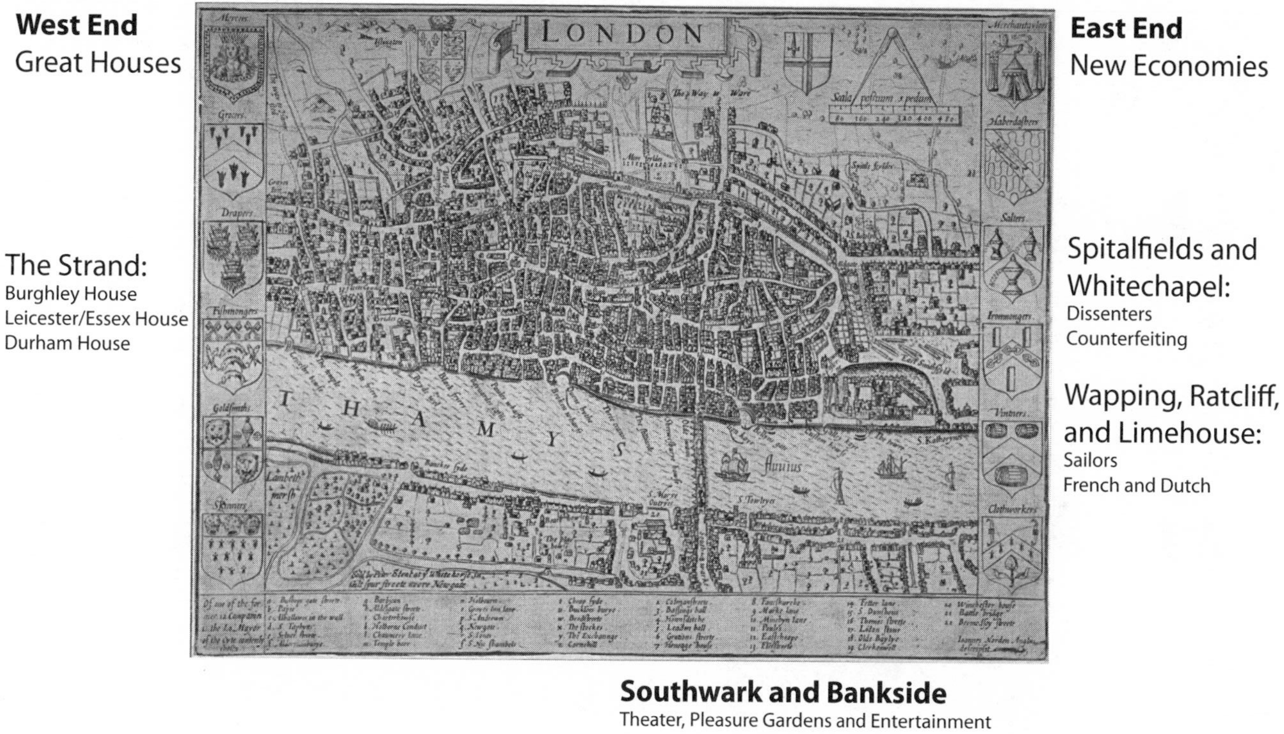
West End
Great Houses
The Strand:
Burghley House
Leicester/Essex House
Durham House
LONDON
East End
New Economies
Spitalfields and Whitechapel:
Dissenters
Counterfeiting
Wapping, Ratcliff, and Limehouse:
Sailors
French and Dutch
Southwark and Bankside
Theater, Pleasure Gardens and Entertainment

Fig. 13. Map showing growth of suburban commercial activity outside of the old guilds and wards of London. Overlaid on John Norden, "Map of the City of London" (London: P. Stent, 1593), based on Agas map ca. 1563. Reproduced by permission of the London Metropolitan Archive.

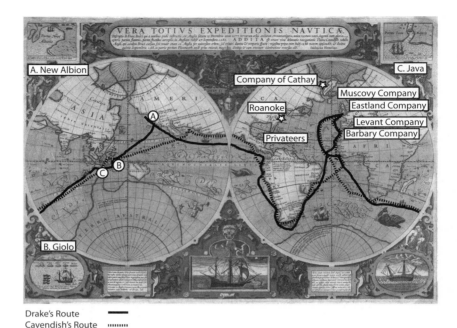

Drake's Route   ━━━
Cavendish's Route   ▪▪▪▪▪▪▪

Fig. 14. Principal London trading companies and routes of Drake and Cavendish
highlighted on Jodocus Hondius, "Vera Totius Expeditionis Nauticae"
(Amsterdam? ca. 1595). Courtesy of the Library of Congress, G3201.S12 1595.H6,
photo by Robert Batchelor.

of 1595 showed not only the routes of Drake and Cavendish but also had
little cartouches suggesting a potential English network of trade in the Pacific connecting Nova Albion above Spanish California with the cluster of
islands around Giolo (including Ternate and Tidore) as well as Java (figure
14). Nova Albion would be the counterpart to the failed Roanoke (Virginia)
colony of Raleigh, which Hondius did not show, having depicted the east
coast of North America as divided between La Florida and Nova Francia.
Likewise the southern routes of Drake and Cavendish replaced the failure of
northeast and northwest passage schemes like that of Michael Lok and Martin Frobisher's Company of Cathay (1577–80), with its claims of Cathayans
and mines on Baffin Island ("Meta Incognita"). In general, the Pacific seems
just that in the Hondius map, a stable and reliable source of commodities,
while civil war in Ireland, dynastic intrigue in Scotland, a vanished colony
in Virginia and the broader privateering efforts against the Spanish made the
Atlantic an uncertain space of war.

Cavendish's ship the *Desire* was brought from Plymouth to London to sail past Greenwich with sails of blue damask and standards of blue silk in an echo of the launch of the Cathay Company voyage, which had never made a proper return.[9] At first glance, Cavendish appears like Drake and the other "privateers," a consummate looter with a talent for terrorizing the coasts of the Spanish Empire. Traveling up the coast of South America, he burned three Spanish towns and thirteen ships, including the returning Manila galleon, the 600-ton unarmed *Santa Ana* captured off Cabo San Lucas in November 1587. It had a large cargo of gold that had been exchanged for silver in Manila, revealing the nature of the trans-Pacific exchange economy developed in the 1570s, in which the Spanish were making profits off of both Asian commodities and differential valuations of gold and silver. He brought back a secondhand report about Manila as "a town unwalled, which hath three or four small block houses, part made of wood, and part of stone being indeed of no great strength: they have one or two small galleys belong to the town." If militarily quite weak, emporially Manila had been a striking success, "a very rich place of gold and many other commodities," that were brought there by twenty or thirty ships per year from China (an accurate figure) in exchange for silver from Acapulco. Most importantly, gold was exchanged for silver "weight for weight."[10] The hundreds of cities marked on the map of China suggested precisely where such silver was going. Cavendish's report portrayed the global economy as emporial rather than imperial—money came from favorable exchange rates on gold and silver and from bringing plentiful commodities to places like Europe where they were scarce.

As Cabot had suggested some forty years previously, engaging in such a world required translation. It was Richard Hakluyt who printed Cabot's "Ordinances" for the first time in 1589. Just before describing the Ming map brought back by Cavendish, Hakluyt tried to emphasize the importance of translation by including a table of words translated from Javanese into English brought back by Drake. If Eden and Cabot produced the great cultural artifacts of the first phase of London's globalization, then Hakluyt's *The Principall Navigations* (1589, expanded 1598–1600) marked the second. Hakluyt, an Oxford humanist and divine, modeled his collection on previous ones by the Venetian Giovanni Battista Ramusio from the 1550s as well as on the Italian-born Spanish historian Pedro Mártir de Anglería's (aka Peter Martyr de Anglería) *De orbe novo*, covering a diverse array of firsthand accounts of explicitly "English" voyages from the year 800 to 1589.[11] But Hakluyt's work was fundamentally different from that of Ramusio and Mártir in trying to define a national sense of sovereignty and to replace the open uncertainties of Cabot's map with a more stable sense of boundaries. It was

also different from the more Orientalist theatrical and poetical models coming out of London itself, whether Christopher Marlowe's dramatization of
the Timurid Empire in *Tamerlane* or Edmund Spenser's claim in the *Faerie
Queene* to emulate Cyrus and the Persians, in which the active policies of
Asian empires seemed to offer enviable models of unity against the fragmented sovereignties of medieval kingship and civil war in Ireland.[12] Hakluyt was more concerned with cartographic limits and sovereign boundaries
defined through technical achievements (navigations, voyages, and discoveries) than authoritarian ones, and it was in terms of these issues that systematically calculated tables of population and military data from the Ming
showed new ways of organizing the subjects of a polity.

Whether because of limited time and resources or a more explicit desire
to avoid reproducing a simple image, Hakluyt did not reengrave the Chinese
map as Samuel Purchas would another Chinese map in the 1620s. Instead,
Hakluyt's translation of the data on Cavendish's map, most likely mediated
by Spanish translations although perhaps made with help from the Japanese
sailors, returned the map to its pre-imagistic state by reproducing the data
and administrative categories established by the Ming census.[13] This was
the most precise accounting of the Ming to have appeared in Europe up to
this point, down to errors of transcription in which some population figures
are off by a factor of ten. So for the Beijing district, the Hakluyt translation
reads, "The great city of Paquin, where the king doth lie, hath belonging
to it 8 great cities, and 18 small cities, with 118 towns and castles: it hath
418,789 houses of great men which pay tribute, it hath horsemen for the
warre 258,100."[14] The figures and categories from the 1579 edition of Luo's
atlas that accompany the *Bei zhili yutu* (Beijing District Map, 北直隸輿圖)
count eight seats of magistrates (*xianfu* 縣府), 17 submagistrates (*shuzhou*
屬州), 115 counties (*xian* 縣) plus two additional counties (*youzhou* 又州)
and one subordinate county (*shuxian* 屬縣) for a total of 118 counties, and
418,789 households (*hu* 戶).[15] Hakluyt's point was to show how different the
premise of Ming mapping was from the Spanish as embodied in Ortelius's
work at Antwerp. Rather than a simple image of the world, Luo's map diagramed the administrative and population structure of the empire through
collected data. "Data," a newly introduced but abstract Euclidean concept
of the "given" (Gk. Δεδομένα, Lat. *datum* and *data*) still remained untranslated into English in the 1570s and '80s by proponents of geometry like John
Dee, but it nevertheless took on a new concreteness in London in a global
and cartographic sense with the arrival of Luo's calculations.[16]

The census and cartographic categories used for numbers were historically and linguistically idiomatic to the Ming Empire and thus difficult to translate. Nevertheless, unlike earlier Portuguese accounts of the closed and inward-looking empire, they suggested a complex political entity based on struggles over the definitions of a series of frontiers. Lineage groupings marked for taxation purposes (*hu* 戶) became "households of great men," like those of aristocratic and wealthy gentry families in England. Administrative districts (*zhou* 州) became in the fashion of Marco Polo "cities," like London itself. The Ming-centered geography also required equivalencies. Luo Hongxian accompanied his maps with a table describing various barbarians and tributaries (*gong* 貢)—Koreans and Japanese to the East, Liuqiu (Ryukyu) to the southeast, Vietnam to the south, Indians and Muslims to the southwest, Turkic people to the West, and Mongols to the northwest. The translated categories in Hakluyt unevenly approximated Luo's—the kingdoms of the "Tartarians" (to the north but also "adjoyning upon Moscovie"), the "Mogores" (Mughals and Uzbeks, to the northwest and stretching to Bengal) and the "Chinian" (a "very deformed" warlord fighting the Tartars and Mogors, perhaps Timur) all "without the circuit of the wall." Luo's map of India and the fact that the Ming considered it a tributary disappeared but so did Cabot's and Dee's Cathayans. The Vietnamese port cities of Qui Nhon ("Champa") and Hoi An ("Cauchinchina"), both identified by latitude indicating a Spanish or Portuguese hand in the translation, were supposedly kept within the Chinese sphere of influence by a navy of hundreds of ships based in Guangzhou ("Canton").[17] Fujian ("Fuckin") garrisoned cities "to keep watch upon the Japans." Guizhou ("Cutchew or Quicheu") manufactured weapons to fight with the "Jawes," Tai (泰) chieftains of the southwest frontier of Yunnan who had been in revolt since 1436. Here was an empire surrounded by dangerous potentials for revolt but ultimately able to maintain itself by carefully keeping track of data about its territory and the status of its frontiers.

Before he had access to Cavendish's map, Hakluyt had commissioned Robert Parke to translate a text published in 1585 in Rome and Madrid by the Augustinian friar Juan González de Mendoza, the *Historia de las cosas más notables, ritos y costumbres del gran reyno de la China*. A moderate Catholic text that conveyed a basic sense of the size and grandeur of the Ming from extensive Chinese sources gathered at Manila, it appeared in English in 1588, ten months before the *Principall Navigations* and right after Cavendish's return. Hakluyt had worried, however, that the book might prove untrustworthy because of Catholic and Spanish obfuscations. At the time, the English Jesuit Robert Persons in Rome supported the idea of a "crusade" against England led by Philip II, which would ultimately take shape as the Armada, and both a pirated Italian edition of 1587 by John

Wolfe and Parke's English translation reworked a complex politics already at play among the London printers after the execution of the Catholic Mary Queen of Scots in February 1587.[18] Parke's translation, dedicated to Cavendish, celebrated the circumnavigation as the culmination of efforts that had begun "five and thirty years" before with Cabot, when "that young sacred, and prudent Prince, king Edward the sixth of happie memorie, went about the discoverie of Cathaia and China," in an effort to escape from regional economic troubles as well as the "mischiefs" and "contempt" directed against London merchants "in the Empire, Flanders, France and Spain." But according to Parke, it was now time for "passing over *Paulus Venetus*, and Sir *John Mandevill*" and thus Cathay, and for the English to begin assembling their own data.[19]

Rather than gathering data, however, Mendoza had argued that Chinese sovereignty rested in the long tradition of the language, the characters understood by mandarin scholar-officials. Chinese characters were likewise used, according to Mendoza, in Japan, the Ryukyus, and Cochinchina, a kind of sphere of Confucian values. The three characters Mendoza chose to transcribe and that Parke carried over into English in a clumsy Gothic rendering are telling in this regard. *Tian* (heaven, 天) and *di* (earth, 地) for "heaven and earth" suggested an alternative cosmology and universalism, while *cheng* (city, 城) implied the importance of cites and the urban fabric more generally in extending language. Mendoza described Ming educational institutions and the bureaucracy itself, which preserved and promoted the language. For Mendoza, even through multiple layers of translators, Ming writers had demonstrated a durable sovereignty through their extension of language, a model appropriate for Latin and more importantly Spanish as well. The idea of the power of the Chinese language popularized by Mendoza in Europe and later alluded to by Bacon helped support the Augustinians and the Jesuits in their emerging strategies for translating Catholicism in East Asia and ultimately by the time of Louis XIV "Confucianism" into Europe.

Instead of trying to set up institutions for translating the entire Chinese language, Cavendish seemed more interested in the need to develop technical skills for navigation and trade in East Asia. Cavendish took captive a bevy of skilled European navigators, including the Acapulco pilot Alonzo de Valladolid, who knew the Pacific routes, and a Portuguese pilot Sebastian Rodriguez Soromenho (Cermeño), who had traveled extensively in China and Japan, knew about the rich silver mines on Honshu, and may have been the one in possession of the Ming map. This was also the explicit reason for returning with the "two young lads borne in Japon, which could both wright and read their owne language, the eldest being about 20 years olde was named Christopher, the other was called Cosmus, about 17 yeeres of age, both of very good capacitie," who would sail with Cavendish until his death. Two boys born in Manila would do the same, while a third became a servant to Francis Walsingham's daughter, the Countess of Essex.[20] They could, as Parke suggested in his introduction to Mendoza, "serve as our interpretors in our first trafick thither."

Cavendish's voyage also revealed with certainty that the Spanish and Portuguese relied upon Muslim, Chinese, and Japanese pilots to find their way around Asian seas. The pastor William Barlow, who had connections with Hakluyt, wrote of interviewing one of the Japanese sailors from "Miaco" (Kyoto) and one of the boys from Manila about East Asian compasses.

> They described all thing farre different from ours, and shewed that instead of our Compas, they use a Magneticall Needle of six ynches long, and longer, upon a pinne in a dish of white *China* earth filled with water: In the bottome whereof they have two crosse lines, for the foure principall windes: the rest of the divisions being reserved to the skill of their Pilots. Upon which report of theirs, I made a present trial howe a Magneticall Needle would stand in water, and found it to prove excellently well; not doubting but that many conclusions of importance in Marine affaires will thereby more readily be performed.[21]

Barlow concluded that the use of the compass had been widespread in the Indian Ocean and East Asia both before the arrival of the Portuguese and before it became common in England, and it thus represented the original networking technology for Asian port cities. He expanded upon this by recounting the stories about Vasco da Gama's use of a pilot from "Melinde" (the Swahili port of Malindi) to navigate to Calicut as well as Ludovico di Varthema's accounts of the use of compass and "carde" (chart) on a Malay or Chinese junk sailing from Borneo to Java, and Cavendish had clearly done the same thing.

Barlow's rhetoric also illustrated how a coherent sense of Protestantism and Englishness could be built out of contact with Asia. *The Navigator's Supply*, dedicated to Robert Devereux, Earl of Essex, and published the year after the latter had turned the war against Spain by capturing and destroying Cádiz, was a paean to God's providential gift of the sailing compass. Barlow claimed that after "considering that the knowledge of languages growth by entercourse and mutuall access," God "ordeyned" the compass, perhaps as a gift to Asia, in order to remedy the confusion of tongues and to help achieve "the Civill or rather Cosmopoliticall union of humane societie." He

explained that despite the fact that many still believed Spanish and Portuguese pilots possessed superior skills for long voyages, English navigators
had now surpassed the Iberians both in the actual practice of navigation and
in instrument design and mapmaking.[22] For Barlow, the encounters of the
1580s and 1590s had enabled English navigators to do this through the recognition that the Spanish and Portuguese in the Indian and Pacific Oceans
had themselves relied on translating the work of Muslim, Chinese, Japanese, and other pilots. Translating Asian techniques into London's navigation and commerce was no mere speculative activity as it had been with
Cabot's Cathay, but a divinely ordained plan to enable the success of the
Protestant English nation and the circumvention of Catholic empire. It was
on a practical level the kind of reinvention of *translatio studii* that Bodley
and James would attempt a few years later in the refounding of the library at
Oxford. By the end of the 1590s, substantial capital would be ready in London as well as the newly independent Dutch provinces to engage in a broad
program of translation and exchange in the "East Indies" as a whole, beyond
the lingering medieval fantasies of Cathay.

TRANSLATING "CHINA" AND "GIAPAN"

Beginning around 1580, after the scandalous failure of Martin Frobisher's
Cathay voyages, a number of Londoners fed up with fiction had begun to
send out calls to obtain an actual Chinese or Ming map, in part to determine whether Cathay and China were the same place and in part to compete with navigators in Spain and Portugal and cartographers in Antwerp,
who were known to have seen maps printed in China.[23] Even before that,
a number of writers had begun to try to clarify language about East Asia in
particular and to refine the process of translating European travel accounts
and geographies into English. One result was that the geographical terms
"China" and "Japan" had begun to supplement and compete with the older
language of "Cathay" and "Cipangu." The shift in language also implied a
shift away from the concept of a massive tributary empire in Eastern Asia
at the end of the Silk Road and towards a broader recognition of a complex
zone of trade and exchange driven by silver, competing merchant networks
and technologies, and a range of types of sovereignty associated with historically dynamic political entities.

A good example of this change appeared in a series of translations of
Spanish and Portuguese writers for aristocratic and gentry women readers
about the Americas and East Asia published by Richard Willes in 1577 as
The History of Travayle in the West and East Indies. The bulk of the book

was dedicated to Lady Bridget, Countess of Bedford, who gave Willes a pension to make translations in order to help publicize Frobisher's attempted voyages to Cathay. Writing in 1576, Willes noted that it had lately become fashionable for people "to discourse of the whole worlde, and eche province thereof particulerly even by hearesay, although in the first principles of that arte, he bee altogeather ignorant and unskylfull."[24] Good knowledge and good translation—he cited the humanist examples of knowing Greek and Hebrew—were essential to good discourse. Beyond the community of interest defined by the joint-stock, he thought it important to have communities of readers educated with a basic framework in cosmography who could then engage with disparate translations in order to help to make decisions about shifts in language. His particular community of readers, the women of the Bedford household, were directly and indirectly invested in the language of Cathay as embodied currently in Lok and Frobisher's new Cathay Company. Women as readers could play a different and important role from that of men as investors, making translation into an activity geared towards the refinement and accuracy of national language.

This was important because in 1577, the accounts coming to London about East Asia were garbled. The polemically oriented Calvinist printer Thomas Dawson, aiming at a London rather than gentry readership, published that year the first translation in English of the Augustinian friar Martín de Rada's activities in Fujian with his edition of the Spanish merchant Thomas Nicholas's *The strange and marveilous Newes lately come from the great Kingdome of Chyna, which adioyneth to the East Indya, translated out of the Castlyn tongue by T.N.* Dawson put out the Nicholas translation at the same time as a translation by Thomas Rodgers of a Dutch news pamphlet, which argued that the double successes of both the Ottomans and the Pope signified the coming apocalypse.[25] *The strange . . . Newes* indeed strangely tried to raise fears either of a Sino-Spanish rapprochement or potential Spanish invasion of China from the Philippines with the story of de Rada's and Jerónimo Marín's diplomatic mission to Fujian in 1575. Plans were supposedly being hatched in Manila by the Philippine governor Francisco de Sande ("Gandi") and by investors in Mexico City for a massive Spanish invasion of the Ming Empire. It also described a fleet of the "Turke King of Brazer" (possibly Brunei, where de Rada subsequently went in 1578, but probably a reference to the pirate "Limahong" or Lin Feng 林鳳), suggesting the reach of Turkish power. In 1577, it would have been hard to tell which translated tales from the other side of the world to believe, especially in the context of the strong debates over religion that still raged in London.

Translated collections like that of Willes thus had a conservative and

humanist aspect to them, geared towards maintaining the careful modera-
tion of Elizabethan policies by the humanist gentry in the 1560s and 1570s.
Because of his interest in East Asia, some have even seen Willes as a crypto-
Catholic, suspicions also aroused by his time studying at Louvain and his
translation of Jesuit missionary works. Most of his book was in fact repetitive of Marian-era knowledge, reprinting selections from Eden's second
book *Decades of the Newe Worlde* (1555), including those of Pedro Mártir
de Anglería as well as of Gonzalo Fernandez de Oviedo on the West Indies,
from the time when as Willes emphasized "K. Philippe was in Englande."
Willes updated these, however, with two new translations about China and
Japan; an account of English navigation and Frobisher's initial expedition
to Newfoundland; a full translation of the travels of Ludovico di Varthema
around the Middle East, the Indian Ocean, and Southeast Asia (1502–7); and
a variety of other pieces that focused primarily on the North Sea, Russia,
and Persia. Against Spanish universalism, he made a sharp critique of the
"generall table of the world" by Ortelius at Louvain, "for it greatly skilleth
not, being unskylfully drawen for that poynt: as manifestly it may appeare
unto any one that conferreth the same." The Duke of Bedford had one of
Cabot's manuscript world maps on the wall at Cheynes, and Willes wanted
to distinguish the supposedly "universall tables" of Ortelius and Mercator
from those of Frisius and Cabot, who made East Asia closer (Frisius) and less
completely understood (Cabot).[26]

In his most original translations in the second half of the book, dedicated
to Lady Bridget's stepdaughter Lady Anne, Countess of Warwick, Willes
pushed the reader towards a more direct engagement with changes occurring
in Ming China and Japan. The translation of Galeote Pereira's account of the
Ming Empire from information gleaned during his imprisonment in Fujian
between 1549 and 1553 marked the first extensive description of China in
English, emphasizing the continuing failure of the Portuguese to make in-
roads there.[27] Willes altered Pereira's account and described "China" as a
"province of Cathay," following Eden's more interchangeable formulation
"Cathay and China."[28] Pereira used administrative geography to give China
more of a presence, listing thirteen "shires" (provinces or *sheng* 省) along
with rough numbers of "cities" in each, as Mendoza would later. Extrapolat-
ing from Fujian, and in an implicit comparison with Europe, he compared
each "shyre" to a "mighty kingdom," describing China in terms of its the
roads, architecture, and bureaucracy as "one of the best governed provinces
in all the world."[29]

For Pereira, the Portuguese relationship with China was problematic,
and the painful punishments inflicted on prisoners described in excruciat-

ing detail stood as a stark warning against attempting to trade with China without permission.[30] But Willes rewrote this Portuguese story of a "closed" China by highlighting the ambiguity between Cathay and China as well as the contrast between China and Japan and then historicizing the question of openness. Cathay had granted "free accesse vnto all forreiners that trade into his countrey for marchandyse, and a place of lybertie for them to remaine in," until the "*Loutea* or Lieuetenaunt" of Fujian converted to Islam, at which point all but a few Muslims were killed or exiled, a story that is perhaps a loose description of the end of the Yuan Dynasty. The Giapans, on the other hand, "be most desyrous to be acquaynted with strangers," including the Portuguese. Moreover, trade more generally in the ports of East Asia was safe, open, and free, plied by "rude Indish *Canoa*," so that "the Portugalles, the Saracenes, & Mores traueil continually vp & downe that reache from *Giapan* to *China*, from *China* to *Malacca*, from *Malacca* to the *Moluccaes*." This mixture of merchants extended to language, and Pereira did explain that "China" was not a term used or even known in Fujian but a Portuguese convention adapted from the Gujaratis (Persian چین). People in Fujian called their political entity *Tamen* (*Da Ming* or "Great Ming" 大明). This remained an open and dynamic world of language, polities, and exchange so long as one did not, like the Muslims and by implication the Catholic Portuguese and Spanish, try to convert the population to a different cosmology, to fix the terms of what was in fact free exchange, or to extend sovereignty into Asia itself.

For Japan, Willes used a translation of a letter from Kyoto written by the Jesuit Luis Fróis dated February 19, 1565, and titled "Of the Ilande Giapan, and other little Isles in the East Ocean" to show how the unification of Japan also served to define the maritime frontier of the Ming. "Giapan" as a name was a convention drawn from Portuguese "Japaõ" or "Japón" by way of the Italian "Giapan," which had entered Portuguese through the Malay "Jepang," a word also used by southern Chinese merchant networks from both Min-speaking Fujian and Wu-speaking Zhejiang provinces.[31] Willes had seen this concept of Japan emerge in printed European sources firsthand. After a few letters from Portuguese missionaries working in the archipelago appeared in the third edition of Ramusio's *Navigationi* (1563), they were printed relatively regularly at Louvain from 1569, where Willes was a student. While there, he befriended the Jesuit historian Giovanni Petro Maffei, who in 1571 published a highly edited and in certain cases deliberately altered translation of both Fróis's letters and Manuel da Costa's history of the Jesuit mission in Asia up to 1568. This was regarded with great suspicion since da Costa himself and later Matteo Ricci described Maffei's book as full of errors and lies.[32] Willes thus painted a contradictory picture of Japan as both highly urbanized and deeply provincial. The islands, which Fróis described as "full of siluer mines," contained sixty-six kingdoms and were plagued by "continual civil wars."[33] In contrast with the Ming, Japan was a space of disorder. But Willes in his prefatory material also used other sources to describe a land of great cities and a destination for trade, with the ninety thousand households of "Meaco" (Kyoto, *miyako* 都 means "capital") making it a much larger city than London.

This uncertainty about Japan had certain similarities to that expressed by contemporary Ming authors. Just before Portuguese Jesuits began to compose such letters, Ming scholars including Luo Hongxian had published a series of books and maps attempting to define Japan. These books from the presses in Zhejiang and Fujian, starting in the 1540s and especially from the 1560s, also tried to define the relationship of the southern coast of the Ming Empire to the archipelagos and territorial kingdoms surrounding it. Using texts and maps to define the nature of conflict-ridden ocean frontiers in East Asia seems to have emerged as a strategy in Zhejiang and Fujian against the *wokou* pirates. In 1547, the new governor of Zhejiang, Zhu Wan, wrote a conventionally titled essay on food and money, arguing that the closing of the Ming tributary office in Ningbo in 1523 after a dispute between two different embassies claiming to represent Japan had resulted in a dramatic expansion of smuggling by creating a series of alliances between coastal merchants and gentry and Japanese merchants, which also came to involve the Portuguese. For Zhu, this absence of law literally created pirates, turning Japan into an entity that could not be dealt with along standard tributary models like the Ryukyus, and in the absence of any overarching maritime policy he tried to institute a system of regional trading passes (*xinpiao* 信票) to verify merchants and restore legality.[34] In 1549, to enforce bans on smuggling and Japanese piracy, Zhu's forces captured and executed the crews of two junks captained by the Portuguese Fernão Borges and none other than Galeote Pereira. Chinese merchant families in the province with investments in these ships arranged to have the governor put on trial, and he committed suicide that same year.[35] Between 1552 and 1554 a series of raids on the southeast coast, led in particular by the merchant fleet of Wang Zhi, saw the emergence of fortified bases on the Ming mainland funded by Chinese and Japanese merchants and populated by Japanese mercenary warriors. The Chinese army suffering major defeats in 1553 and 1554 retreated into the walled cities. Coupled with high demand for silver this created even more complex relationships, including permission for the Portuguese to set up at Macao (1555–57), tighter relationships between some Japanese lords on Kyushu and the Portuguese and many Japanese merchants moving down to central Vietnam to reach Guangdong and Fujian indirectly, as the descriptions on the map brought back by Cavendish suggested.[36]

By the mid-1550s, both Chinese mapmakers like Luo Hongxian and Portuguese cartographers like Lopo and Diogo Homem at Lisbon were using ancient Nara-era maps by the monk Gyoki (行基 668–749) to try to define Japan as a coherent entity. These were maps largely made and copied by Buddhist monks, placing Japan at the center of the purity of the world and depicting Japan as a thunderbolt of the Universal Buddha.[37] One of the Portuguese maps ended up as part of a folio atlas of the world originally intended for Philip and Mary but presented by Diogo Homem to Elizabeth upon her ascension in 1558.[38] Luo's work (figure 15) aimed at a more comprehensive and accurate picture of Japan, also containing a detailed new survey of the Ryukyus (琉球國, Liuqiu Guo). Unable to extend his grid system to Japan because of disparities of technique between his method and Gyoki's system, he nevertheless used an opening in the ocean waves to imply the need to do so. Then in 1561, an ambitious new atlas project appeared from a press in Zhejiang that relied upon Luo's work and a large number of other maps—Zheng Ruozeng's coastal atlas of China (figure 16).[39] This was literally an "ocean plan," the word *chou* (籌) in the title having the active and passive connotations of both a survey and a strategy. Zheng wanted to build a comprehensive knowledge of both the Ming coast and Japan to drive a wedge between the Japanese and Chinese who had been collaborating in pirate raids and smuggling. His principal map rotated Luo's by ninety degrees, making east into north and literally reorienting the Ming towards Japan and the Donghai or Eastern Sea. Ortelius would subsequently use this convention in an inverted form for his map of China and Japan published in Antwerp in 1584. Other maps in the atlas, like that of *wokou* routes, depicted in diagrammatic fashion an almost abstract struggle with Japan as a concept on the eve of the so-called Unification Wars (1568–1603). The tributary barbarian model and the broader zone of Chinese culture and writing (*wen*) could no longer alone adequately explain the processes of transcultural conflict coupled with political and linguistic consolidation occurring in East Asia.

In this sense, Willes's effort in London to open a discussion about supplementing tributary "Cathay" and "Cipangu" with a more precise geographical language of "China" and "Japan" mirrored in a limited sense a process going on in East Asia. In 1580 John Dee wrote in his "Instructions"

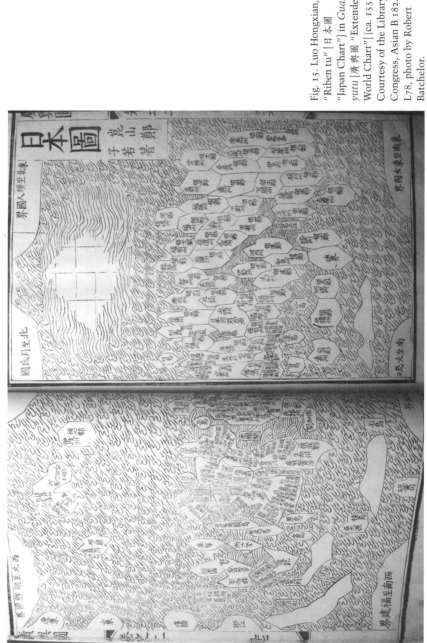

Fig. 15. Luo Hongxian, "Riben tu" [日本圖 "Japan Chart"] in *Guang yutu* [廣與圖 "Extended World Chart"] [ca. 1555]. Courtesy of the Library of Congress, Asian B 182.07 L78, photo by Robert Batchelor.

Fig. 16. Map of pirate routes from Japan. Zheng Ruozeng (郑若曾), *Chouhai tubian* [筹海图编 "Ocean Plan: Compilation of Maps"] ed. Hu Zongxian (胡宗宪), juan 1 (1624, originally 1561]. Japan (*Riben 日本*) is the circle at the top of the map. The Ryukyus or *Liuqiu guo* (流球国) are the circle to the right. Courtesy of the Library of Congress, Asian B145.8 H86, photo by Robert Batchelor.

for the expedition of Charles Jackman and Arthur Pett that "You may take opportunity allso to sayle over to Japan Island where you shall finde Christen men, Jesuits of many Countries of Christendom (and perhaps some Englishe men) at whose hands you may have great instructions & advice for our affaires in hand."[40] Such efforts at "accommodation" between Christianity and Japanese beliefs had officially begun only recently with the 1579 arrival in Japan of Alessandro Valignano, who had been disturbed in Macao the year before by the failure to establish any inland mission in the Ming Empire. Even though Nagasaki became a Jesuit enclave, with the last Buddhist and Shinto temples torn down during the 1580s, Valignano's broader policy involved learning the Japanese language and accepting various aspects of Buddhist ritual and dress to make Catholicism look less alien. At the same time, a similar process of translation and language learning with Chinese and Tagalog took place in the new Spanish colony on Luzon.[41]

By the mid-1580s, as England's undeclared war with united Spain and Portugal began in earnest, accommodation appeared dangerously successful. The keystone to Valignano's accommodation project was to send a Japanese "embassy" to Europe in 1582; it stayed from 1584 to 1586.[42] Rather than an official embassy from the "capital" at Kyoto, it was basically a student exchange trip involving four fourteen- and fifteen-year-old converts to Christianity from Kyushu. The young gentlemen were carefully supervised so that they did not see religious divisions in Europe.[43] After their arrival in Rome, Pope Gregory XIII had European-style clothing made for the group under the assumption that even under the terms of accommodation, the Japanese would eventually undergo a kind of assimilation into a broader Catholic universalism.[44] A newsletter to Elizabeth's court from Madrid dated November 12, 1584, relayed these ideas: "They are white and of very good intelligence, and when they return to their own land it is hoped they will be of much benefit to Christianity, because being Christians and such great men, they may easily convert all those Indies by the respect and authority they have amongst them."[45] This news came just one month before the formation of the Catholic League of Spain and France, which resulted in a dramatic shift in English policy favoring intervention in the Netherlands against Spain. Between March and August 1585 almost biweekly reports arrived from Italy detailing the entertainments of the Japanese ambassadors at Florence, Rome, Venice, Milan, and Genoa and their final return to Barcelona on August 16. Elizabeth at Walsingham's urging signed the Treaty of Nonsuch on August 10, committing England to Dutch independence and de facto war with Spain. With war certain, the news about the Japanese abruptly ended.[46] It must have seemed to some in London as if the joint monarchy of the Iberians stood on the verge of taking over the entire world.

But Christopher and Cosmus, the Japanese converts who came to London with Cavendish in 1588 with news of "the state of their Easterne habitiations," offered a very different version of Japan and East Asia than did the "embassy." Instead of a barbaric Japan conceived of as a launching point for the Christian conversion of East Asia, they represented a more cosmopolitan and networked world of sailors, cities, navigation, trade, translation, and exchange. For them, despite working for the Iberians and then the English, Ming China remained the culturally definitive entity producing and circulating representations of itself and indeed of the Japanese themselves. They had been as happy to engage with Cavendish and learn English as they had been with the Spanish, and in their second and last voyage with Cavendish in 1591 they denounced a Portuguese pilot as a potential traitor.[47] The story of Japan thus indicated to Londoners how contested and fluid, at a global level, languages of sovereignty had become, with the complex struggles over the Japanese archipelago and the line between piracy and commerce distantly mirroring those over sovereignty and unification in Ireland, Scotland, England, and the Netherlands.

EXCHANGING CHINESE MAPS

If the Japanese and Filipinos who came with Cavendish subverted Spanish and Portuguese narratives of closure and empire, the possession of a Chinese map on a Spanish ship suggested that the Ming, itself only founded in 1368 with the collapse of the Mongol Empire, had gone a long way towards solving the problem of sovereign autonomy. During the 1560s, the Ming, which defined and then closed down direct trade with Japan, also helped create a vast sphere of indirect trade by officially reopening trade with Southeast Asia on the request of the governor of Fujian.[48] In addition to renewing its tributary relations with Ryukyu (1561, 1579), it legitimized trade routes with overseas Chinese merchant communities in the Siamese cities of Ayutthaya and Patani, the ports in Vietnam, northern Java, and from 1573 Spanish Manila. Because of this, maps printed on Ming presses in southern China took on a new purpose in East and Southeast Asia during the 1570s; Luo's atlas itself was reprinted and augmented in 1579. Rather than attempts to show a closed, regulated, and bounded empire secure from the lawlessness and piracy of Japan, maps became gifts that overseas Chinese merchants would use to help establish trade relationships. Population and administrative data that accompanied such maps became ways for merchants from Zhejiang, Fujian, and Guangdong to promote the potential for trade with the mainland. This not only worked, it worked as far away as London, producing the initial momentum for the formation of the East India Company.

The first recorded transfer of such maps by merchants occurred at Manila in the 1570s. The Spanish governor general, Guido de Lavezaris, wrote in a letter to Philip II dated July 17, 1574, that the Chinese "continue to increase their commerce each year, and supply us with many articles as sugar, wheat, and barley flour, nuts, raisins, pears, and oranges; silks, choice porcelains and iron," and "this year they gave me a drawing of the coast of China, made by themselves, which I am sending to your Majesty." A second letter sent with another ship on July 30, 1574, added that Lavezaris was also sending Spanish charts of the Chinese coast and of Luzon as well as "another paper which I received from the Chinese, upon which is printed a map of the whole land of China, with an explanation which I had some Chinese interpreters make, through the aid of an Augustinian religious [probably de Rada] who is acquainted with the elements of the Chinese language." This map was most likely the large sheet map usually attributed to Yu Shi called the "Gujin xinsheng zhi tu" [古今形胜之圖 "Ancient and Modern Adventageous Map"] published in 1555 by a local academy called the Jinsha Studio in Longxi County, Fujian.[49] It contained short notes about terrain and geographical changes, including around one thousand place names as well as extensive notes about the history and geography of Ming frontiers, but unlike Luo's atlas, it offered nothing substantial about the demographics of the Ming Empire. Yu's map came out of an early Ming tradition, borrowing from the *Da Ming hunyi tu* (Amalgamated Map of the Great Ming Empire, ca. 1390), the *Ming yi tong zhi* or "Unified Gazetteer of the Ming Dynasty," as well as the ancient *Shanhai jing* (Classic of Mountains and Seas). The synthesis resulted in a somewhat fanciful world picture, not unlike medieval European maps, in which monstrous foreigners lived on the edges of the known world. It nevertheless indicated the desire of Fujianese merchant networks to place themselves in a more global context, and Lavezaris wrote that the Fujianese promised to bring more and better maps next season.[50]

Even in Europe, the Philippines were called the "Islands of China" because as the first stop of the eastern trading route down to the Malukus, they became a focal point for Fujianese and other East Asian merchants to rework ideas about the Ming Empire. The English merchant Henry Hawkes, who lived in Mexico for five years (1567–72), had used the phrase "Islands of China" for archipelago making up the eastern route to the lawyer Richard

Hakluyt (the cousin and guardian of the editor of the *Principall Navigations*). Hawkes also described the trans-Pacific route from Luzon to Barra de Navidad, including the silks, gold, and cinnamon that the Spanish could get from Chinese merchants at Manila in exchange for Mexican and Peruvian silver.[51] Traders from Luzon, Mindanao, and Borneo would claim to be "Chinese" as an indication not of their ethnicity but their access to mainland markets from these islands. Plenty of Fujianese as well as Guangdong merchants traded directly at Manila from 1573, but larger junks built in the Ming Empire would not travel among the smaller islands, as the shoals posed too much of a risk.[52] Maps were convenient signifiers of authenticity in such situations, indicating who truly had merchant networks linked with the Ming and who was just claiming to be Chinese. If the Selden Map was produced in Manila, which is a strong possibility given the detailed port and navigational instructions around Luzon, then it would be one in a line of several attempts to rethink the broader conception of trade among a range of merchants in this period, who all saw themselves associated in one way or another with the Ming economy. The Jesuit Matteo Ricci's famous combining of the work of Ortelius and Luo should be seen as competing with the earlier work coming out of Manila, trying unlike Hakluyt to subsume Ming cartography into the broader context of Ortelius's Iberian worldview.

Because of these exchanges, a small but well-chosen library and map collection developed at Manila, largely thanks to Fujianese merchants and printers as well as the remarkable Augustinian friar Martín de Rada. Having learned Visayan and Chinese after his arrival in 1565, de Rada's two trips to Fujian in 1575 and 1576, for which Chinese provincial authorities extended an invitation after the Spanish repelled the renegade Guangzhou merchant pirate Lin Feng from Manila in 1574, opened the dense world of commercial Chinese printing and mapmaking centered on Jianyang in Fujian to the Spanish.[53] De Rada brought gifted copies of Luo's atlas along with a number of other books back to Manila, supplementing what merchants had given independently. Because of their Fujianese connections, for a brief period in the 1570s it looked like Augustinians and the Franciscans of Manila might take the lead in both Chinese language translation and establishing missionary connections with the Ming. In Seville in 1577, the Spanish priest Bernardino de Escalante, who had lived in London during the 1550s as a member of Philip II's entourage, put together a synthetic account of China drawn from the Portuguese Dominican Gaspar da Cruz's 1569 book and other sources. It highlighted the uniqueness of the Chinese written language and would quickly be translated into English and printed at London and drew awareness to the translation project.[54]

In 1585, Mendoza included in his account of the Ming Empire a list of the Chinese books that had been sent from Manila to Madrid, a list duly translated into English by Parke for his London edition of January 1589. Mendoza, who was part of a failed embassy to the Ming sponsored by Philip II in 1581, returned from Mexico to Spain just as de Rada's papers arrived there on a separate ship. With the help of Augustinians as well as possibly some Chinese in Madrid and Seville and Escalante himself, Mendoza transcribed a catalog of twenty-eight different categories of publications, beginning with texts of geographic and administrative interest.[55] The descriptions suggest a startlingly good collection that most likely included Luo's work or derivatives of it as well as Zheng Ruozeng's *Chouhai tubian* (1562) along with standard Ming administrative manuals like the *Ming shilu* (1574, 1577) and the *Da Ming huidian* (1576).[56] What Mendoza himself gleaned from this material was in many ways quite limited, although he did have tables for the number of cities and towns in each of the fifteen provinces, military units as well as tribute-paying subjects in each area, using population figures for the householder category.[57]

By the late 1570s, the exchange of maps and conflicts among missionary and merchant activities in East Asia had encouraged a breakdown of information control at both the Spanish and Portuguese Casas in Europe. Printing of missionary reports, mutual jealousies, and increasing amounts of East Asian books and maps made it necessary to engage in battles over geographic information at a public level both on a global scale and through Antwerp in particular. Juan Bautista Gesio, the cosmographer to Philip II who had been recruited away from Portugal by the Spanish ambassador in 1573, used the Portuguese cartographer Luis Jorgé de Barbuda (Ludovicus Georgius, ca. 1564–1613) to gather maps in Lisbon, and Gesio ultimately smuggled Barbuda out in 1579 to work with the Spanish on Ming maps.[58] In 1577, apparently working with Escalante in Seville, Barbuda produced a map of the "China regnum" for Ortelius's competitor in Antwerp, Gerard de Jode's *Speculum Orbis Terrarum* (1578), from sources that included the maps of both Yu Shi and Luo Hongxian available in the archives there.[59] Ortelius, whose 1570 *Theatrum Orbis Terrarum* had earned him the post of geographer to Philip II in 1575, used his courtly contacts to have this map of China suppressed.[60] The Bible translator Benito Arias Montano, who knew Ortelius through the publisher Plantin, obtained a version of the Barbuda map and then sent it to Ortelius in 1580, but Ortelius waited to publish a modified version in the supplements to the new version of his *Theatrum* in 1584 (figure 17).[61] The publishing and political power of Antwerp was being leveraged both in Europe and in East Asia in an attempt to control the more

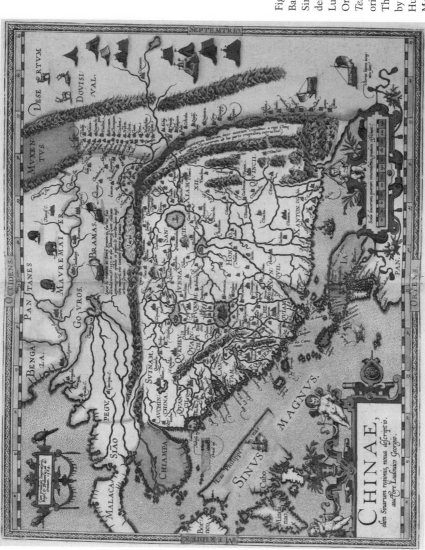

Fig. 17. Luis Jorgé de Barbuda, "Chinae, olim Sinarum regionis nova descriptio, auctore Ludovico Georgio," in Ortelius, *Theatrum Orbis Terrarum* (Antwerp: 1595), originally published 1584. This item is reproduced by permission of the Huntington Library, San Marino, CA.

complex translations of mapping occurring in both places, producing a kind of stagnation that alternative networks of translation were actively trying to circumvent. Ortelius merely included this map in his atlas and did not alter his original 1570 world map to reflect its findings until well after 1589, when Hakluyt had used a version of it as an emblem of current understandings of the world.[62]

Ortelius's willingness to publish in 1584 what he had been responsible for suppressing in 1578 came out of the revelations of Drake's circumnavigation and the treasure he brought to London in October 1580.[63] Ortelius had visited London in 1577, meeting with the Westminster schoolmaster William Camden, whose four-kingdoms vision of *Britannia* (1586) Ortelius subsequently encouraged at a time when the Dutch were courting Elizabeth to become their sovereign.[64] In December 1580, Camden had written to Ortelius about Dee's attempts to locate Japan on a map he was drafting.[65] That same month, Mercator wrote from Duisberg to Ortelius congratulating Ortelius about receiving Barbuda's map and suggesting that Drake's treasure indicated that the English had achieved a northeast passage and reached China. Once it became clear that Drake's efforts meant little in East Asia and the simplified map of China would not significantly help the English, Ortelius became more interested in publishing it as a piece of propaganda about Iberian dominance of East Asia.

It was also the case that Spanish Jesuits had begun to aggressively supplant Augustinian efforts at accommodation from 1583, and a distinct cadre under the Jesuit Father Alonzo Sanchez made common cause with certain provincial Spanish officials to argue for an invasion of the Ming.[66] The map captured by Cavendish on the *Santa Ana* in November 1587 probably was designed to intervene in such arguments, and the English read it to suggest the impossibility of such an invasion. Manila's Governor Gonzalo Ronquillo thought a force of eight thousand men and twelve galleons would be sufficient, in addition to similar numbers of forces recruited from Christianized Japanese and Filipinos. Ricci himself confirmed this bizarre idea when he wrote to Macao in a letter accompanying his first map of China that despite sixty million people registered by the census, "the power of China rests rather upon the great number of towns and the multitude of inhabitants, than upon the valour of the people . . . They have no more spirit than women, and are ready to kiss the feet of anyone who shows his teeth at them."[67] If the Portuguese came up with the cliché about the cruelty and isolation of the Chinese, Ricci and the invasion planners introduced a new claim, repeated for the next several hundred years, that the Chinese were effeminate and an easy target for takeover. This kind of posturing went over poorly in Madrid, where Philip II had more pressing concerns on his mind, and orders went out from the new Jesuit superior, José de Acosta, to recall Sanchez and curtail the dangerous fantasies coming out of Mexico City, Manila, and Macao. The Augustinian Mendoza's 1585 book on China dedicated to both Pope Sixtus V and to Fernando de Vega, president of the Royal Council of the Indies, tried to provide arguments to support the moderates in Madrid and Rome. Using census data, Mendoza explained that the Ming army contained 5,846,500 footmen and 948,350 horsemen, who were not as "valliant" as those in Europe, but nevertheless implying that a force of even 16,000 was rather small.[68] Ricci's militant letter would be published in Italian, French, and German in 1586–87 to compete with such claims.[69] Cavendish's capture of the *Santa Ana* and the subsequent collapse of the Armada assembled that spring to attack England spelled the death of this highly unlikely invasion plan.

Conversely, Cavendish began the process of what would later be symbolized by the Selden Map, integrating London merchants into what was perceived as a Chinese-dominated merchant sphere in East Asia. When Cavendish arrived in southern Java in early 1588, map in hand and following Drake's route, he lied and told the local "King's secretary . . . that wee had been at China and had had trafique there with them, and that wee were come thither to discover, and purposed to goe to Malaca."[70] Pretending to be merchants involved in the Chinese trade had the dual advantage of keeping their identity as privateering pirates hidden and linking them with a large legitimate empire that was neither Spain nor Portugal. Cavendish's interpreter was a "Mestizo" Malay-Portuguese, who told the English that Portuguese factors on Java buy slaves, cloves, pepper, sugar, and "many other commodities." Cavendish told the Portuguese factors that their true king was now in exile in London (another lie; the pretender Antonio had moved to Paris before Cavendish left London) and that he had sailed to Java on orders from the Portuguese pretender and Elizabeth to defend them against the Spanish. The Portuguese seemed to accept this, as tensions with the Spanish over the Philippines had never been resolved, going so far as to accept a gift of three cannons from Cavendish. Cavendish had thus turned the "large map" of the Ming into a tool for undermining Iberian plans for empire, entering into the struggle for definition of the seas around the Ming. On his return, Cavendish described the powerful Ming in a letter to Lord Hunsdon, the Lord Chamberlain, explaining that he needed the map as evidence, for he otherwise feared his reports "should not be credited," admitting that even he was "incredulous."[71]

The fact that the map could undermine Iberian claims to empire came out of the particular techniques used in its composition. Connecting mapping to the administration and fiscal-commercial organization of the empire had been emphasized starting in the early Ming, when the Hongwu emperor (1368–98) had compiled a series of population registers (the Yellow Registers or *Huangce* 黃冊) as well as land-survey and tax handbooks (the Fish-Scale Maps and Books or *Yulin tuce* 魚鱗圖冊).[72] The atlas was a tool developed in the late Ming to help systematically reorganize the complex and frequently dysfunctional census, taxation, and administrative categories of the empire (inherited from the fifteenth century) from a local level as part of a movement to rebuild *xiangyue* (鄉約) or village covenants.[73] Wang Yangming (1472–1529) and his followers like Luo had used the covenants to aid in the reconstruction of Jiangxi province after a long rebellion from 1516 to 1522. The idea was to rehabilitate the more elite level of "bandits" and "rebels" into what Wang called *xinmin* (新民) or "renewed people."[74] The map was literally a tool for building a network of subjects. Luo in particular, through his reading of Wang Yangming's *Chuan xi lu* (傳習錄, "Instructions for Practical Living," 1518), was interested in the connection between the scholar and his place of residence (*ju* 居), and his atlas attempted to develop techniques for mediating between local networks and knowledge and broader imperial administrative structures.[75] The atlas itself appeared not as an imperial but as a local text printed as Luo himself engaged in a process of Buddhist and Daoist spiritual discovery in his home prefecture of Ji'an in southwest Jiangxi, where he subsequently made a community covenant for Tongshui, Jishui, in 1562.[76] Woodblock printing, however, made the atlas an object that could circulate much more widely. Luo's comprehensive map, the *Yudi zongtu* (輿地總圖 "Overall Geographic Map"), was printed separately as a larger map, while the atlas *Guang yutu*, after the initial 1555 printing, seems to have gone through five more editions (1558, 1561, 1566, 1572, 1579) over the next quarter century.

Luo Hongxian's atlas is usually remembered for reviving the use of the graticule from the Song and Yuan Dynasties and for including a legend or key to map symbols, which appears at first glance to be a kind of top-down courtly administrative strategy for rationalizing the state. But analogies with European maps can be misleading here, as the grid and the key were used not to accurately represent a set of distances but to organize information in a diagrammatic or topological fashion, as a *tu* (圖), in order to show relations.[77] Each symbol on the key to Luo's atlas was literally a starting point (*cong* 從) from which one would move to further information on the same map or go deeper and begin investigating a province and its subdivisions. The map envisioned and idealized through geographical data a developing, open, and populous and urban empire, well-integrated with travel routes and the growing global trade in silver, well-governed in its administrative divisions, data collection, and military garrisons. But it was designed to be used in localities to understand the broader connections of the empire rather than at the court in Beijing itself. In contrast to Münster's *Cosmography* with its elaborate imperial emblem or Cabot's first map with its prominent imperial crest, the Ming maps inspired by Luo conveyed sovereignty through the grid, geographic data, and cities themselves. Rather than calling the Ming an "empire" or a "kingdom" as Mendoza did, Cavendish, who had experienced China only through a map, referred to it simply as a "country," one of great "stateliness and riches." It was, unlike the popular image of the Khan's Cathay, Cyrus's Persia, or Timur's empire in 1580s London literature and theater, a model for building a rationalized and unified state out of urban commerce, based on data and relationships rather than a simple assertion of sovereignty embodied in the figure of an emperor.

THE STATE AND SOVEREIGN SPACE

The term "state" became fashionable in England in the 1570s and 1580s as a replacement for "commonwealth" in the struggle to realign institutions of divinity and law, relationships that had been shaken by the Reformation and growing dynastic tensions. In the first decades of Elizabeth's rule, the crown and the Anglican Church, along with the traditional London guilds, had tried to retain a somewhat neutral position with regard to Spanish sovereignty, papal supremacy, and economic dependence on Antwerp. The emerging idea of the state remained tied to medieval and Roman ideas about dynasty and *imperium*, securing and maintaining sovereign authority rather than rationalizing it. In 1571, Edmund Plowden's *Reportes* chronicled the claim (from an early Elizabethan legal case concerning the problematic reign of Edward VI) that the king had "deux corps," one "naturall" and one "politike," in an effort to abstract some of this authority.[78] Even more abstract was Jean Bodin's classic "absolute" formulation from 1576, "La souveraineté est la puissance absolue et perpétuelle d'une République." For Bodin divine and natural law limited such sovereignty, except in the case of "le grand Roy de Tartarie," who held "puissance absolue" once elected by his subjects after the death of a previous sovereign.[79] But neither in early Elizabethan case law nor in Bodin's theories, which both carried forward Catholic and Thomist assumptions about universality, was the abstraction of the sovereign territorial state based on the rationalizing techniques of the state itself. Luo Hongxian had developed the concept, within the framework

of Wang Yangming's neo-Confucianism, of making data collection coherent
and comprehensible from the local level upward in order to consistently
renew the broader framework of the state. The basis of the Ming state in the
imperial cult and rituals of Beijing may not have been rational, but its durability and extension, including the technical image it projected as far away
as London, represented a more coherent model of sovereignty than those
put forward by the lawyers of European monarchs.

The concept of "political arithmetick" and strengthening government
through data is usually associated with seventeenth-century thinkers like
William Petty, whose own understanding of the state seems influenced by
Bacon and Hobbes as well as his experience in the Downs survey in Ireland during the Civil War and Restoration. Although the example of the
first English Bills of Mortality, compiled intermittently from at least 1528
and more regularly after 1603, suggested that the city or at least the parish
should keep track of deaths as well as births, there was no indication before
the seventeenth century that such activities were guided by an overall rationality of the state (or church) let alone designed to produce a coherent sense
of the national population.

In the late sixteenth century, the ex-Jesuit Giovanni Botero published
a series of books developing the concept of "reason of state" out of certain
arguments in Machiavelli's writings that became popular both in Italy and
in London in a series of English translations. He was the first political theorist to appreciate the significance of Mendoza's writings on China.[80] In his
Della cause della grandezza e magnificenza della città (1588, Eng. 1606),
Botero outlined the factors contributing to population growth and urbanization. Reading in Mendoza that even small Ming commercial cities like
Guangzhou were "greater than Lisbon, which yet is the greatest city that
is in Europe except Constantinople and Paris," Botero composed a paean to
metropolises around the world as well as a critique of Italian provincialism
and the failure of its more limited urban ethos. China had been so successful in its urbanization that it seemed "one body and but one city" to him.[81]
Botero tied the success of cities to the natural environment, economic resources, prosperity, and the resulting urban population, and thus began to
develop a "universal" methodology for comparative urbanism. Such themes
about the objective rationality of the success of states were also developed
in Botero's second and more famous book, *Delle raigon di stato*, published
in Venice in 1589, which described how states could found, preserve, and
extend their dominion.[82] Finally, in his extremely popular *Relazioni Universali* (1591–93), a standard textbook for English university students after

its translation in the early seventeenth century, Botero proclaimed China the model state. "There is no countrey moderne or ancient governed by a better forme of police then this Empire: by this government have they ruled their Empire two thousand yeeres: And so hath the state of Venice flourished 1100 yeeres, the kingdom of France 1200."[83]

It is hard to gage Botero's popularity in London before 1601, when his works started appearing in translation and when the successful separataion of the economically and increasingly globally successful United Provinces from Spain had demonstrated the power of his arguments. Empire was the principal theme of Botero's most ardent English admirer, Sir Walter Raleigh, whose 1614 *History of the World*, written while he was in the Tower shows a strong influence.[84] Certainly with the return of Cavendish, a substantial faction of the Levant or Turkey merchants began to argue that relations with the populous, highly urbanized, industrious, and commercial Ming Empire and Japan were the only way to ensure London's economic prosperity and England's autonomy from Spain. This echoed previous strategies supported by Elizabeth in the 1670s and early 1680s to link London with Islamic enemies of Spain like Morocco or the Ottomans through chartered companies.

Thanks to Cavendish, arguments about promoting and establishing trade with East Asia could rationally rest on data about urban networks as well as the clear definition of independent sovereign entities over which the Spanish and Portuguese had little if any influence. A memorial of English Merchants to Lords of Council dated October 1589 offered a survey of the history of the Portuguese "settlements in India," their occupation of Melaka, the Bandas, and the Melukus, at the same time suggesting other places where the English could establish themselves. The complexity of trade at Sumatra, Java, and "Sayam" all presented interesting opportunities, but most important were the densely populated and urban Ming China and Japan, "ready to entertayne all strangers that resorte unto them."[85] But in March 1593, Cavendish's ship *Leicester*, destined for China for a second voyage, returned to Plymouth without its captain. Cavendish, who died at sea perhaps by his own hand, had tried to convince his crew after failing to round South America to go to China by way of the Cape of Good Hope, but when he spread out his charts, "all manner of discontents were unripped amongst themselves, so that to goe that way, they plainly and resolutely determined never to give their willing consents."[86] The mere juxtaposition of Spanish and Ming maps had convinced London investors but not sailors, and the failure of data strategies on this voyage became the principal focus of its two most prominent survivors—Robert Hues and John Davis. Both

Hues and Davis published books in 1594 supporting comprehensive and literally "global" strategies for data collection in London as well as making direct trading links with the Ming as part of this effort.[87]

In general, the problem with Botero's conception of reason of state was that it relied more heavily on the concept of the "interest" of the state rather than the actual data about "the greatness of cities" and their relations between one another that Luo Hongxian's map suggested it was possible to quantify and measure. In London, there had been a number of failures based on imposition of fantasies of state "interest"—the outright fraud of Frobisher, the increasing reliance on angelic communication by Dee to support his "British Empire," the poetic veneer of Spenser applied to fraught Irish colonization. While surveying Munster, Sir Valentine Browne as surveyor general of Ireland encouraged utopian urban projects, geometrically and mathematically organized, that had clear defenses and standing garrisons supported by taxation revenue. The almost total collapse of Munster in 1598 during the Nine Years' War suggested how little had been accomplished through this approach.[88] Raleigh's efforts were also decidedly mixed. In October 1584, hoping to find a passage to the Pacific, he brought back Manteo of Croatan and Wanchese from Roanoke to work with astronomer and mathematician Thomas Harriot in Durham House until April 1585 on translating between Algonquin and English and then later with John White on mapping the coastline.[89] But no new passage to Asia could emerge from these projects, and the two attempts at settlement between 1584 and 1587 in Virginia ended disastrously. The ad hoc character of exploration and measurement, the general problem of records "lying scattered and neglected" and data collected haphazardly and unsystematically were key motivations for Hakluyt when planning his travel collection in 1587. He continued this theme in his actual publication, suggesting that the supposedly "universall cosmographie" outlined in Continental books had never actually proceeded from unified premises but was in fact "most untruly and unprofitablie amassed and hurled together," unlike the more coherent if less comprehensive Ming map.[90]

Along with Hakluyt, Hues and Davis, and Harriot perhaps the clearest advocate for the collection of actual geographic data in the service of building up the autonomy of the state in London during this period was Edward Wright. Wright is most famous for what Shakespeare in *Twelfth Night* (1601) called his "Augmentation of the Indies," in a map made around 1600.[91] As Shakespeare noticed, the detail ("more lines") supported by new data techniques was extraordinary in comparison with the globes of Emery Molyneux or contemporary maps by the likes of Hondius and Ortelius. Often found attached to the third volume of Hakluyt's *Principal Navigations* (September 1600), Wright's map included China, Cochinchina, Korea, and Japan as well as references to recent Dutch activity at Banten on Java (figure 18).[92] Like Cabot's map, it left "Tartary" above China blank, but Wright now made no reference to Cathay or the Great Khan. The blank space was a new kind of unknown, not speculative or mythic but lacking data while having a framework for its future placement. According to Wright, the effort to fill such spaces with the extended royal body of the sovereign, whether or not through the centralized institutions of Iberian pilots and German-Dutch cosmographic traditions, had created "inextricable labyrinth of error," a "false projection," "faulty to the very groundwork thereof."

Radically different from the contemporary maps of Hondius or Ortelius, Wright's map only showed coastal regions about which data had been collected, thus enabling certainty in terms of routes as well as providing a scalable framework and methodology for gathering and organizing new data, doing for the oceans what Luo Hongxian did for the territorial state. It is often noted that Wright relied on Emery Molyneux's globe for his chart; nevertheless, although the cartography had certain similarities, there were substantial differences in terms of East Asia from both Molyneux's surviving original globe at Petworth House (1592) and the Hondius revision (1603). For Japan, Korea, and the Chinese coast, Wright in 1599, Molyneux for his 1592 globe, and Hondius for the "Christian Knight" map of 1596 all seem to have had at least indirect access to a copy of a map similar to the Spanish manuscript "Sinarum Regni Alorusq Regno Rus et Insularus Illi Adiacentium Descriptio," which synthesized Yu Shi's sheet map, Luo Hongxian's atlas, and a Gyoki-style map of Japan.[93] Wright was willing to extend his map into parts of Ming China and Japan that were known only through the work of East Asian cartographers, and in this sense he had more clearly "augmented" the Indies than Hondius.

Wright himself did not begin with Shakespearean ambitions to shift Londoners' understandings of either the "Globe" or the "Indies." His 1599 book only contained a map that showed the English coast west of Portsmouth, Spain, and the Azores, the first usable chart based on the Mercator projection derived from data on magnetic variation. The data had been collected in 1589 on an experimental voyage Wright made to the Azores with Hues and Davis right after Cavendish's first voyage. It was sponsored by George Clifford, Earl of Cumberland, who later became influential in founding the East India Company and provided its first ship, the Deptford-built *Red Dragon*.[94] Tables of data formed the basis of Wright's actual method, and he used them to help Molyneux finish the globe project before Davis

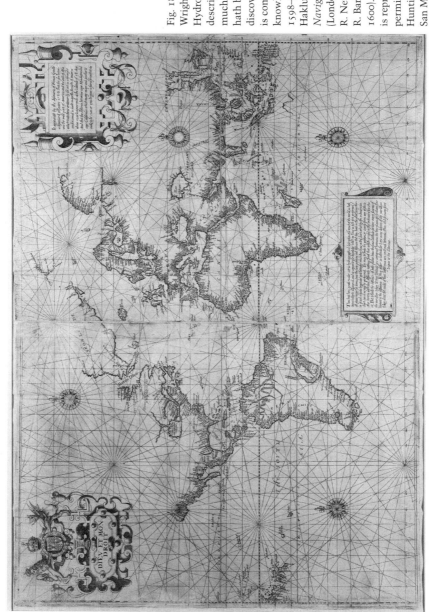

Fig. 18. Edward Wright, "A True Hydrographical description of so much of the world as hath beene hitherto discovered, and is comme to our knowledge" (ca. 1598–9) in Richard Hakluyt, *Principall Navigations* 3 (London: G. Bishop, R. Newbery, and R. Barker, 1599–1600). This item is reproduced by permission of the Huntington Library, San Marino, CA.

and Hues returned from the second Cavendish voyage. Later, he claimed this made calculation automatic so that average sailors did not actually have to learn astronomy, geometry, and trigonometry. It also meant that "navigation might by Arithmetical calculation onely, be performed without Chart or Globe." Mathematicians would make navigational tables, and once made the actual tables "black-boxed" the mathematically created and thus authoritative navigational space.[95]

The idea that empirical data rather than representations alone should serve as a mediating force in making authoritative claims about space was the great scientific realization of the 1590s in London. It came out of the interrelated failure of data strategies in Ireland and Virginia and apparent success of Ming data collection. Thomas Blundeville's *Exercises* (1594), a work on mathematics, geography, and navigations that played an important role in popularizing Wright's ideas even before he made his map, emphasized the new geographical order of the "seniories" of empires in Asia beginning with China. This new and strikingly diverse interest in London in geography and mathematics attempted to define ways of navigating a world of multiple sovereignties, which held their authority independently and at least in part through the mathematics of census, cartography, and navigation. It contrasted with Elizabeth's apparent disinterest in maps and focus on a more limited sense of the British Empire confined to extensions of power across the "narrow seas" surrounding the British Isles.[96] It was the efforts of London mathematicians and mapmakers and increasingly their Amsterdam counterparts strengthening sovereign claims through the collection of data that most effectively undermined the universal and abstract cosmographic efforts of the Spanish and Portuguese Casas.

This context saw the birth of the London East India Company, like Cabot's Cathay Company a private joint-stock venture but one that explicitly had the comprehensive translation of data rather than the simple creation of trading relations as part of its chartering. The first subscription list appeared on September 20, 1599, with 101 names ranging from Sir Stephen Soame, the Lord Mayor of London, to several aldermen, to more apparently humble names like Thomas Cutteler, grocer, Nicholas Pearde, clothworker, and John Harbie, skinner. They pledged amounts generally between £100 and £500, to raise a total of £30,133 6s 8d.[97] On September 24, they resolved as an "assembly" to appoint fifteen directors and formally petition the queen. Factors or officers were appointed "by a generall Assemblie of the adventurors and there elected by the consent of the greater number of them assembled."[98] The next day's meeting resulted in talking points—the Dutch were preparing a new voyage following a successful return from Java in 1597, a "joint and united stock" was necessary for distant trade, export of foreign silver would be required because of the nature of trade with Ming China.[99] The committee delivered a petition to the Privy Council in October and received a negative answer on October 16. The council claimed the Company might endanger peace negotiations with Spain, arguing that "the generall state of merchandize" rather than the particular interests of this joint-stock would be better helped by peace with Spain.[100]

Hakluyt's dedication of volumes two and three of the *Principall Navigations* to Robert Cecil in 1599 and 1600 were certainly an effort to persuade both patron and readers that following Portuguese routes to independent Asian sovereignties would be the best way to challenge Iberian claims to universalism and to relieve pressure at home.[101] English sailors like John Davis and William Adams were already signing on to Dutch voyages from mid-1598, and in September 1598, Cecil received correspondence that two English ships had intercepted two Portuguese carracks on their way from Goa to Macao.[102] Rumors swirled around London in August and September 1599 that the new Spanish king Philip III had readied another armada of 220 ships and 30,000 soldiers backed by French, Danish, and Scottish allies to aid Tyrone in Ireland.[103] Spanish efforts foundered in the Azores in September, but that same month the Earl of Essex, fighting a losing war in Ireland against the Earl of Tyrone, called a truce. The assembly for the East India Company first met in London four days before Essex left Ireland on September 24. Essex appeared before the Privy Council on September 29, confronting Cecil and Raleigh over the issue of the Irish war and was placed under house arrest on October 1.[104] By mid-October, the queen had warmed to Essex's position, and a stalemate ensued as Cecil tried to force a trial of Essex, which finally began on June 5, 1600. Meanwhile, the Dutch were pioneering a successful commercial strategy with English sailors, and in August 1600 the Earl of Northumberland wrote to Cecil from Middleburg that John Davies himself had returned after a voyage of twenty-eight months under the command of Cornelius Houtman with a load of pepper and spices from Aceh in Sumatra.[105]

In this context, Cecil wanted to know whether England could develop trading relations with autonomous sovereignties other than the Ottomans and the Persians if it pursued an anti-Spanish and anti-Portuguese strategy in Asia. There was no room here for speculative Cathay.[106] Cecil and Elizabeth asked Fulke Greville, from December 1598 the treasurer of "Marine Causes" and a semi-neutral client of Essex, to investigate. Greville replied on March 10, 1600, beginning, "You demand of me the names of such kings as are absolute in the East and either have warr, or traffique, with the King of Spaigne." Beginning with Africa—from Fez and Morocco—Grenville gave a comprehensive list but saw the biggest opportunities in Sumatra ("or Taprobuna"), where the kings of Aceh and Tor were enemies of both the Portuguese and the Spanish and gathered in commerce from both East Asia and the Indian Ocean. He noted that the "Philippinas" had been abandoned by the Ming government, thus allowing the Spanish to use them as a point of access to the China trade, and that ships come to Goa from "Arabia, Armenia, Persia, Cambaia, Bengala, Pegu, Siam, Malacca, Java, Molucca, and China," in both cases suggesting that the Iberian powers relied heavily on networks of non-European seaborne merchants to bring goods to them. It was a rushed document, Greville "having neither meanes nor tyme to seak other helpes," relying on a handful of sources.[107]

The Company's equivalent to Greville's document, drafted most likely by Hakluyt as a response or supplement to Greville, was entitled "Certayne Reasons why the English merchants may trade into the East Indies especially as to such wise kingdoms and dominions as are not subjecte to the kings of Spain and Portugal; together with the true limits of the Portugals conquest and jurisdiction in these Oriental parts." They gave another list of places in Asia that had their own distinct sovereignties, highlighting among other kingdoms and islands "the most mightie and wealthy empire of China," as well as Bengal, Pegu, Siam, Cambodia, Cochinchina, Sumatra, Java, Borneo, New Guinea, Mindinao, the Liuqius, Japan, and Korea as well as the multitude of islands in the "Malucos and the Spicerie," all of which Greville had only touched on indirectly. They also demanded that as part of peace negotiations the Portuguese and Spanish draw up a list of ports, cities, fortresses, and islands over which they claimed sovereignty. The concluding argument prefigured that of Hugo Grotius a decade later about the freedom of Christian princes and states to "the vaste, wyde and infinitely open ocean sea" and access to the numerous "territories and dominions of so many free princes, kings and potentates in the East, in whose dominions they [the Spanish] have no more soveraign comaund or authority, then wee, or any other Christians whatsoever."[108] It ended with a long list of sources and people interviewed, raw data on sovereignty collected by those interested in forming a trading company.

By late September 1600, the queen and Privy Council indicated that a charter would be forthcoming, officially granting it on December 31, 1600, and the committee began to purchase supplies and Spanish *reales*. In the charter itself, the language of a world of independent sovereignties was framed in the domestic language of nation, people, and commonwealth rather than reason of state: "We greatly tendering the Honour of the Nation, the Wealth of our People, and the Encouragement of them, and others of our loving Subjects in their good Enterprizes, for the Increase of our Navigation, and the Advancement of lawful Traffick, to the Benefit of our Common Wealth." Both the development of a language of nations for China, Japan, and other Asian polities along the lines pioneered by Willes and other translators of the 1570s and 1580s as well as the collection and mapping of data about sovereignty had been necessary elements for creating the new company. The charter granted status for fifteen years for "one Body Corporate and Politick" made up of the "Governor and Company of Merchants of London, Trading into the East-Indies," and it also required the London-based Company to serve as a supplement to the Royal Navy, agreeing to let its ships be fitted for war and royal service in case of attack.[109] The first voyage left in 1601 and returned successfully in 1603, having established relations with Aceh and a factory in Banten in the same part of the city as the houses of overseas Chinese merchants. It would be in Banten that the English captain John Saris in 1612 would seize a second map of the Ming Empire, a printed sheet map entitled *Huang Ming yitong fang yu bei lan* [皇明一统方與備覽, "Comprehensive directional view map of the Imperial Ming"], revealing even more directly the ways that Chinese merchants from Fujian in their own personal collections valued their relation to the empire and its maps.

The actual sovereignty of the Ming and the emerging sovereignty of Japan, as well as the methods developed by Ming cartographers for comprehending their own urban networks, allowed for substantial redefinitions of sovereign and cartographic space by urban merchants and intellectuals in both coastal China and London in the late sixteenth century. As Luo's atlas suggested, it was no longer the mere drawing or chart that defined claims to sovereignty with crests and emblems. The method of assembling the data itself and translating it offered a much more substantial basis for creating spatial authority that was globally recognizable. Translating Ming strategies for assembling territorial data along with broader East Asian strategies for exchange and navigational knowledge outside of imperial borders allowed Londoners in the 1580s and 1590s to take a key step towards a global reach in terms of trade, most notably with the founding of the East India Company.

There is no question that the East India Company was still a kind of organ of state, understood as part of the queen's virtual body and rooted in older strategies of world pictures and corporate identity. Nevertheless, unlike Cabot's Cathay Company, the East India Company relied on real data about the precise locations of sovereign space. At least one of these sover-

eignties, Ming China, considered its population relatively equally as households, using a census to compile an aggregate national population rather
than identifying corporate groups with specific privileges. This notion of a
collection of householders, which gave clear evidence of Ming sovereign independence from Spanish and Portuguese claims to imperium, had arrived
in London with Thomas Cavendish on a Chinese map in 1588 along with
witnesses of its success from Japan and the Philippines. In the early 1590s,
Barlow having met these sailors could still write that the broader population of London still had little if any knowledge about the extent and nature
of global trade that sustained their city. "I knowe right well, that among
the vulgar people, there are many thousands, which going unto market,
doe bring home with them a peniworth or two of Cloves or Mace nowe
and then; and yet never asked the question on where they grewe, nor by
which way they were transported from their Native country unto us."[110]
By 1604, however, Shakespeare was making jokes about the desirability of
"china dishes" in *Measure for Measure*, and his audience increasingly knew
whereof his characters spoke.[111]

CHAPTER THREE

The Value of History:
Languages, Records, and Laws

The law speaks by record, and if these records remain, it will to poster-
ity explain the law.
—John Selden, speech to the House of Commons, February 12, 1629[1]

1619: JOHN SELDEN, HUGO GROTIUS, AND EAST ASIA

In the early seventeenth century, two additional Chinese maps made their
way to London. These in many ways symbolized the way developments in
East Asia were beginning to shape a kind of global or international politics
in London, a politics that increasingly had implications for how laws as well
as languages and history were understood. The map captured by John Saris
in Banten in 1612 and reprinted by Samuel Purchas in his voluminous 1625
Hakluytus Posthumus, or Purchas his Pilgrimes, showed a bounded and
distinct Ming Empire that, following Luo's atlas, was organized according
to a hierarchy of municipal locations and connected through a series of river
systems. Conversely the Selden Map, which was probably drawn around
1619 and became an object of display in John Selden's "museum" or library
in the early 1650s, suggested an empire integrated into a regional system of
trade, thriving in many ways because of the navigational and economic successes of its overseas merchants. The map that Selden collected rather than
the one printed by Purchas ultimately reflected in a more concrete manner
the kind of "empire" that the Britain of the Stuarts and the Commonwealth
would become, one in which claims to dominion over the seas rested largely
upon the technical achievements of mariners and the contracts with diverse
sovereignties that made exchange possible. In a traditional history of England, the Petition of Right (1628) that revived the claims of Magna Carta

and in which Selden played an important role would be understood as the great milestone in the advancement of English rights during this period. But the Selden Map reveals a different dynamic in which both the developing autonomy of the state and domestic resistance to the kinds of martial and maritime claims being made in the name of the monarch could be reconceptualized both legally and historically because of maritime networks, technical achievements, and autonomous relationships emerging on a global scale.

Like the more celebrated 1588, 1619 was a watershed year in which a number of debates about legal sovereignty and historical rights came to a head at key nodes along the routes of the silver trade and in which the two different concepts of East Asian trade embodied by the Purchas and Selden maps began to diverge. In London that summer, as the Dutch and English concluded a mutual defense treaty in June against the Spanish and Portuguese in Asia, John Selden submitted the manuscript of his *Mare Clausum* to James I as a response to Hugo Grotius's *Mare Liberum* (1609). To James's dismay, Selden advocated a British closed seas policy against the Dutch goal of open seas, which as Richard Hakluyt knew could directly justify blockades of Iberian shipping in Asia.[2] But the Company went ahead and sent a ship from London, the *Elizabeth*, to support the Dutch blockade of Manila from Hirado in Japan. Likewise in 1619, the Verenigde Oost-Indische Compagnie (VOC, established 1602) established a new headquarters on western Java at the old Sundanese capital of Sunda Kelapa or Jayakarta that they would rename Batavia. After two previous years of war with the Dutch and struggles to balance accounts with Chinese merchants in neighboring Banten, the London East India Company's "Presidency" would have to work uncomfortably in Batavia between 1620 and 1628. Also, around 1619, in the Chinese parian of Manila or perhaps even in Quanzhou, Fujian, the Chinese mapmaker who drew the routes on the Selden Map was trying to devise a way of defining legitimate shipping routes across East Asia in the context of competing imperial ideas of sovereignty, most likely for an independent merchant pirate like Li Dan of Hirado, Japan, who was building up capital networks and simultaneously leveraging the Dutch, English, Spanish, Tokugawa, and provincial officials in the Ming. The mapmaker did not think the English worthy of mention nor did he know about changes in distant Batavia, which until that year was still Jayakarta (*Yaoliuba* 咬留吧), but he was aware that the Dutch (*hongmaozhu* 紅毛住) and the Spanish (*huarenzhu* 化人住) were pursuing distinct strategies in East Asian trade from fortified positions in Ternate (*Wanlaogao* 万老高).[3] Dutch blockades, along with competition to find a new southern route across the Pacific past New Guinea, were disrupting both the traditional eastern route of Chinese traders and the

supply of silver making its way to the Ming Empire. In turn, Chinese traders were responding with innovative strategies for balancing silver between
different sovereign entities that would have profound effects on the future
development of the London East India Company's trade in East Asia and
the Indian Ocean as well as the emerging "Atlantic World" itself, even if no
single power or merchant network was able to issue a kind of Navigation
Act for East Asia.

In a European context, the developing autonomy of the English and
Dutch "states" in the late sixteenth and early seventeenth century might
be described as a process of devolution from Spanish and imperial authority
that inverted the attempt by the three Philips (II, III, and IV) to coopt the
Portuguese monarchy and trading networks. More tangibly, this national
autonomy was reflected in and sustained by their own devolutions of power,
notably the creation of corporations for trading in Asia, extensions of dominion or property paralleled in East Asia by the Japanese "Red Seal" system between 1600 and 1635. Perhaps because of their novel forms of organization and the distances involved, the London East India Company and
the VOC both displayed a unique sense of autonomy from the theoretical
national source of their sovereignty. Even more than the process of state
formation in Europe, their development as a result was entangled in processes concerning trade and sovereignty emerging from Asia.[4] In the 1610s
and 1620s, there was even talk of merging the Dutch and English companies as a way of pooling capital in order to challenge the joint-monarchy of
the Iberians and to bargain more coherently with various Asian merchant
networks or emerging state formations. The fact that this merger did not
occur has much to do with the different situations of London and the united
Dutch cities, with their divergent relations with Asian trade, and most importantly with their different attitudes towards both techniques and strategies for building relationships in East Asia.

Asian information networks required and indeed were pioneering new
kinds of contractual relations and through them new kinds of histories—
from the *Hikayat Aceh* (ca. 1613) commissioned by Iskandar Muda to the
Tokugawa annotated edition of the *Azuma Kagami* (1605).[5] As chapter 2
suggested, sixteenth-century history writers in both Europe and China had
tended to emphasize centralizing developments in the territorial empires
like the Ming to the neglect of more decentralized maritime efforts. But
during the seventeenth century, devolution of authority required by shifting patterns of exchange, multipolar and overlapping trading networks, and
extension of exchange relations beyond the distance of easy communication
caused a proliferation of objects ranging from manuscript histories to more

technical achievements such as popular printing or instrument and map-making designed to address such problems. Significant numbers of these made their way from East Asia and the Indian Ocean to London, eventually passing into collections like those of the Bodleian Library. Rather than knowledge disseminated from a central authority, each of these artifacts provides evidence of attempts to establish new kinds of authority over time and space distinct from, supplemental to, or more broadly devolved from the imperial center. They appear to be efforts by overseas merchants in particular both to comprehend the broader extraimperial world in which they operated and at the same time to find an epistemological grounding for their own activities. London's relationship to these more diffuse activities, rather than to centralizing processes in imperial capitals like Beijing, proved crucial to shaping conceptions of how translation and cultural interaction occurred.

The person who perhaps best understood this in early seventeenth-century London was John Selden, and by the late 1640s he had his own house and library at Carmelite in Whitefriars, London. The map with its system of routes became an emblem of "closed" and indeed measured seas in East Asia. It suggested the successful resistance and possibly even indifference to both Iberian and Dutch imperialism by Asian and especially Chinese merchants as well as polities from the Bantenese sultanate on Java to the Tokugawa in Japan. Selden wrote a codicil to his will dated June 11, 1653, requesting specifically that his map of China, in some ways the centerpiece of his library in Whitefriars, be placed with the rest of his collections into what would have been London's first public library and museum. Like the first Navigation Act in 1651, the map was also a retort to the ambitious Dutch, who in 1662 received a physical demonstration of the limits of their own power by the Ming loyalist army of the merchant-pirate Zheng Cheng-gong (aka Koxinga), who forced them to leave Taiwan.[6] Selden's *Mare Clausum*, translated into English in November 1652 and republished in 1663 with a new dedication to Charles II, became central to thinking in London about trade and law until Adam Smith's attacks on mercantilism in the *Wealth of Nations* (1776). It provided the principal justification for the new state-building policies of the Rump Parliament and Cromwell in the first Navigation Act of October 1651, "the first parliamentary statute that in any comprehensive way defined England's commercial policy."[7] Selden's book in tandem with his Chinese map revealed how the creation of successful economic, legal, and even military strategies by the early seventeenth century in London and the Atlantic had come to rely upon forms of capital accumulation dependent on legal agreements but largely independent from direct state control; these techniques were developed on a global scale but particularly in East Asian waters. They also provided a basis for alternatives to new kinds of absolutism and imperialism such as those practiced by the Qing from Beijing and the Dutch from Batavia.[8]

In the substantial recent scholarship about Selden, his role in placing London's mercantile activities in a comparative global context has been neglected.[9] Selden's first draft of *Mare Clausum* responded not only to the Dutch propagandist Grotius, who had come to London in 1613, but to Grotius's English translator Richard Hakluyt (d. 1616) and the picture of English navigation that Hakluyt had composed in the aftermath of Cavendish's return that depended purely on national achievements.[10] As an Elizabethan thinker, Hakluyt had responded to the image of closure presented by Philip II as joint monarch of Spain and Portugal, by suggesting a nation composed through the process of gathering data, a process subsequently articulated by a number of Londoners from Harriot and Wright to Raleigh and Bacon. But such data gathering also revealed what seemed in Christian terms to be a radically open world of sovereignty in Asia, constrained only by the presence of territorial empires like the Ming with substantial bureaucratic, contractual, and textual apparatuses that generated *imperium*. For Selden, the seas were not as open as Bacon's famous image of the Pillars of Hercules in the *Instauratio magna* (1620) suggested. To address this conceptual failing, Selden linked the old Latin concept of *dominium* (property, as opposed to sovereign *imperium*) to mutual technical and contractual recognitions, which he believed occurred on a global scale despite having examples drawn mostly from Latinate Europe (figure 19). Law and agreements were never simply matters of assertion. Global claims had to respect sovereign entities like the Ming on the basis of *imperium* as well as contracts, treaties, and scientific and even technical achievements on the basis of *dominium*. Strategies of collection, as Selden engaged in with historical manuscripts and data-oriented and technical objects ranging from the Selden Map to North African astrolabes and Mexica painted books, would demonstrate and show how to engage in these kinds of relations (figure 20).

Ideologically, Grotius's initial treatise on the free sea had attempted in the broadest possible terms to justify a singular historical act that occurred off the Malay Peninsula in 1603. That year a Dutch company, which because of the delay in transmitting information across the world was not yet operating as part of the newly united VOC, seized the Portuguese ship *Santa Catarina* as it approached the Straits of Malacca, supposedly as part of an alliance with the Malay sultanate of Johor. Instead of justifying the legality of this through the company's history of relations with Johor, Grotius ab-

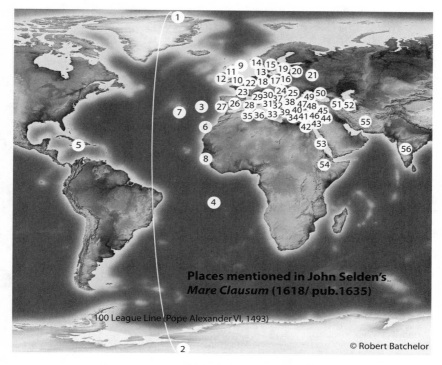

Places mentioned in John Selden's
Mare Clausum (1618/ pub.1635)

100 League Line (Pope Alexander VI, 1493)

© Robert Batchelor

Fig. 19. World map with locations mentioned in *Mare Clausum* (1618/1635).

1. Arctic Pole	15. Sweden	29. Genoa	43. Israel
2. Antarctic Pole	16. Poland	30. Venice	44. Tyre
3. Atlantic (Al-Andalus Sea)	17. Pomerania	31. Rome	45. Syria
4. South Sea (South Atlantic)	18. Germany	32. Naples	46. Syrian Sea (Al-Shem Sea)
5. West Indies	19. Baltic	33. Sicily	47. Constantinople
6. Canary Islands	20. Lithuania	34. Crete	48. Ottomans (Turks)
7. Azores	21. Muscovie	35. Algeria	49. Black Sea
8. Guinea	22. Holland	36. Carthage/Tunis	50. Sea of Azov (Maeotis Sea)
9. British Sea	23. France	37. Adriatic	51. Caucasus
10. England	24. Austria	38. Macedonia	52. Caspian Sea
11. Scotland	25. Hungary	39. Greece (several locations)	53. Red Sea
12. Ireland	26. Spain	40. Aegean	54. Ethiopia
13. Denmark	27. Portugal	41. Cyclades Islands	55. Persia
14. Norway	28. Mediterranean	42. Egypt/Nile	56. India

stracted the concept of free seas from natural law, removing from the ocean any claims to *imperium* or *dominium*. Despite the subtitle "De Jure quod Batavis competit ad Indicana commercio," his immediate motives were so obscure that in Scotland and the western coasts of England, where fishing was a key source of income, many thought the text was an attack on British rights to North Sea cod or even whaling off the Svalbard Archipelago.[11] However, Hakluyt's manuscript translation, made in London a few years before his death in 1616 and passed among the intellectual and legal circles

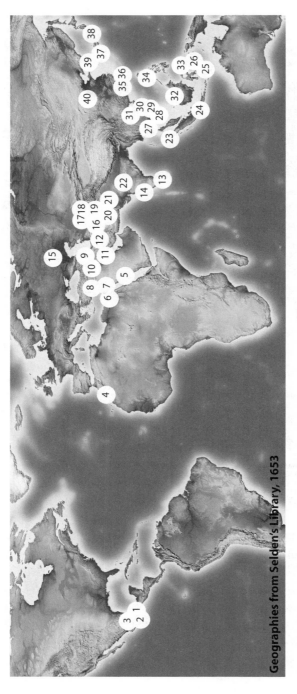

Geographies from Selden's Library, 1653

Fig. 20. World map with locations from astrolabes, the Selden Map, and Mexica manuscripts in Selden's library in 1653. The Laud astrolabe shows fifty locations. Selden's astrolabe only marks Marrakesh but includes several latitude ranges. The Selden Map shows over sixty East Asian ports.

Regions from Selden's Mexica books:
1. Jaltepec, Mixteca Alta
2. Coixtlahuaca, Mixteca Alta
3. Tenochtitlan, Aztec Triple Alliance

Selected cities from Laud's and Selden's astrolabes:
4. Marrakesh
5. Mecca and Medina
6. Misr (Cairo, Egypt)
7. Dimashq (Damascus)
8. Halab (Aleppo)
9. Tabriz
10. Baghdad
11. Basra
12. Isfahan
13. Sarandib (Sri Lanka)
14. Kanbayat
15. Volga Bulgar
16. Herat
17. Bukhara
18. Samarqand
19. Kabul

20. Qandahar
21. Lahore
22. Delhi and Agra

Selected places from Selden Map:
23. Aceh
24. Sunda Kelapa (Batavia)
25. Timor
26. Banda
27. Siam
28. Cambodia
29. Champa
30. Hoi An
31. Tonkin
32. Brunei
33. Ternate
34. Manila
35. Quanzhou
36. Taiwan
37. Nagasaki
38. Edo
39. Korea
40. Beijing

of the Inner Temple, appeared as the East India Company engaged the VOC in extensive negotiations in London (1613) and The Hague (1615), where Grotius also served as a delegate, bringing the question back to East Asia. Both companies in 1615 still supported the idea of free seas that had since the 1580s helped create a united front for making Asian alliances against the Iberians, although skeptics like Selden were emerging, especially over the question of a merger.[12]

With Spanish and Portuguese claims to papally sanctioned *imperium* tottering during late sixteenth-century wars, Grotius and Hakluyt had observed that Catholic claims to *dominium* or state property seemed to rest increasingly on opportunistic Machiavellian assertions of power, frequently made in the context of those wars. Spain had recently, in 1606, staged a massive invasion of Ternate, with thirty-six vessels and two thousand soldiers, out of fear that the Chinese, the Japanese, the Dutch, and the English all had designs on the Malukus. Ternate was well known in London and Amsterdam, in part because an inset in Hondius's 1596 map of the world had shown Drake's *Golden Hind* being towed in 1579 by galleys into its harbor. The struggle between the Dutch and Spanish to establish a route to Ternate around New Guinea from Banten and Peru respectively had been closely watched by the English, and the substantial efforts by the Spanish to fortify Ternate and thus create a potential southern silver route across the Pacific were well known in both Europe and as indicated on the Selden Map among Chinese merchants. The Dutch decision to send a massive contingent of 1,900 men in 1607 to Ternate, which arrived by way of Banten in 1609 and proceeded to build fortifications and blockade Manila, was a dramatic maneuver that the London East India Company, with fewer resources, could never have emulated.[13]

Clearly both the Iberian and the Dutch models required substantial investments in military force and defenses, a Machiavellian confidence in the relation between military and state power. The neo-Machiavellian Botero in his 1598 *Relationi del Mare*, first published in English in 1635 as a kind of alternative theory to the first edition of *Mare Clausum*, had followed other jurists from Genoa and Venice in arguing that a degree of dominion over the sea existed through reason of state to the extent that a city needed to control its ports or its people on surrounding islands. Possession of "Mediterraneans" (from the Greek Μεσόγειος or "middle earth") had the effect of concentrating traffic, as against vaster and more dispersed "oceans." For Botero, Spain's actual power rested on dominance of the Mediterranean as a special kind of *mare nostrum* that created leverage over the Atlantic, the Caribbean, and through Mexico and Panama the Pacific. Likewise, Spain's

claim to the Malukus rested, as Bartolomé Leonardo de Argensola argued in
1609, on their relation to the lower part of the chain of islands making up
the Philippines and the connection with the trans-Pacific routes.[14]

But Grotius's dedication of *Mare Liberum*, "To the Princes and Free
States of the Christian World," rejected the idea that legal issues of right
and wrong were determined in Machiavellian terms by popular "opinion
and custom" as well as "will," "fortune," and "profits." As an Arminian
going against the sectarian grain of Dutch Calvinism, Grotius appealed to
the mutual recognition of natural law by the Christian community (not
Catholic or sectarian Protestant) of princes and independent states. This
limited Spanish imperial claims without losing ground on older Catholic
visions of a world governed by Christian cosmology. It also elided key issues
in Asia. The Zeeland branch of the VOC had actually commissioned *Mare
Liberum* to support the 1609 Treaty of Antwerp. While this ultimately ini-
tiated a twelve-year truce between the Dutch and the Spanish in Europe, it
was rejected for Asian waters at Ternate in 1611. Grotius's pan-Christian
claims in print ignored the more complex problems of negotiating relations
among differing Christianities and with non-Christians in Asia, Africa, and
the Americas.[15] In private, he remained more equivocal. His manuscript
De Jure Praedae Comentarius gave a precise account of the particular "just
war" alliance between the VOC and the sultan of Johor against the Portuguese, which had led to the initial capture of the Portuguese ship in 1603
and to formal alliance in 1606.[16] But publicly, because he relied so heavily
on the Christian notion of natural law, he had no way to frame such alliances with non-Christian and especially Asian princes, and in *De Jure Praedae*, he fell back on a different and much looser Christian argument about
treating "neighbors" well.[17]

Selden's *Mare Clausum* certainly borrowed concepts from both the
natural law tradition and the neo-Machiavellians like Botero, but it rested
upon more comparative, historical, and indeed legal foundations than Grotius's work. The simplified Latin "Argumentum" from 1635, translated on
the English title page in 1652, reduced the claims to two basic points for the
benefit of diplomats, courtiers, and Parliamentarians.[18] First, according to
either the law of nature or the law of nations (*jure naturae seu gentium*),
whichever framework one preferred, the sea could be parceled into dominion (property). Second, Britain was historically an independent empire, a
narrow concept borrowed from John Dee and other Elizabethans involving
dominion over the seas surrounding the islands, and this could be historically reconstructed through traditional claims to property supported by documentary record. This simplified argument might have satisfied the court

of Charles I, with its limited ambitions for "ship money" and desires to please ambitious venturers like the "association" of William Courteen, but neither claim addressed the larger and unprecedented problems faced by the new Virginia or the East India Companies of maintaining relations over time in the absence of strong state power. Thus in the text itself, Selden had to offer a more attractive global alternative than Dutch free trade.

Claims to abstract rights of free trade were haunted by the question, as Gerard Malynes put it at the time, of "whether a merchant may trafficke with Turkes, Heathens, Barbarians, and Infidels, and performe promise with them."[19] So Selden first cut through a number of the standard arguments about the supposed universal "Law of Commerce and Travel (by them styled the Law of Nature)." He then challenged the technical claim that surveying the sea was not possible in the same way as the land, finding an actual quote about ancient Thalassometricians (sea-measurers) who "measured according to the rules of Geometrie, no less than the land," with machines for measuring distances at sea.[20] This verified the importance of new strategies for the collection of instrumental data through the sea compass and trigonometry pioneered by Harriot and Wright and inspired by the voyages of Drake, Raleigh, and Cavendish.[21] But in place of universal commerce or the purely English navigations articulated by Hakluyt, Selden returned to the tradition of mutually recognized transcultural exchanges and contracts first defined in London by Cabot in his "Ordinances." In return for abandoning abstract ideas of free trade or natural law, Selden offered London merchants real and sustainable relationships of dominion, mutually defined through contracts and the careful collection of data about the customs surrounding the routes and ports of Asia. Selden's Chinese map would later demonstrate that such relationships had a long and still active legacy outside of Europe.

Selden qualified his technical emphasis on measurement by arguing that all claims to measurement, whether made by Christians or non-Christians, were valid only if acknowledged, especially in a contractual sense. In the early sixteenth century, Manuel of Portugal, for example, had claimed dominion by conquest over the western Indian Ocean as part of his title—"per grace de Deos Rey etc. Senhor de Guinee et da conquista et Navagacum et commercio d'Ethiopia, Arabia, Persia et da India." After making arguments why both Queen Elizabeth and the Ottoman sultan might refuse to "acknowledge" this claim, Selden also quoted de Góis's translations of the letters sent to Manuel by the emperor of Ethiopia, Dawit II, who politely repeated the "hyperbole" of Portugal's claim to the Red Sea in diplomatic correspondence but made sure to emphasize in his reply that, "Title doth not intrench upon the Law of Nature or Nations, any more than this."[22] In tandem with this, Selden explicitly refuted the idea that the Spanish had the lawful right to subdue through "conquista" the inhabitants of the Americas "because they denied commerce and Entertainment," that is because they were not sociable, something that Grotius understood as a key element of Christian natural law. Selden instead emphasized the need for formal treaties, a process theoretically occurring in Virginia with the Powhatans. Rather than "natural law," as Grotius would have it, commercial rights for Selden without exception came out of mutual recognition and the universal law of "keeping faith," a religious concept no doubt rooted in Jewish tradition and divine law but one far more ecumenical than natural law.[23] As in Jewish law, contracts could be verified through the process of writing and archiving, and thus relied upon mutual translation and subsequent reenactment through the archive itself.[24] The law spoke through record.

The great Commonwealth scholar J. G. A. Pocock once dismissed Selden as having a "common law mind," alluding to his reluctance to abstract principles from the archive and indeed from language. But this also made Selden a good translator. Selden understood that any rigorous set of translations aiming for universalism required a reenactment of the particulars of history. In the context of the debates over the Petition of Right in 1628, Selden argued that royal prerogative had to come from an actual *act* of Parliament, a position markedly different from his contemporary Sir Edward Coke, who claimed that the king was bound only by ancient usage and custom.[25] And if the exertion of authority always came from particular acts rather than vague tradition or abstracted arguments, then for London to establish legitimate exchange relations, sovereignty could not be simply universalized or naturalized from Europe or the legacy of Roman law. Moreover, any attempt to write history from a purely European perspective (let alone a national one) was fundamentally flawed. For Selden, not only did the study of history as interrelated acts, translations, and agreements as well as reenactments of those translations through the laws create, in an institution like Parliament or the London East India Company, a different kind of post-Machiavellian civic consciousness; it also allowed for the emergence of languages and translation practices appropriate to a global age in which sovereignty had devolved onto the microprocesses of *dominium* embodied in exchange networks.[26]

LEGAL RELATIONS: OPENING LONDON TO ASIAN TRADE

Unlike earlier companies tied to a particular empire, the Russia Company with the czar or the Levant Company with the Ottoman sultan, the two East India companies were the first successful joint-stock corporations to make claims that required a multitude of trading relationships around the world. Prior to this, complex trades tended to fall apart because of the chains of relations involved. The experiments with Persian trade through Russia, the first Virginia efforts, and the various "Guinea" (West Africa) enterprises involving multiple, decentralized, or simply misunderstood sovereignties are cases in point. The Dutch solution was *koophandel met force* (commerce with force), using military power to stabilize relationships when negotiation failed. But this was an expensive and ultimately destabilizing strategy that alienated merchants and states. The merger discussed in the Anglo-Dutch conferences in 1613 and 1615 was designed to help defray the costs of potentially endless military expenses, and the VOC proposal of mutual defense of "Indian Kinges and people" against the Spanish effectively ended negotiations at The Hague in 1615. When renewed plans for the merger were drawn up in 1619, the year the Dutch took Batavia, Dutch authorities in Java were angry that the English received too great a share of the trade. Relations in Asia drove the breakup, as the London East India Company's factors suffered a series of humiliations, as well as a massacre in Ambon, until returning to Banten in 1628.[27] In this sense, it was not the institutional form of the joint-stock company itself or even ambitions to Iberian-style *imperium*, but the legal and contractual interaction among various networks of Asian trade that set London and Amsterdam on separate paths.

Even in the case of the Dutch, historians have long recognized that the concept of "imperialism" provides an inadequate framework for studying Asian and especially Southeast Asian commerce in the seventeenth century.[28] Forming successful relationships in each case required a substantial learning curve, negotiations often among several merchant networks and local elites (like the Chinese and Javanese in Banten), and usually a willingness to buy into markets with large quantities of precious metals. In the early seventeenth century, North America and the Caribbean and even Ireland did not offer a simpler alternative to this. Virginia and the other new colonies in the Americas admittedly could appear to be Roman-like *coloniae*, especially if their seizure could be legally justified, but practically there were dramatic struggles. In the abstract, "plantations" had a certain appeal for a gentry and Parliamentary politics rooted in ideas about natural law. This often drove aggressive efforts to acquire land, whether in the efforts in Ireland that exploded in 1641, or the 1619 gentry coup led by Sir Edwin Sandys and others that ousted the London merchant Sir Thomas Smythe as treasurer of the Virginia Company to push for a similarly disastrous plantation policy.[29] Zealous gentry soured Virginia, and the crown had to take over after the newly reorganized company caused the disastrous

1622 uprising by Powhatans. The exertion of authority by the monarch was in many ways appreciated by the more imperialist voices among the gentry, who wanted to maintain authority in decision making but feared too much devolution of power. Writing as part of the East India Company's Select Committee, Sandys noted that because the Dutch government was "popular" and "our[s] monarchical . . . our King treats them upon very great disadvantages, because the States-General have no powers over their Subjects to command them, nor will the Bewinthebbers [the VOC directors] obey them if they should."[30]

Yet, unlike the ventures in America and Ireland in which the crown was brought in to directly intervene, a large degree of autonomy can be seen in the East India Company's first efforts in building relations in response to conditions in Asia. Still beleaguered by Tyrone's Rebellion in Ireland (the Nine Years' War, 1594–1603), Elizabeth's strategy in giving the Company a charter seems to have been a continuation of her earlier policies of loose alliances with Muslim states like the Moroccans and Ottomans to challenge Spain. The first targets for trade relations were Aceh (a cosmopolitan sultanate connected with the Ottomans), various sultanates in India, and China and East Asia more generally, as well as the Portuguese and Dutch. Sultan Alauddin Riayat Syah al-Mukammil (1589–1604) had put his seal on a letter that the Stadtholder Maurice of Nassau wrote to him in Spanish in December 1600 and knew that the English had successfully fought off the Iberians.[31] Sir James Lancaster and Henry Middleton brought with them to Aceh a Jewish interpreter from Barbary who knew Arabic, but Elizabeth's January 1601 letter to the "King of Achen in the Island of Sumatra" was generic in its address, requesting merely friendship and "peaceable traffic," in distinction from the Spanish and Portuguese who tried to claim sovereignty over Asian waters.

Two versions of the letter to the sultan survive, the January 1601 version as a first draft and a much more precise one reprinted by Purchas that Lancaster in his journal claims to have delivered, which seems to have been written by Company representatives. In that second letter, the ghostwriters for Elizabeth referred to the unsuccessful Achenese attack on Portuguese Malacca in 1575 and the valiant Captain Ragamacota (Rajah Mahuta) in order to engage the sultan with both a localized history and a heroic narrative.[32] Having refused the first actual letter from Elizabeth, Alauddin agreed to take the revised version. The ensuing negotiations at the Achenese court enabled Middleton to receive a permit for trade, a plain document in a green silk envelope written in Malay Jawi script and stamped with the lampblack circular seal of Alauddin Syah. This document was clearly modeled upon the more translatable seals or *chaps* used in the Indian Ocean and *farman* documents of the Mughals and Ottomans, but it was only useful locally at the port.[33] A formal signed letter in Arabic to deliver to Elizabeth only came in October 1602, after Company representatives captured a 1200-ton Portuguese carrack off Malacca coming from the Coromandel Coast. Lancaster had a generic privateering commission from the English lord high admiral, but rather than keeping the prize to themselves, Lancaster and Middleton split the cargo with the sultan and the Dutch captain Joris van Spilbergen, who was working for the Moucheron Company.[34] At this point, the sultan produced a long and flowery letter in Arabic to Elizabeth as well as a ruby ring and two pieces of clothing woven with gold thread in a Chinese lacquer box.[35] The East India Company had become a translator between sovereign entities, itself a "go-between," and in Arabic the sultan recognized the language of a "joint Company," "one Corporation" with "common privileges" as an "absolute society" with "liberties" and "patent privileges." This is precisely what Selden would later say a world of *Mare Clausum* required— not merely mutual diplomacy between states but the development of mutually recognized and translatable contractual languages, archived as documentary records.

The Dutch, still acting as independent firms before news arrived of the incorporation of the VOC in March 1602, learned a strikingly different lesson from this event. The joint actions of the East India Company and the Moucheron Company embarrassed the representatives of the United Amsterdam Company, like the Moucheron Company one of several precursors to the VOC formed in the late 1590's. The Amsterdam Company was based at Patani, up the Malay Peninsula from Johor, and their captain Jacob van Heemskerck resolved that to maintain their position in Patani was "necessary and desirable to defy the enemy and show the natives that we do not fear Portuguese power." He convinced the Portuguese crew of a small ship coming from Champa to abandon their vessel off Tioman Island on the east side of the Malay Peninsula. This had the desired result, for almost immediately Raja Bongsu, "the young King of Johor," sent Heemskerck a letter and embassy suggesting that he await the larger carrack from Macao in the straits and use Johor as a base.[36] After capturing the rich galleon the *Santa Catarina* in February, the event that subsequently inspired Grotius's treatise, Heemskerck took his prize to Banten in June, where his Moucheron competitor Joris van Spilbergen noted in his log the far larger prize of seven million guilders that Heemskerck had secured.[37] Not only were individual Dutch captains in conjunction with local elites on the other side of the globe now improvising policy that could have major implications for the United Provinces, but they also did not secure in the process letters acknowledging either formal sovereign relations or develop a mutual legal language to comprehend such policies.

The East India Company seizure made under the aegis of Acehnese sovereignty was largely forgotten, normalized by the new legal relationship and terms of agreement with Aceh that subsequently served as a model for defining relations with Banten after 1603; conversely, the *Santa Catarina* incident and its aftermath became well known across Europe. Images of ensuing battles with the Portuguese showed the Dutch cooperating with the "young king" Raja Bongsu.[38] In response to the scandal, Grotius from 1604 composed the longer *De Jure Praedae* for the Amsterdam Chamber, which represented the leading investors of the newly united VOC, in an attempt to define a comprehensive theory of just war allowing seizure of Spanish and Portuguese ships and bases. To justify this action as defensive and even liberating, Grotius falsely claimed that the sultan of Johor, Abdulla Hammayat Shah, himself participated in the *Santa Catarina* incident. By 1607, he was explaining Dutch policy in draft letter to the sultan of Tidore—"Our purpose was not just to protect ourselves against the Spanish and Portuguese [deleting: *who have unjustly sought to proscribe free trade throughout the world*], but also to be most diligent in liberating East Indian princes and nations from Iberian tyranny"—warning the Sultan that Londoners would soon come trying to trade but not actually spending money to fight the Spanish.[39] Then when Grotius switched to work for the Zeeland Chambers in 1608, he edited down the twelfth chapter of his unpublished *De Jure Praedae* into *Mare Liberum* in order to articulate a more abstract policy based on natural law. At Zeeland, Balthasar de Moucheron's old company had broad interests in Russia, the Americas, and East Asia. He wanted from Grotius a comprehensive theory of trade that also encompassed the Atlantic World, where there was no question of cooperating with non-European maritime powers.[40]

In London, by contrast, an increasing divergence emerged in the first half of the seventeenth century between more gentry-oriented "British" Atlantic plantations ("the Atlantic World") and London's Asian strategies of engagement, including those of the Levant Company, or to put it more starkly, between colonization and translation efforts. With a smaller initial investment and competition from the Dutch, the London East India Company had to shift and expand its trading networks rapidly. It used new investment to disperse risk, quickly leaving expensive Aceh, and establishing over twelve new factories between 1600 and 1620.[41] They also constantly sought out new European markets, leveraging other London companies trading with the Mediterranean and Baltic in the process—as most of their primary commodity pepper was reexported. The company increasingly used domestic consumption of commodities to convert the gentry residing at least part time in London as well as the broader populace to the cause of Asian trade. The bridge for gentry investors based in London was the fashionable new arcade on the Strand, "Britain's Burse" or the "New Exchange" as it came to be called. Owned by Robert Cecil, the Earl of Salisbury, the Lord High Treasurer and Secretary of State, it opened to great fanfare in 1609 with James I, Queen Anne, and Prince Henry all in attendance. Cecil was directly involved in approving silver exports and also held the customs farm for silk from 1601.[42] Thus like Gresham's Royal Exchange, the New Exchange played a role in bullion and currency management until a more open market in silver emerged in the 1630's in relation to Spain. The opening "Entertainment," written by Ben Jonson, featured a shop master who proclaimed that the Burse was better than all the "divers china houses about town" and indeed to "all the magazines of Europe." He cries, "What doe you lacke?" and answers, "Veary fine China stuffes, of all kindes and quallityes? China Chaynes, China Braceletts, China scarfes, China fannes, China girdles, China knives, China boxes, China Cabinetts," along with a range of Dutch optical instruments and "Indian mice, Indian ratts, China dogges, and China cattes." This was a new kind of appeal to desire, reversing Cabot's old formula for success in trade by intoxicating wealthy Londoners rather than unsuspecting natives.[43]

These differences suggest some of the reasons why negotiations over the merger of the VOC and EIC fell apart. As London became increasingly intertwined with Asian trade and the bullion trade that supported it, the VOC from 1619 became increasingly independent of policy-making decisions coming out of the cities of the Dutch Republic, becoming self-financing after the initial investment and acting almost like a state or independent republic. The Dutch shifted their main base of operations from their large fortress at Ambon (1605–19) to Batavia, which became a nodal point attracting stocks of commodities and bullion from Europe but especially from Japan and to a lesser extent Persia.[44] The VOC made much of its money bringing silks from Tonkin or later Bengal to Japan in exchange for silver, but the EIC, lacking such forts for the storage of bullion (with the exception of Fort St. George, Madras, est. 1644), largely sent bullion from Europe and then reexported goods from London to the Continent, Africa, and the Americas. London was the English Batavia.

Tensions between James and the gentry forced a short-lived cooperation between the Dutch and English companies, which ended disastrously. Cooperation came about only for a brief period in 1619 after the king, wearing his other crown, tried in 1618 to grant Sir James Cunningham a patent to create a Scottish East India Company, a move the London Company prevented through a forced loan to the crown. Under pressure, the Company responded to anti-Spanish factions among the gentry led by Sandys and Coke, who wanted joint-defense fleets operating against both the Portuguese in the Western Indian Ocean and Spanish Manila from the factories at Hirado, Japan, as a way of pressuring the Spanish in the Atlantic.[45] An early joint action was against the sultanate of Bijapur's shipping in the Red Sea at the port of Dabhol because the sultan's ships employed Portuguese *cartazes* (passes) from Goa. The East India Company also captured Ormuz from the Portuguese in 1622 in tandem with a Safavid force.[46] The biggest hauls were made in the eighth and ninth Dutch blockades of Manila, jointly made with English ships for the first time in 1621–22, which mostly plundered Chinese merchant ships coming out of Fujian and Guangdong. Selden's map of China, taken by an English commander, was most likely captured at the very beginning of these operations. But aside from obtaining the map, the disruption of Chinese merchant networks, unlike the joint efforts in the Persian Gulf with the Safavids, worked largely against London's interests, causing them difficulties from Banten to Hirado and doing little to help normalize the flow of silver coming from New Spain and Peru.[47]

In East Asia, the more the Company pursued common cause with the VOC (and at home the anti-Spanish gentry), the greater became the encouragement for London traders to break away from the Company and become interlopers, giving them more freedom to make particular agreements to amass commodities and silver. On the small islands (*pulo*) of Run and Ai in the Bandas southeast of Ambon between 1617 and 1621 (strategic locations in the Malukus between Celebes and New Guinea near key spice producing centers), Company factors acting semi-independently during the collapse of the Banten presidency even built forts as a defense against the Dutch and Iberians.[48] In a kind of mutual inflation of titles, they claimed to have made deals in the name of James I with local elites managing the nutmeg and mace harvests, who themselves styled their households as Malay *orang kaya*.

A pamphlet battle in English and Dutch over these events raged in Amsterdam in 1622 designed to sway opinions on both sides of the Channel.[49] Then in June 1624 news reached The Hague during treaty negotiations over war with Spain of what came to be called the "Amboyna massacre," which involved the torture and execution of ten Company employees from London, who had been acting semi-independently with one Portuguese and nine Japanese to supposedly try to take advantage of the Dutch shift from

Ambon to Batavia. A bilingual slew of pamphlets followed.[50] Confused due to James I's illness (he died in March 1625), as well as the efforts of George Villiers, Duke of Buckingham, and his protégé the future Charles I to provoke war with Spain, the English court had few constituencies who wanted to challenge the Dutch directly. James offered to become an East India Company shareholder and let the company fly his new Union Flag (1606), having just revoked the Virginia Company charter and turned it into a crown colony. The Company rejected this offer and maintained its own flag, suggesting that commercial "partnership" was below the dignity of a king but primarily worrying that if the king became a shareholder, then London merchants would lose control to courtly patronage in relation to its officers and to the monetary demands of the patrimonial state. The Company received a clear message of Charles's indifference to its difficulties in July 1625, when the Dutch East India fleet, for whose capture James had issued a warrant the previous September, was allowed to pass Dover unharmed within sight of the English navy. A joint Anglo-Dutch war against Spain began that November with an assault on Cádiz.[51]

The lesson for the London East India Company was that cooperation with either the Dutch VOC or the English court ultimately undermined its effectiveness, and by 1628 it had abandoned operations in Japan (1623) and returned to Banten, cooperating with the local pangeran, who was increasingly interested in fashioning himself a sultan. Beyond the limited circles of merchant investors in London, the East India Company had to look to both Asian merchant networks and to a broader London public of consumers and shareholders for support. The Company's director Thomas Mun began to formulate a neo-Aristotelian theory of "traffic" and "accommodation" tied to temporal change in an effort to define London's relations with both Asia and Europe, although as a member of James I's Standing Commission on Trade from 1622 he would not publish it. "Thus by a course of traffick (which changeth according to the accurrents of time)," wrote Mun in 1623, "the particular members do accommodate each other, and all accomplish the whole body of the trade, which will ever languish if the harmony of her health be distempered by the diseases of excess at home, violence abroad, charges and restrictions at home or abroad."[52] Using the 1623 massacre in Ambon as a focal point, pamphleteers also made explicit ideological contrasts between London merchants and the Dutch, suggesting a more precisely legal difference. They depicted the East India Company working cooperatively with Asian sovereigns as well as pilots and soldiers like the Portuguese and Japanese killed at Ambon, "answerable only to a course of commerce and peaceable traffick; not expecting any hostility, neither from the Indians nor especially from the Dutch." The Dutch conversely engaged in a violent colonial strategy, reminiscent of the Spanish and Portuguese and predatory upon it, "from the beginning of their trade in the Indies not contented with the ordinary course of a fair and free commerce, invaded divers Islands, took some Ports, built others, and laboured nothing more, than the conquests of Countries, and the acquiring of new dominion."[53] Dutch free sea policy thus became the opposite of truly free commerce, which respected the legal limits of dominion, regardless of whether that dominion was held by a Christian or non-Christian. In tandem with such claims, Selden's concept of "closed seas," adopted even by Charles I after 1635, following the signing of the Cottington Treaty with Spain in 1630, increasingly made more sense in London than Hakluyt's translation of Grotius, which tellingly remained unpublished.

ASIAN LIBRARIES IN LONDON, OXFORD, AND CAMBRIDGE

The problem with Selden's model and indeed the evidential underpinnings of *Mare Clausum* itself was that London lacked the libraries of texts and experience in translation to comprehend such a program on a global scale. While the Cottington Treaty solved the problem of obtaining Spanish silver in London, it did little that could offset the growing Dutch military presence in Asia and monopolization of important nodal points for trade. In revising his *Titles of Honour* (1614, 2nd ed. 1631), Selden made manuscript transcriptions of and notes on correspondence from princes around the world to the English court, gathered with the help of his new ally Archbishop Laud and the East India Company. This included letters from Algeria, Morocco, Tidore, Ternate (labeled "Molucco"), Banten, Aceh, Persia, the Mughal Empire, Ethiopia, and Japan. Selden's onetime collaborator at the Virginia Company and then ambassador to the Mughals, Sir Thomas Roe, sent him a copy of the Mughal seal, and Selden corresponded with the Leiden scholar Thomas van Erpe (Erpenius) on Muslim titles, collecting as much as he could from various sources. It would have been part of a globally comparative work on notions of aristocratic authority and sovereignty, something Selden could ultimately only make gestures towards.[54]

To the extent that London did have such linguistic materials, Selden trained and then lived amidst them at the Inns of Court, which played the role of creating a "public" akin to the Elizabethan theater or the Restoration coffeehouse. Ben Jonson called the Inns "the noblest nurseries of humanity and liberty in the kingdom"; in a dedication to the 1614 *Titles of Honour*, a book that attempted to include "Mahumedan States" in a comparative study of the sources of aristocratic authority, Jonson wrote of Selden, "You, that have been/Ever at home, yet have all Countries seene;/And like a Compasse, keeping one foot still/Upon your center, do your circle fill/Of generall knowledge."[55] Even more than the new Gresham College (1597), where the lord mayor provided for lectures in geometry, astronomy, music, and divinity, a full course in cosmology, and the Mercers' Company underwrote the more practical law, physic, and rhetoric, the Inns of Court were, as John Stow wrote, "in and about this city a whole university," suggesting how the informational aspects of the city were woven into the urban fabric itself.[56]

As sons of gentry as well as those educated in the new grammar schools of the sixteenth century moved into the area around the Inns, rediscovering it and bringing in new kinds of texts, this became the first true locus of the gentrification of London. Emerging from and circulating within these circles, the histories and topographies of William Camden and the travel collections of Samuel Purchas framed Selden's efforts, as did the efforts by merchants and the court to make legal engagements with Morocco, Virginia, the Ottomans, Persia, the Mughals, Sumatra, and Java in the early seventeenth century.[57] Selden's training meant that he could employ sources in a range of languages, making him a formidable critic for Grotius, whose use of Greek and access to a Greek font paled in comparison to Selden's Greek, Arabic, Samaritan, and Hebrew sources and lettering. Surrounding the Inns of Court were the collections of books, manuscripts, and antiquities like those of Sir Robert Cotton, the Earl of Arundel, and later Selden himself. Selden actively used these collections, Along with the libraries at Lambeth Palace and the Tower of London, in the case of Arundel's Greek marbles from the Ottoman Empire even cataloging them, a process that connected him to a dynamic nexus of legal scholars, writers, publishers, and aristocratic collectors in London.[58]

By the 1640s, Selden himself had put together one of the best collections of books and manuscripts in England. His library included books in a staggering array of languages, which at his death in addition to numerous Latin, Greek, and Hebrew works included 117 in Arabic, 42 in Persian and Turkish, 9 in Chinese (along with the map and compass), 1 in Japanese, and 3 postconquest Aztec texts. This was possible because especially from the 1620s, large numbers of manuscripts, books, and objects from Asia ranging from Ottoman manuscripts to Tokugawa printed editions began to fill the various private libraries of London as well as the university libraries at Oxford and Cambridge. These archives also pulled the universities into the city's concerns, and in the case of Bodley's library and Bacon's science, they gave Oxford's and London's intellectual circles a kind of European respectability they had previously lacked. Aside from the cases of scholar-buyers like Edward Pococke and Thomas Greaves in the Ottoman Empire, most manuscripts were not acquired for collections directly but indirectly through the book markets of London as well as Amsterdam and Leiden. Distinct from the chinaware and other commodities of Britain's Burse, these manuscripts and books reframed the basic linguistic, historical, and technological framework for understanding exchanges with Asia in particular. As such, they formed a material legacy for translation in London and the universities that could be built upon in the future, one that by and large still remains intact to this day.

Among the wealthy in London, having a library of both ancient and non-European manuscripts as well as collections of instruments, ancient and modern artworks, and natural history specimens became important signs of status by the 1620s, and the collectors wanted their books compared, contrasted, researched, and examined. George Villiers, the new Duke of Buckingham from 1623, broadened his interest in collecting in 1625, asking John Tradescant to enlist the help of the navy, the Virginia Company, the Somers Island Company, the Newfoundland Company (est. 1610), the Levant Company, the East India Company, and later the newly chartered Guiana Company (1626–27) to supplement his own collections.[59] Buckingham also acquired ninety-three manuscripts in 1625 that were supposed to go to Leiden University from Erpenius's collection. Buckingham used them in his role as the new chancellor of Cambridge to transform the university's library, which as of 1632 had only had one Arabic manuscript, a Qur'an donated by William Bedwell the previous year.[60] The Cambridge Arabic scholar Abraham Whelock could now do actual research, the manuscripts including what is still the oldest Persian manuscript at Cambridge, along with texts in Arabic, Coptic, Javanese, Hebrew, Syriac, and Malay as well as one Chinese book.[61]

Books and technical objects like maps, compasses, and astrolabes had once been curiosities, mere emblems of a lack of understanding of true causes; by the 1620s, targeted collection of these items had become part of an effort to build archives for active research. Raleigh had called his ship an "ark," and outright plunder of Spanish and Portuguese ships by privateers had been the basis for much of the late Elizabethan collection of Sir Walter Cope in Kensington (ca. 1595–1614) as well as some of the choicest pieces of Hakluyt's collection. But already in the 1580s, in an effort to preserve the archival nature of what had been captured on such ships, Cope advised Hakluyt that they would be better understood as a collection rather than as individual "curios."[62] Defining time, space, and language were the hallmarks of the new wave of collecting in the 1620s. The most remembered legacies of this collecting process are the monumental set of travels published by Samuel Purchas in 1625 as a "posthumous" Hakluyt and Tradescant's museum in Lambeth (ca. 1628),which included parts of Cope's collection and subsequently became the core of the Ashmolean at Oxford. In 1634, the well-heeled Peter Mundy claimed in his journal that by visiting Tradescant's museum a man could in "one daye behold and collect into one place more Curiosities than hee sould see if hee spent all his life in Travell."[63] Purchas, who included at the beginning of his *Pilgrimes* a series of combinatorial and comparative tables of scripts, noted that writing had produced not only a diversity of alphabets for diverse languages but also a diversity of alphabets within one language over time, "as is most easie to bee seen in well furnished Libraries" like that of Sir Robert Cotton. Amidst this broader process of collecting, the singularly unique object actually took on a new importance as a way of gathering together joint translation efforts. In 1626, the physician and astronomer John Bainbridge came across an "Arabic book of astronomy" (Ibn al-Banna's *Minhaj*) in London. Bainbridge appealed to Archbishop Ussher's interest in chronology, Selden's in geographic measurement, and William Bedwell's in translating Arabic to encourage the bringing of more such books of astronomical tables to London and to sponsor further translations.[64]

The first major divergence over the role of such books and objects emerged between Francis Bacon and Thomas Bodley in the 1600s. Both entered Parliament in 1584 and were in many ways late Elizabethans—Bodley began work on his library in 1598, while Bacon published his *Essays* in 1597 and began to seriously think about revising Aristotelian science after the trial of Essex in 1601. Unlike Bodley, who explicitly preserved Chinese and other books whose titles could not even be read in England, Bacon held suspicions about the benefits of library collections in themselves as well as of knowledge drawn from Arabic or Chinese sources. Bacon certainly used the great libraries of early seventeenth-century London and, on the orders of King James, oversaw the cataloging of the Lambeth Palace library from 1610.[65] But for Bacon, who knew very little about the subject aside from what he read in Mendoza and Escalante about Chinese, Asian languages, like ancient European languages, lacked a coherent program for developing new instruments and knowledge. In the *Advancement of Learning* (1605), a copy of which he sent to Bodley for the library, Chinese characters as well as Egyptian hieroglyphics merely provided evidence for disproving Aristotle's understanding of spoken language as a necessary mediating force between letters and thoughts. Bacon toyed with the idea of Chinese as using "Characters Real," that "express neither letters nor words in gross, but Things or Notions" enabling communication across divergent spoken languages, but the "vast multitude of characters" seemed an insurmountable barrier and indeed historical error.[66] This still merited keeping such books, and Bacon included with his presentation copy a letter about how books are "shrines" and Bodleian was an "Ark" to save knowledge from the deluge of barbarism.

For Bacon, modern arts were substantially different. The *Novum Organum* or "new instrument" of 1620 (part two of the *Instauratio magna*), celebrated printing, gunpowder, and the compass as the keys to European success, "For these three have changed the whole face and state of things throughout the world . . . whence have followed innumerable changes, insomuch that no empire, no sect, no star seems to have exerted greater power and influence in human affairs than these mechanical discoveries." In these cases, however, he sidestepped the question of *translatio studii*, the Chinese invention, or even East Asian usage of all of these, let alone the idea that the technical basis of European modernity rested on a broader engagement with Asia.[67] By the time he wrote his "New Atlantis" (ca. 1624, published 1626–27), the idealized site of collecting knowledge in the eastern Pacific Ocean about causes, development of instruments, and translation between knowledge systems that he called "Salomon's House" conspicuously lacked a library.[68]

Between the poles of Bodley's archive of preservation, which could not comprehend Chinese, and Bacon's desire for new instruments and languages that could circumvent it, Londoners engaged in a more complex process of collecting and exchanging. Each new book or manuscript helped to redefine networks, encourage further collecting, and find collaborations that might lead to a more coherent method. Some collections highlighted the difference between EIC and VOC strategies as well as the cross-fertilization going on between them. Three of the five early Malay manuscripts at Cambridge acquired by Buckingham bear the signature of Pieter Willemszoon van Elbnick (aka Pieter Floris). Floris had transcribed them in Aceh in 1604 while with the VOC before he joined the East India Company's expedition on the *Globe* in 1611 to explore trading possibilities on the Coromandel Coast.[69] Floris disagreed with the VOC's restrictive trade policies and limited efforts towards translation. One book he obtained was the popular Arab-Persian story the *Hikayat Muhammad Hanafiyyah*. Famously one of two *hikayat* (romance stories) read to warriors before the Portuguese attacked Malacca in 1511, it told the story of resistance against the tyranny of the Umayyad ruler Yazid I by the sons of the martyred Husayn ibn Ali under command of the fantastic hero Muhammad Hanafiyyah. Another of his manuscripts included a Telugu alphabet for the Coromandel Coast and a Dutch-Malay vocabulary, which Floris called a "Magasin" or "warehouse," "wherein are assembled diverse words in Dutch and Malay."[70] Floris's translations suggested a world of trade, stories, and languages that reached across political boundaries to moderate monarchies, if necessary by fraternal resistance to tyranny. They thus formed a little archive of resistance to centralizing policies by the Achenese sultan and by the VOC in Asia as well as those like Robert Filmer who supported patriarchial monarchy in England and Holland, ironic given their donation to Cambridge by Buckingham's widow after his 1628 assassination.

Many of the manuscripts obtained were themselves collections or "assemblies" of texts and languages, indicating at a popular level how the linguistic "cosmopoli" of the Indian Ocean and Southeast Asia worked. Edward Pocock obtained a fragment in Malay of the *Hikayat Bayan Budiman* from Borneo. The stories, which ultimately derived from the Sanskrit *Sukasaptati* ("The Parrot's Seventy Tales"), had great prestige in the Mughal Empire in its Persian form, the *Tutinama* (1330), and the popular Malay version derived from a 1371 translation by Kadi Hassan.[71] Laud's *Hikayat Seri Rama*, the Malay version of the Sanskrit *Ramayana*, came to the Bodleian in May 1635 as part of a much a bigger donation from his London library classified vaguely as "rolls" or records.[72] The palm-leaf "menak" *Caritanira Amir* in Javanese given to the Bodleian in 1629 was a similar kind of donation, in this case a shortened performance-oriented version of the *Hamzanama*, a collection of stories popular among the Persians and Mughals about the picaresque adventures of Mohammed's uncle against secular enemies of Islam.[73] This particular collection of stories, usually performed as theater, was given at the death of William Herbert, Third Earl of Pembroke, the recent patron of another collection in folio, *Mr. William Shakespeares Comedies, Histories and Tragedies* (1623). In each of these cases, almost a century before the Arabian Nights became popular in France and England, collections of stories and plays that circulated widely in Asia in multiple languages were arriving in London from the port cities of Southeast Asia, suggesting transcultural models of linguistic exchange. New strategies of collecting and new kinds of history writing in Southeast Asia in part responding to English and Dutch activities—the *Sejarah Melayu* written in 1612 in response to the ship seizures of 1602–3 and subsequent events at Johor; the *Hikayat Aceh*, commissioned by Iskandar Muda in either 1613 or 1615 after his capture of Johor; the Javanese *Sejarah Banten* (ca. 1660) in response to the expanding power of the Dutch at Batavia and elsewhere in Java; and the broader tradition of the *Babad Tanah Jawi* beginning in the seventeenth century—all suggested not an imposition of European models in Asia but complex emerging practices of history writing and archiving. This interaction of textual traditions and active historical writing in Southeast Asia implied the need as far away as London for comparative assessments, recognition of historical bodies of law and sovereignty, openness to other interpretations, allowances for incompleteness, and ultimately awareness of the possibility that future techniques of translation might open the door to deeper understandings of language, space, and time.[74]

Because of this complexity, in early seventeenth-century London, following Bacon's lead, the search for historical and linguistic models capable of comprehending global interactions and change could often lead to abstraction. Even though he published primarily in Latin, Selden shared some of the goals of those like John Wilkins and Francis Lodwick who were working on "universal" language projects in Oxford and London in the 1640s and 1650s, aiming to replace Latin as medium for the exchange of information. A note left among Selden's manuscripts at Lambeth Palace reads, "To make a perfect lexicon of the harmony of all languages is not the work of one man." A collaborative "universal character if it were contrived as it might be would serve for universall commerce of all nations tho they did not understand one anothers language so a Catholick lexicon of all the primitive and veary scant words in all languages with alphabeticall tabbes to each perform'd by lexicall men would be much to the ease of all learners."[75] This passage on comparative linguistics, dating from after the mid-1630s, in its ambiguous use of "Catholick" in relation to commerce, echoed the religious project for a Polyglot Bible sponsored first by Archbishop Laud and then completed in the 1650s in London under the editorship of Brian Walton.[76] Organizing language and concepts in tables went back to Selden's earliest interests in the knowledge trees created by Porphyry of Tyre to clarify Aristotelian logic, which in the sixteenth century Petrus Ramus developed into an abstract framework for organizing knowledge, a standard tool in English education from the 1570s.[77] Such a language would be a "contrived" method designed to produce "universal commerce," a technical achievement. But despite Stoic-like claims to a "harmony of all languages," translation or *translatio* was not for Selden a kind of mystical preexisting phenomenon that could be simply uncovered or activated, as it had been for Renaissance thinkers like Botero. Selden's particular formulation required the work of translation, open-ended investigation into fundamental linguistic difference and especially historical legal practices and institutions, a process sidestepped by the Stoic "Οικειωσιν," a more abstract and simplified notion of man's innate sociability and domesticity, popular with both Grotius and many universal language theorists.[78]

This is why of all the collectors of early seventeenth-century London, Selden was perhaps the most engaged in grappling with these challenges of rethinking time, space, and language that had been raised by new practices of collecting and translation. After the publication of the Latin edition of *Mare Clausum* in 1635, Selden's printed works seem to indicate a turn to more secure foundations of divine law and providential Hebrew history, rather than the path of a universalism rooted in a comparative science of measurement and history. But Selden's growing collection of manuscripts, from which eight thousand books and manuscripts went to the Bodleian in 1659, told a different story.[79] Second only to Cotton's library, which Charles I dispersed in 1629, Selden's library became the most significant private collection of books in London during the period of the Civil War and Interregnum. In 1639, one of Selden's patrons, Henry Gray, the Eighth Earl of Kent, died; somewhat scandalously, Selden moved into Gray's house with his widow Elizabeth. Sometime during the drama surrounding the capture and execution of Charles I between 1648 and 1649, Elizabeth changed her will to give Selden substantial amounts of property including Carmelite, a large old monastic house near the Inner Temple that was named after the dissolved Fleet Street priory; Selden inherited it upon her death in 1651. Like John Pym's house for the Puritan radicals, Carmelite became an important gathering site for moderates in Parliament, and Selden was already living there by the mid-1640s. After the Civil War, Cromwell wanted Selden to directly link his research work to republican projects related to sovereignty, but Selden rejected or ignored requests to defend the regicide, to write the first comprehensive English constitution, and to respond to the royalist *Eikon Basilike* as a false emblem of divine translation. For the last one, Cromwell turned to his second choice, John Milton, the Council of State's new "Secretary for Foreign Tongues," himself a great admirer of Selden's "exquisite reasons and theorems almost mathematically demonstrative."[80] In the early 1650s, Selden did use his prestige to defend research into translation, promoting Arabic and Hebrew printing at Oxford and countering suspicions that scholars studying such languages had Catholic or Laudian motives.[81] At Carmelite, Selden also wrote the great summation of his research into Hebrew law, the three-volume *De synedriss et praefecturis juridicis veterum Ebraeorum* (1650–55), as well as his response to Theodor Graswinckel's caustic attacks on *Mare Clausum* before his death in November 1654.[82] He also set up a kind of modern revision of the monastic library, what he called

"Museum meum Carmeliticum," displaying his Chinese map and compass, his Greek marbles and inscriptions, along with precious stones and crystals in cases, all surrounded by shelves of manuscripts organized along lines similar to Cotton's library.[83] The coupling of Hebrew studies with Asian science and instruments suggested potential research in comparative universalisms, something that the medievalism of Cotton or the classicism of Arundel could not claim.

Many have seen Selden's work in this period as similar to that of his contemporaries Milton, Harrington, Hobbes, and even Grotius himself, gravitating towards the idea of the "Hebrew republic" as a central concept as monarchy collapsed.[84] Selden certainly collected Hebrew manuscripts, and to encourage study along these lines, he convinced Parliament, with the help of John Lightfoot at Cambridge, to put forward a £500 grant to acquire the collection of the Italian rabbi Isaac Faragi, imported in 1648 by the bookseller George Thomason, which included one Persian and ten Hebrew manuscripts and over four hundred Hebrew printed books.[85] For Selden and others, the idealized and literally reorienting Hebrew republic addressed the Roman republican problem of the diversity of laws that required an empire to manage. Both Grotius in his manuscript *De republica emendanda* and the Leiden scholar Peter van der Cun's 1617 book *De republica Hebraeorum* had recapitulated the Ciceronian argument from *De Legibus* that an art for defining natural law was the necessary solution to such diversity. Selden himself in a series of works culminating in *De synedriss* claimed that Jewish institutions, contracts, and law could be put forward as universals because of their durability over time, something that derived from their historical and archival rather than their natural (or pre-Babel or antediluvian) basis.

If this turn to Hebrew dominated much of Selden's later published work, his library demonstrated historical legal and technical achievements outside the Jewish, Christian, and Islamic traditions that like Hebrew seemed to carry with them the potential for translation. Selden's acquisition of the Chinese map and compass is one example of his eye for the diagrammatic—objects and books that could give both visual and technical depictions of space while at the same time conveying a sense of movement over time and history. His two Arabic astrolabes—one of which he gave to Laud along with a gazetteer for towns in Syria, Persia, and India at the time of the publication of *Mare Clausum*—highlight the broader theme of collecting objects that revealed differing attitudes towards history, measurement, and geography.[86] On the Laud astrolabe, six concentric circles (five changeable) have the names and latitudes of fifty towns, including Serendib (Sri Lanka),

Khambat, Lahore, Delhi, and Agra, going up to Volga Bulgar and westward all the way to Cairo, Aleppo, Damascus, and Aden (figure 21). The astrolabe kept by Selden was made in the Eastern Mediterranean and gave locations from Fez to Aleppo.[87] These were readable instruments as Selden, along with many others in London and the universities, knew Arabic, and they indicated that sea-measuring "Thalassometers" had mapped and still did map the Mediterranean and the Indian Ocean, as the Selden Map showed the Fujianese doing in East Asia.

The more explicitly spatial navigational materials like the astrolabes or Chinese map and compass were juxtaposed to books that tried to diagram a sense of time. Selden's unique twenty-juan edition in six volumes of Luo Guanzhong's *Sanguozhi yanyi* (三國志演義, "Three Kingdoms Romance"), a Ming historical novel, is a good example of this. A reprint of the illustrated edition by the publisher Liu Longtian, it had serial woodcut images from the novel at the top of many of the pages. This popular novel cheaply printed in Fujian and obtained most likely outside of China through Hokkien merchants was a way that such merchant families maintained their links to Chinese history, albeit in a manner that was explicitly fictionalized.[88] Selden of course could not read Chinese, but the serial, "fully illustrated" (全像 *quanxiang*) character of the woodcuts at the tops of pages would have visually suggested a kind of history to Selden, temporal rather than hieroglyphic. Indeed, such images contrasted with the Polyglot Bible of Brian Walton (1657), which took as its sole exemplar of Chinese printing the conventional lotus-leaf Jianyang colophon from the 1592 edition torn from the first *juan* of the novel. It must have appeared to Walton as a kind of emblematic stele.[89]

The illustrations of the *Sanguozhi yanyi* also suggest an oblique connection to Selden's remarkable collection of Mexica pictorial narratives and a kind of rereading of standard narratives of the ancient originality of their "hieroglyphic writing." The most famous of these, the Codex Mendoza, already had a long history of readings in London, previously translated by Michael Lok and then published by Samuel Purchas, who called this "Mexican History in Pictures . . . the choicest of my jewels." Selden stored it with the *Sanguozhi yanyi* in the middle of the Asian books in section F of his library along with a "rotulus" and "liber."[90] Purchas, during the abortive war with Spain that was the Anglo-Dutch Cádiz expedition (1625), had read the codex as an indication of the incompleteness of Spanish sovereignty and the willingness of the indigenous Mexica to engage in translation projects. In an elaborate translation in his *Pilgrimes*, he explained that the book showed movement in space and time through "rundles or pricks"

Fig. 21. Laud brass astrolabe, Persia, before 1636. © Museum of the History of Science, Oxford, 47063.

(calendrical notations) as well as "partitions" showing time.[91] Connecting this "accompt[ing]" with "history," Purchas argued that until this book, which began in 1324, "a Historie, yea a Politicke, Ethike, Ecclesiastike, Oeconomike History, with just distinctions of times, places and arts, we have seen neither of theirs nor of any other Nation," including the ancient Egyptians and the modern Chinese and Japanese.[92] Selden had the postconquest Codex Mendoza as a kind of Rosetta Stone, which in its indications of historicity, genealogy, and place revealed the inadequacy of a purely pictorial "hieroglyphic" reading that Bacon and some of the artificial language theorists tried to employ. As Selden's label itself suggested, it was "Cum figures, quasi hieroglyphicis."

The "rotulus" and "liber," which Selden obtained separately, came from Mixteca-speaking areas to the south of Tenoctitlan on the Pacific side of Mexico in northwestern Oaxaca. The roll, a preconquest document on amatl paper from Coixtlahuaca in the far north of the Mixteca areas, told a story of four chieftains who descended from heaven on a journey, using a footprint design for "deeds done" to create a kind of narrative map.[93] The screenfold "liber," which also used footprints and a kind of snaking "ligature" for narrative mapping, was a 1560 manuscript from the Nochixtlán Valley dealing with a dispute over the ownership of the town Zahuatlán that used depictions of marriage alliances and alliances between communities to make its claims (figure 22).[94] Selden had also probably seen the twenty-two-foot-long pre-Columbian and Mixtec Codex Bodley, which had similar genealogical sequences and representations of temporal travel using trails of footprints, and the Codex Laud, a divinatory almanac for offerings, rain, eclipses, death, and marriage.[95] If Purchas was largely collecting around the Baconian trope of language, Selden was more interested in history, time, and space.

Kept in reserve, all of these texts represented the rather faint possibility of reactivating Mexica and Chinese mathematical models of time and space in the face of Spanish and Catholic universalism. The Jesuits at the Ming court had been pushing Christopher Clavius and Pope Gregory's controversial calendar reform of 1582, especially from the 1630s, and these reforms were ultimately accepted by the Qing dynasty after 1644 (and in England in 1752). The Jesuit Martino Martini would publish his *Atlas Extreme Asiae Sive Sinarum Imperii* in 1654, and the next year it would become volume six of Joan Blaeu's sequel to Ortelius, the Amsterdam-published *Theatrum orbis terrarum*. Dedicated to the Archduke Leopold Wilhelm of Austria, the governor of the Spanish Netherlands, Martini's volume argued that divinely inspired Catholic science was being exported to the world, with the Jesuits

Fig. 22. Codex Selden (ca. 1560), bands running from bottom to top, right to left, showing from year 12 Flint 7 Deer to 3 House 10 Deer, including several marriages and the transitions between the first and second dynasties of Jaltepec as keepers of the tobacco ritual at the temple altar or house (top center). © Bodleian Library, University of Oxford, 2009, MS. Arch Selden A.2, f5.

imagined as enlightened emissaries and swift messengers, "angeli veloces," who used technical data to rationalize space and "open the closed" ("clausa recludo"). Conversely, the books and instruments in Selden's remarkable collection suggested the linguistic, historical, and technical challenges to universal claims. They also suggested the limited sources that were currently being used in compiling sacred chronologies closer to home, in projects like the universal timelines of John Lightfoot and James Ussher.[96]

THE SELDEN MAP

In the codicil to his will written on June 11, 1653, Selden requested that all of his Hebrew, Syriac, Arabic, Persian, Turkish, Greek, and Latin manuscripts as well as the Greek marbles now at his house in Whitefriars to go to "some publick use," whether a "convenient library public" or "some college in one of the universities." He then made particular reference to his "map of China made there fairly, and done in colours, together with a sea compass of their making and divisions, taken both by an English commander, who being pressed exceedingly to restore it at a great ransome, would not part with it."[97] He singled out this valuable and practical piece for description along with the Arabic editions of Hippocrates, Galen, and Ibn Sina that he gave to London's College of Physicians. Just a month and a half before Selden decided that his collection should be made public, Cromwell had dissolved the Rump Parliament, and his Council of Officers drifted towards an almost theocratic "Sanhedrin" as a solution to the problem of governance. Selden in response composed a massive historical critique.[98] He had already been skeptical about the Rump Parliament's use of *Mare Clausum* to justify both the first Navigation Act and the war with the Dutch that had been declared in July 1652, technically over whether the English could call themselves sovereign "lords of the seas" in the Atlantic.[99] The English edition entitled *Of the Dominion or Ownership of the Sea* appeared in 1652 through the efforts of the pamphleteer and newspaper publisher Marchamont Nedham, who argued that it "vindicated your [Parliament's] Right of Soveraigntie over the Seas."[100] Selden's idea of a public library was meant to be a moderating force in the political turmoil of the early 1650s, checking the messianic and apocalyptic tendencies in the Commonwealth with a sense of the complex global history of the development of law and science. The Chinese map revealed the hubris of the Commonwealth's activities as well as those of the Jesuits and in its technical dimensions gave a sense of what kind of approach to international relations could result in the creation of good positive law.

Obtaining the Chinese map had been fortuitous for Selden, as it demonstratively expanded on the efforts of Purchas, Laud, and Bodley. In 1637 or 1638, Laud had acquired a unique manuscript copy of the *Shunfeng xiangsong* (順風相送) or "Escorting Picture for Favorable Winds," which like the Selden Map describes both the eastern and western trade routes through the Philippine archipelago and the Vietnamese coast respectively, although there is no evidence he knew what it contained.[101] A decade earlier, however, Purchas, in his 1625 *Pilgrimes*, had been the first person in Europe to attempt to reprint a Chinese woodblock sheet map, "The Map of China, taken out of a China Map printed with China characters, etc. gotten at Bantam by Capt. John Saris," with the Mandarin title, 皇明一统方舆備览 (*Huang Ming yitong fang yu bei lan*, "Imperial Ming unified square world prepared for viewing"), faithfully reproduced at the top (figure 23). Purchas claimed to have engraved the map "to express my love to the pubicke in communicating what I could thereof," but also to show "China Characters (which

Fig. 23. Samuel Purchas, "The Map of China, taken out of a China Map printed with China characters, etc. gotten at Bantam by Capt. John Saris" [皇明一统方舆备览, *Huang Ming yitong fang yu bei lan*, "Comprehensive directional view map of the Imperial Ming"] from *Hakluytus Posthumus or, Purchas his Pilgrimes: Contayning a History of the World in Sea Voyages and Lande Travells by Englishmen and Others* [London: W. Stansby for H. Fetherstone, 1625], 400–401. Reproduced by permission of the Huntington Library, San Marino, CA.

I thinke, none in England, if any in Europe understands)." Because he could not translate these, Purchas decided to remove most of the characters from the map except for the title, leaving only the abstract marks for provinces, cities, and rivers, adding descriptions and latitude and longitude as he understood them, since "I could not wholly give it, when I give it; no man being able to receive, what he can no way conceive." Like many of his other engravings, the map was for Purchas primarily a writing sample, representative of a kind of script he defined as "New Worlds" because "the Ancients knew [them] not."[102]

But in taking away most of the text, Purchas also abstracted the map so that what remained was a far more accurate transcription of the river systems and the location and hierarchy of the seats of provincial and county officials than even the main comprehensive maps in Luo Hongxian, giving an image of the Ming Empire very similar to the Ming *Yu ditu* (輿地圖, "World earth map" ca. 1512, surviving copy 1526). Purchas had obtained the sheet map from Hakluyt, during the period in which he was translating the *Mare Liberum* of Grotius, and in many ways it confirmed what Hakluyt had published in 1589 from the lost Cavendish map. Hakluyt had in turn received it in 1613 from the East India Company captain John Saris, who had somewhat forcibly obtained the map in exchange for a debt owed by a Fujianese merchant while he served as chief factor at Banten between December 1608 and October 1609. The map Purchas printed, unlike the Selden Map, imagined China as a closed empire surrounded by a kind of lawless space, and it gave far more detail about the administrative structure and riverine topography of the Ming than did the Selden Map. For the merchant in Banten who had it, the map was almost certainly a connection with both home and civilization in the world of foreigners (fan 番 or yi 夷), while on the Selden Map maritime East Asia (including Banten) was known space that contrasted with the regions north of the Great Wall, where there lived "Huangwali fan, Henanli fan, ju zaicihou" (黃哇黎番呵難黎番俱在此後 "only Huangwali foreigners, Henanli foreigners, after this point").

That Saris would seize such a map seems typical of the situation at Banten in the early seventeenth century as described by the Company's first factor Edmund Scott, with competing houses of Chinese, English, Dutch, and Javanese that traded, loaned, gifted, extorted, and stole money, commodities, and objects from each other.[103] Generally, however, Saris had disapproved of outright theft, as when Sir Edward Michelborne seized a Chinese junk going from Banten to Timor in January 1605. This action apparently upset the Javanese pangeran and courtiers, interested in establishing a reputation as a stable and trustworthy port like regional competitors at Aceh or Johor, more than the local China "captain" or "admiral" and the Chinese port-master or shahbandar, who saw the situation as more fluid compared with the stability of the Ming. Purchas also seemed interested in depicting the region as a frontier. He published the account of Michelborne, formerly an Irish adventurer with Essex, had received special license from James in June 1604 after being expelled from the East India Company to trade on his own authority with "Cathay, China, Japan, Corea and Cambaya" (Cambodia), causing no end of consternation not only among the Chinese and Gujaratis but also among the Dutch, whom he also looted. But Michelborne was also concerned with collecting data, and along with his navigator—the famous John Davis, who was killed by a group of Japanese pirates near Pahang in 1605—another John Davis (also Davy, d. 1621) took notes that he would later expand into a full rutter on the ninth East India Company voyage of Edward Marlowe from 1612 to 1615.[104] For Saris, after his own return in 1610, such data quickly attracted more investment for the eighth EIC voyage in April 1611, which he personally commanded and which saw the establishment of the English factory at Hirado in 1613 with the help of the "China Captain," there Li Dan. In 1625, Purchas published Saris's extensive accounts of Gujarati, Tamil, Malay, and Chinese shipping coming in and out of Banten, which contained detailed data on prices, commodities, and routes, as well as the Davis rutter from the Michelborne and Marlowe voyages.[105]

Thus Purchas tried to remain faithful to the Chinese sheet map's Sino-centric concept of a square or measured world (*fangyu* 方與) with a single sacred, political, and economic center, rather than the accounts Saris was developing out of Banten or Li Dan out of Nagasaki, which involved much more cosmopolitan networks of trade. Purchas also explicitly used the map to critique the "erroneous conceits which all European Geographers" have employed in translating maps of the Ming, explaining that a square was more appropriate than the "heart" shape used by Ortelius and Hondius derived from Yu Shi's map and later used for the basic shape on the Selden Map rather than the square shape of the *Yu Ditu* or Luo's atlas.[106] The errors by Ortelius had the paradoxical effect of both elongating China—making Beijing at 50° rather than 40° latitude—and of creating a distinct kingdom of Cathay. Purchas's map served as an emblem of the Ming as a contained, unified, and central commercial *imperium* with a distinct language system, an iconic approach probably not all that different from the reasons the Fujianese merchant in Banten had kept the map in a wooden case in his house, a sense of the true center of East Asian trade.

Although Beijing is placed near the vertical axis on the Selden Map, which was not the case with the Purchas map, Selden's shows a much broader world of trade and ocean space centered on a point in the South China Sea near Hoi An. One could argue that this image of East Asia as a whole was a European rather than a Chinese conception. Selden himself had seen a copy of a manuscript map in Madrid in Robert Cotton's library that had a similar framing; the manuscript map was presumably copied for the English envoy Charles Cornwallis in 1609 as an emblem of Spanish resurgence after their success against the Dutch in Ternate in 1606 (figure 24).[107] In fact a number of maps from this period tried to depict the ocean in East Asia, such as the "Indische Noord" (1621) of the VOC's Hessel Gerritsz and the "Ko Karuta" (ca. 1613) of Kadoya Shichirojiro, which shows in a series of pinholes a trunk route from Nagasaki to Hoi An similar to that in the Selden Map.[108] If Purchas's map looked back to the kind of stability desired from the Ming in London in the late sixteenth century, Selden's map depicted ocean space of the early seventeenth century, which was being re-imagined by a number of different groups contesting those routes. Although "Chinese" in both its language and its design, notably through a particular technique of drawing navigational routes that differed markedly from European portolan or grid approaches, the Selden Map was also historical in the sense that it was responding to changes in both cartography and trade in East Asia as a whole and attempting to leverage a variety of sources of knowledge to visualize a system of Fujianese trade.

The one definitive claim by Selden about his map and compass was that they were not exchanged but "taken both by an English commander," paradoxical emblems as loot of the "open sea" policies that both he and Saris rejected. But the map itself also was a product of the multipolar situation of the early seventeenth century, in which Tokugawa Japan had unified, the size and significance of the Manila silver trade had grown, a series of Southeast Asian polities had solidified their territorial presence, and on the islands trade had increased substantially and the use of Islamic law, courts, and education as part of statecraft had become more prominent. The seizure of the Selden Map could have occurred during a number of possible encounters, and Timothy Brook has argued that Saris himself took both maps; there is also some evidence that it was taken during the short period of the ambitious Anglo-Dutch joint-defense fleet (1621–22), when the English and the Dutch cooperated in sending ships out of Japan to blockade Spanish Manila. Certain aspects of the map itself, including the visibility of two landings on Taiwan, make it likely that the map was made in the later 1610s, while the detailed navigational instructions around Manila and ports on Luzon, including what the map labels as the "jiawan men" (甲万門 "Ten Thousand

Fig. 24. Map copied from Spanish original in Madrid in 1609, Cotton collection. "The
kingdom of China." © The British Library Board, Cotton Augustus, II.ii.45.

Shell Passage") or the Mindoro Strait and Apo Reef that are the gateway to
a route leading down to Ternate and Tidore, suggest that the map may have
been made in Manila or by someone very familiar with navigation around it.

Given the Company's general policy of working with Chinese merchants, indeed relying upon them for help with political alliances, loans of

silver, and access to more extensive networks of trade, it was not a usual practice after the first decade of the English presence in East Asia to capture Chinese ships or interfere with merchant houses. During debates over the union of the two East India Companies in 1615, the Dutch had proposed the joint-defense fleet as a measure against both Spain and competition from indigenous shipping:

> this union would impede the traffic of the Chinese, Malays, Javanese and others who trade in the Moluccas . . . On the contrary, while the two companies remain separate, the Malays, Chinese, Javanese and others will more and more take over all the traffic of the Indies and above all that of the Choromandel, of the Moluccas and elsewhere.[109]

The East India Company did not agree with such strategies, which endangered their status in both Banten and Japan, and in June 1618 the VOC even had to pay off five or six captured Chinese merchants who were working for the Hirado Chinese shipping magnate Li Dan with a *tael* each for fear of being called out and evicted as pirates in a Japanese court.[110]

The two Anglo-Dutch blockade operations in 1621–22 were recorded and regarded as legitimate in Japan, where the goods were off-loaded, but the very first English one in 1620 was not, making it a potential candidate for the location of the map's seizure. In August 1620, the East India Company ship *Elizabeth* seized an armed frigate on Taiwan. Onboard were a Sakai (Osaka) merchant named Jojin as captain, a mixed Chinese and Japanese crew, two Portuguese sailors, a pilot named Diego Fernandez, and two Spanish priests, Pedro Manrique de Zuniga and Luis Flores, disguised as merchants trying to smuggle themselves into Japan. The ship had left Manila in June to avoid the Dutch blockade, traveled to Macao, and stopped at Taiwan on July 22 to get water, wood, and other provisions for the final leg of the journey to Japan.[111] The *Elizabeth*, which had come from London by way of Batavia and had by chance stopped on Taiwan as well, took the opportunity of finding the priests to make the first English seizure in the new anti-Spanish policy.[112] The captured ship was full of silk, ginger, and cotton that had been bought on the mainland with Manila silver to be sold on the Japanese market, all of which was eventually confiscated by Japanese authorities. The *Elizabeth*'s captain, Edmund Lenmyes, was rumored to have taken a chest of gold; if the map and compass were aboard, they possibly fell into the hands of Gabriel Tatton, the ship's chartmaker. None of this can be proven definitively. Both Lenmyes and the head of the Japan factory, Richard Cocks, died in disgrace on their way back to London, while Tatton left only a half-finished volume of charts after falling drunk off a ship in Hirado and drowning. All of their personal goods were confiscated and dispersed by the Company, including Tatton's charts as well as Cocks's remaining journals and some of his collection of Japanese books, without clear provenance records and ending up in a variety of collections.

If the map had been seized by the *Elizabeth*, it would have been originally intended for a Hirado mansion near the spot on the map marked by two large red chrysanthemums, a house owned by one of the principal investors in the seized ship—the smuggler, merchant, pirate, and "China Captain" or "Rey de China" Li Dan (d. 1625), popularly known to the English as Andrea Dittis. Li and his brother "Augustine" have a shadowy past in the historical record, and some have argued that they may even be composite figures. Most accounts have Li being born in Quanzhou, setting up as a merchant in the Manila parian where he spent time on the galleys as a prisoner for debts, and subsequently moving to Japan to build a commercial empire. In terms of the English record, however, it is clear that Li leased Saris a house for a factory in Hirado in 1613 and subsequently taken some 1,000 *tael* (ca. 37.5 kg) of bar silver in January 1618, supposedly to give to his family contacts in Quanzhou to open up trade with China. He took 1,500 more after telling Cocks in January 1621 that the Taichang emperor, an illiterate fifteen-year-old who came to the throne in October 1620, had given permission for two English ships to go to Fuzhou in Fujian each year, and that they were only awaiting formal permission from the Zhejiang military governor.[113] But if the rumors of Chinese merchants are to be believed, this represented only a small percentage of the 30,000 *tael* investment fund ("capitael" in the Dutch account) that Li had assembled from Osaka and Nagasaki merchants, the Bakufu, and no doubt a host of other sources for opening up the silk trade, at the time limited to smuggling by ships like the one caught by the *Elizabeth*.[114] Li had a vast and active merchant network involving shares in perhaps eighty ships that mirrored that of the Selden Map, ranging from Hirado in Japan, through Taiwan to Manila, and down through Quang Nam to Tonkin, Siam, and Banten, and because of this he had important relationships with the Tokugawa, the Spanish, the Dutch, and the Portuguese.[115] He needed both Japanese and Spanish silver to make the China trade work, buying Chinese goods with Spanish silver and selling them at higher rates for Japanese silver and gold. When all Spaniards were exiled from Japan in 1624 in part because of the two priests secretly aboard the ship captured by the *Elizabeth*, Li had to move his operations, some of the Chinese community on Kyushu, and with them the Dutch VOC itself to the more central location of Penghu and then Taiwan, something that may have even been contemplated when then map was made.

The back of the map indicates that it was drawn around the routes and measured according to a scale, which Selden (as well as Tatton) would have been able to see. The draft of the trunk route follows a line parallel to the Fujianese coast drawn from a point off Quanzhou that extends northeast to the Goto islands (五島 *Wu dao* or five islands) off Nagasaki (*Longzishaji* or "Languesaque" 笼仔沙机) and Li's home in Hirado (*Yulin dao* or "fish scale island" 魚鱗島) and southwest to the islands off Quang Nam (廣南), marked on the map as Hoi An (會安) and the Da Nang Peninsula (在峴港), the launching point for Southeast Asian trade (figure 25). Letters from James I for the king of Cochinchina carried by Tempest Peacock and Walter Caerwarden on a Japanese ship followed this precise route in 1614 on a voyage that ended disastrously; this was the route followed by the Japanese sailors who killed John Davis in 1605 when he seized their ship on the Michelborne expedition; and English sailors were also catching rides on junks to Siam branching off the same trunk, usually ending up lost if they tried without the help of a good Chinese pilot, as the Portuguese had done before them.[116] With Gabriel Tatton making soundings, the *Elizabeth* had followed this coastal trunk route as far as Taiwan, but relied on the captive pilots and perhaps the map itself to avoid the Ryukyus and cross the East China Sea from Taiwan to Nagasaki.

Building from this trunk route, the Selden Map envisions a "T" structure off Quanzhou and Zhangzhou that branches southeast towards Manila, making Taiwan an ideal location for arbitrage of bullion and transmission of Chinese goods to other markets. On the front, the scale and an orienting square box reappear along with a compass displayed as technical emblems (figure 26). The compass rose at the top of the map uses the word *luojing* (羅经), or geomantic rather than navigational compass, and the map, according to Selden's will, was paired with a separate and more elaborate geomantic compass used for drafting directional angles, which are labeled on the routes themselves as well (figure 27).[117] The textual instructions on the western edge of the map for going from Calicut (古里国) to Aden (阿丹国), Zufar/Salalah (法兒国), and Ormuz (忽鲁謨斯) use the word *ji* (計), meaning plot rather than what are often in the literature called *zhenlu* (針路) or "needle paths." The emphasis on physical draftsmanship is also linked in the map's text to the ocean-based system of measuring time through *geng* (更) or ship's watches that involved burning a standardized length of incense from particular starting points (yong 用), which would use a drafting

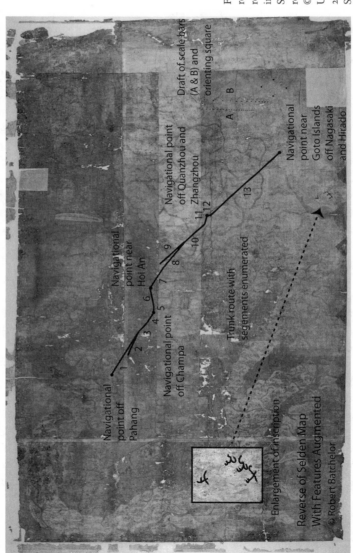

Navigational
point off
Pahang

Navigational
point near
Hoi An

Navigational point
off Champa

Navigational point
off Quanzhou and
Zhangzhou

Draft of scale bars
(A & B) and
orienting square

A   B

Navigational
point near
Goto Islands
off Nagasaki
and Hirado

Trunk route with
segments enumerated

Enlargement of inscription

Reverse of Selden Map
With Features Augmented
© Robert Batchelor

Fig. 25. Selden Map
reverse, with the trunk
route, scale bars and
inscription highlighted.
Segments of the trunk
route are numbered.
© Bodleian Library,
University of Oxford,
2009, MS Selden
Supra 105.

Fig. 26. The compass rose and scale bar on the Selden Map. © Bodleian Library,
University of Oxford, 2009, MS Selden Supra 105.

compass to measure distances taken from the scale bar.[118] Tree-structures
of routes marked a approach distinct from either the graticule-strategies of
Mercator and Luo or the drafting compass circles of Portuguese and Japanese portolans. The significance of the map's structure to Selden, an admirer of visual trees and charts of relations since his schooldays, could not
have been lost.[119] As a potential display object, the map had parallels with
a genealogical tree or *zongzhi tuben* (宗支圖本, ancestral chart) in showing the outward-reaching nature of the lineage group with its ties to the
mainland. But the technical and mathematical visualization of routes in
the Selden Map indicated the much broader rise in mathematical literacy
in the sixteenth century in fields from carpentry and navigation to accounting, and the sense that traders like Li Dan were increasingly involved in
tasks that required more complex forms of mathematics.[120] If it was population data compiled by the empire that made Luo's map translatable, it was
navigational data about the ocean compiled by merchants that allowed the
Selden Map to translate. Moreover, the credit system of exchange, which
the East India Company had to rely on because of the seasonal nature of the
monsoons dictating the arrival times of both silver and commodities like
pepper, rested upon this kind of popular mathematical knowledge.

Fig. 27. Selden's geomantic compass. China, early 17th century. © Museum of the History of Science, Oxford.

As Stephen Davies has pointed out, the map appears to have a magnetic declination of about 6° or 7°, and for those who showed it to Edmund Halley in 1705 it represented a possible source of such data for East Asia, where the English had made few empirical observations of differences in magnetic declination from true north.[121] In the seventeenth and eighteenth centuries, neither European nor Chinese mapmakers knew how to predict declination, which, because magnetic zones shift across both space and time, can only be extrapolated from historical data. It appears that English navigators were the first to collect enough reliable data to notice aberrations over time and space, so that when William Baffin measured declination at the northern end of Baffin Bay off Greenland at 56° west on his fifth voyage in 1615, he found it "a thing almost incredible and matchless in all the world beside."[122] By 1634, Henry Gellibrand noticed variation of magnetic declination over time at London shifting from 11° east in 1571 to 4° in 1634, and its measurement became a regular activity of the Royal Society from the 1660s.[123] Apart from regular measurements in London, most English data came from the Indian Ocean and the South Atlantic, following East India Company routes from London to ports in India and Banten on Java.[124] Regular shipping routes in the South Atlantic and Indian Ocean required consistent remeasurement because of significant and changing declinations. English sea captains starting around 1615 began to notice problems with variations in old charts in the Indian Ocean, and the VOC and the English sent chart makers like Gabriel Tatton out to East Asia in the late 1610s as well after finding the Mercator system of plane charts defective.[125] Having various new theories on the earth's magnetic field like those of William Gilbert but no actual solution to predicting variation, ships would generally carry multiple charts and keep logbooks to estimate variations on particular routes.

Joseph Needham argued that unlike the English, who had been measuring declination for only a few decades, the Chinese appeared to have grappled with the question for centuries, largely in relation to geomantic compasses and not having enough regular data to make definitive claims about variation over either space or time. Nevertheless, serious debates over declination appear to have arisen in China only in the late Ming, although not as the result of European contact. The key problem was that mariners who ventured into the South China Sea noticed "needle confusion" (針迷 zhen mi) as Fei Xin, writing in 1436, suggested happened when passing the Seven Islands in the north (near Hainan) and Kunlun Island (Pulo Condor off Champa) in the south.[126] This was possibly a reference to the line of 0° declination running across the South China Sea in this period. It was a subtle problem, given that most Chinese and Japanese compasses could only measure intervals (*jian* 間) as small as 3.5°.[127] Needham found two "Chinese" measurements of declination in the early seventeenth century, both for Beijing. Xu Guangxi confusingly claimed 5°40′ *east* and Mei Wending scoffed at Adam Schall's claim to have found over 7° of western declination by sundial measurement, believing it closer to 3° west.[128] This revival in interest was probably an effort to buttress geocentric concepts both there and in Rome by suggesting that magnetism does not shift but instead maintains order in the universe by connecting terrestrial and celestial poles.[129]

Both maritime East Asia and the North Atlantic had particularly fortunate conditions for compass navigation circa 1600 because the declination was close to 0°. Regional trade in these areas could manage without any comprehensive approach to declination. However, like European ships entering the South Atlantic and Indian Oceans, the Chinese and Japanese merchants plying the East Asian trade that ranged from Japan to Siam beginning in the late sixteenth and early seventeenth centuries also required new approaches. Declinations over this area ranged from approximately 5° west at Siam and Sumatra to 5° east in southern Japan. The Selden Map suggests that in seventeenth-century maritime China, declination had for the first time become important enough to be represented in Chinese cartography, which had previously dealt with the problem as one of the differing systems of finding directionality in order to follow rutters like the *Shunfeng Xiangsong*.

Given that some of the corrections on the Selden Map indicate knowledge in flux, notably an erasure of island chains between Timor and Ambon in the Banda Sea, the system it uses for arranging data seems well suited to change over time. The drafted system of the routes and the use of a modifiable orientation by declination allows the use of a much looser approach to representational cartography, one that appears to assemble and rearrange previously printed sources into a coherent image of East Asia structured by a system of mathematically defined routes relying on compass navigation data. The part showing the Ming Empire is copied from a map from a tradition of Fujianese encyclopedias linked to Jianyang publishers like Yu Xiangdou, possibly the one entitled *Ershiba su fenye huang Ming gesheng di yu tu* [二十八宿分野皇明各省地輿圖, "Twenty-eight mansion, field-allocation, imperial Ming, all provinces terrestrial world map"] in Wu Weizi's 1607 Fujianese encyclopedia, the *Bianyong xuehai qunyu*, which survives in a copy brought to Leiden from Batavia by the Dutch missionary collector Justus Heurnius in the late 1620s (figure 28).[130] The map thus shows strong connections to the world of Fujianese popular printing familiar to Chinese overseas merchants and well-represented in English and Dutch collections

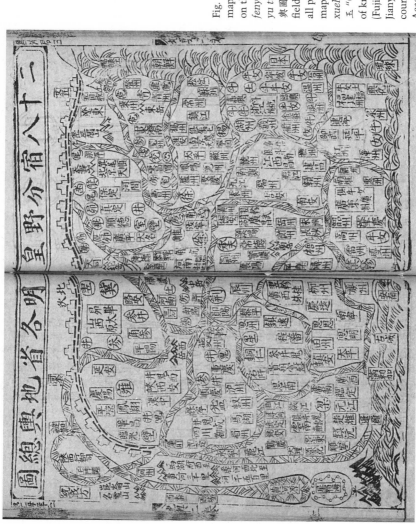

Fig. 28. Chinese encyclopedia map used for the Ming Empire on the Selden Map. *Ershiba su fenye huang Ming gesheng di yu tu* [二十八宿分野皇明各省地輿圖 "Twenty-eight mansion, field-allocation, imperial Ming, all provinces terrestrial world map"], in Wu Weizi, *Bianyong xuehai qunyu* [便用學海群玉 "Convenient to use: Seas of knowledge, mines of jade"] [Fujian: Xiong Chogyu from Jianyang, 1607], juan 2. Image courtesy of Leiden University, Acad. 226, v.1.

of Chinese books and maps like that of Purchas. But the printed encyclopedia used for the Ming section included a series of other maps connecting the
square world "under heaven" to the sun, moon, and circular heavens, including *fenye* markings for relating each province indicating how to match
terrestrial fields (provinces) with the twenty-eight mansions (constella-
tions). On the Selden Map, although the *fenye* are copied, they are relevant
only for the provinces of the empire itself, while the sun and the moon on
the map can allude only to a broader cosmological order defined by origin
myths like that of Fuxi and Nawa.[131] In other words, not only is the image
of the Ming on the Selden Map a cliché that contains far less detail than the
original of the Purchas map, but the Ming no longer appears to be cosmolo-
cally central as in the encyclopedia tradition, which would have appealed to
overseas merchants longing to make a connection with home.

     This does not mean that the Selden Map indicates a move towards
European models of cartography, which seems to be more the case with the
Japanese portolans of the late sixteenth and early seventeenth century. It is
possible that the mapmaker had access to a European-style printed or manuscript map, in particular the Spanish one copied for Cotton (figure 24),
which used Beijing in a similar way as a location for orienting true north
and marking the horizontal middle of the map itself. However, the Selden
Map gives no indication of adopting longitude and latitude systems. Neither
the encyclopedia map nor the Spanish map, emblematic of the Ming and
Spanish Empires, can be said to have been important to the basic structure
of the Selden Map, which relied on the routes themselves. They were used
only as visual clichés and framing devices, a way of displaying what the
system of routes connected. As a result, the elements on the Selden Map
related to the land are relatively uninteresting in terms of cartographic data,
but those related to the ocean—ports, islands, reefs, currents—take on a
greater significance than in most traditional Chinese depictions of the seas
because they are located relative to offshore navigational points defined by
the routes. By breaking out of and abstracting from traditional cosmology,
the Selden Map revealed the way in which the East China Sea (東海, *donghai*, labeled twice on the Selden Map) and the silver cycle more generally
were reshaping the old paradigm for navigation based on the South China
Sea (南海, *nanhai*, absent from the Selden Map).[132]

     The Selden Map was thus a kind of mathematical model for creating a
logic of possession and exchange that no imperial map—whether Chinese,
Spanish, Portuguese, Japanese, Dutch, or English—fully comprehended.
Jane Burbank and Frederick Cooper have argued that in the seventeenth
century, pace Grotius, "innovations in what would later become known as

'international law' took place at the interfaces between empires and their various legal traditions."[133] They are referring to formal and informal agreements about how to create interfaces between systems of writing, courts of law, systems of justice, and, as the Selden Map suggests, cosmological understandings of space and time. Both Selden's *Mare Clausum* and the Selden Map suggest how this emergence of rules of law and a space of contractual exchange actually occurred. Selden did not merely place a Chinese map at the center of his collection as a symbol; he highlighted a unique kind of map that demonstrated the use of technology by Chinese merchants to define sea routes and achieve *dominium* outside of the strict sphere of *imperium*. The result, as Selden perceived so incisively, was not so much a social contract but a technical contract that formed the basis of the rule of law, agreements about how things worked. Certainly European law at times served as a device for translating "native" claims, but the broader notion of the rule of law as a translation device had to be abstracted from a wide variety of exchange and translation practices rather than from "nature" itself. While Bacon, Grotius, Descartes, Hobbes, and Harrington may have written more canonical and European solutions to these issues of language and sovereignty, Selden's writing and archiving projects and the Selden Map itself put forward the concept that law and language, territory and ocean, space and time, needed to be radically rethought in the context of translation of highly heterogeneous historical texts, devices, and exchange relations from around the world. Moreover, by suggesting that under conditions of mutual recognition the measurement of time and space was a universally valid form of experience, the movements of the stars, the tides and currents of the ocean, the directions of the winds, the pull of magnetism, and the claims of mathematics all became paths towards the universal rather than the cosmological. The maker or makers of the Selden Map were pioneers in this regard.

CHAPTER FOUR

The Image of Absolutism

And, while this famed emporium we prepare,
The British ocean shall such triumphs boast,
That those who now disdain our trade to share
Shall rob like pirates on our wealthy coast . . .
Thus to the Eastern wealth through storms we go;
But now, the Cape once doubled, fear no more:
A constant trade-wind will securely blow,
And gently lay us on the spicy shore.
—John Dryden, "Annus Mirabilis" (London: 1667)

1661: TAMING THE REBELLIOUS EMPORIUM

In the world described by Selden and the makers of the Selden Map, one of overlapping dominions with complex translations and negotiations over contracts, sovereignty increasingly became a question of image. This was something that Selden's friend Thomas Hobbes and the rising star John Dryden understood very well. For the 1661 coronation of Charles II, the city, in imitation of imperial Rome, erected a marble arch for the new king's procession from the Tower to Whitehall, a spectacle of Rebellion vanquished by Monarchy and Loyalty. The translator and theater impresario John Ogilby, who had come to London from Scotland with James I, received the commission from the Common Council of London to create three additional arches—the naval might of England, the return of Concord, and the coming of Prosperity. The second arch, with a relief depicting global trade, dedicated to the navy and placed next to the Royal Exchange, provided a vision of how monarchy would "prepare the emporium."[1] *Emporiki*, the exchange or translation of goods from city to city, protected by a stable monarch would

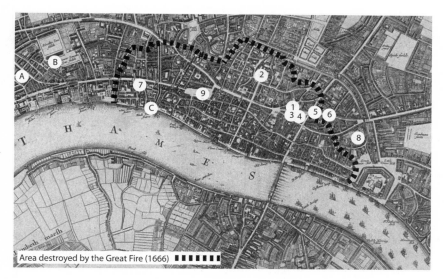

Fig. 29. Post-fire cultural and economic institutions overlaid on John Oliver, "Map of the City of London, City of Westminster, Lambeth, Southwark and the River Thames as it is now rebuilt since the late dreadfull Fire" (London: John Hill and John Seller, 1680). Reproduced by permission of the London Metropolitan Archive, K1268437.

1. Royal Exchange, Cornhill (rebuilt 1669)
2. Royal Society, Gresham College (1660)
3. Garraway's Coffee House, Exchange Alley (ca.1668)
4. Jamaica Coffee House, St. Michael's Alley (1668)

5. Royal African Company, Leadenhall Street (1672)
6. East India Company, Leadenhall Street (1658)
7. John Ogilby's House (1667)
8. Samuel Pepys's House (1660's)
9. Christopher Wren's St. Paul's (under construction 1675-1697)

New theaters:
A. Theatre Royal: Drury Lane (1663)
B. Duke's Theatre: Lincoln's Inns Field (1661-1671)
C. Duke's Theatre: Dorset Garden (1671)

become the ideological keystone of a commercial empire that surpassed those of the Spanish and the contemporary Dutch.[2]

This vision came out of the decision among some Londoners that accepting a certain image of absolutism would be a useful tool in the expansion of commerce, the kind of consent that the title page to Hobbes's *Leviathan* (1651) suggested. Along with support from the Restoration Parliament, this image making had general support at least initially from many involved with the new theaters, coffee houses, reworked institutions of exchange, and printers and engravers of London (figure 29). The image of the king remained, however, ambiguous and indeed a subject of conflict and compromise.[3] With relative stability in England, Scotland, and Ireland after the turmoil of the Civil War, much of both the dynamic and the conflict during the Restoration in London stemmed from divergent approaches to exchange and translation in the Atlantic as pioneered by Cromwell's Navigation Act and Caribbean policies and the East India Company's trade in the Indian Ocean and East

Asia. The latter required increasing amounts of flexibility and translation amongst different exchange and sovereign strategies, and so the image of empire in the East was indeed an image, one that was frequently used as a political tool by the Company, interloping traders, and Asian merchants to build their own networks outside of any substantial imperial authority.

The tributary aspirations of the coronation pageant were ultimately just that, aspirations, but the emporial concept, derived from past and ongoing experience with Asian trade, increasingly shaped London's trade and growth. Developing relations with Asia, which in the late sixteenth century had provided models for national separation from Spain and in the early seventeenth century had given London both economic and intellectual dynamicism, became more problematic by the late seventeenth century, as the contradictions between dynastic and corporate sovereignty and dominion became more apparent. During the Civil War, both Royalists and then Cromwellians had tried, with mixed success, to build an Atlantic empire based on the "sugar revolution" on Barbados in the 1640s—first at Suriname (1650), then with the failed invasion of Hispaniola (1654), and finally through the capture of Jamaica (1655). Additionally, after the failure of Courteen's Madagascar experiments, Cromwell in 1657 offered the South Atlantic island of St. Helena (granted 1657, settled 1659) to the East India Company as a secure base for resupplying ships. Along with the first Navigation Act, these seem to have been efforts to create a coherent Atlantic World that could support the ambitions of the Commonwealth in a manner paralleling that of the Dutch.

If the Cromwellian effort was ambitious and unsustainable, the Restoration lacked the resources that had been mobilized by Parliament during its experiments with sovereignty in the 1640s and 1650s. Charles II did not maintain the kind of army Cromwell had used with mixed success to establish empire in Scotland, Ireland, and the Caribbean, and he began with limited naval resources. Even before plague, fire, and Dutch invasion rattled London in 1665–66, there was a sense that the monarchy was fragile. Yet by the 1670s, fears arose that England might be sliding towards Catholicism and French-style absolutism, suggesting a dramatic reversal of events. Much of this was image making. Ogilby proved instrumental in the late 1660s and 1670s in the effort to create an image of a London-centered global absolutism embodied by Charles II that could be the focus of general appeal. Theatrical world-image making was not new, having played a prominent role in the quite distinct pageantry of Elizabethan and Jacobean theater, court masque, and London's mayoral pageants.[4] Ogilby's efforts also paralleled those in the 1650s by Innocent X and Alexander VII to remake Rome, especially in using engravings to produce a new vision of urban and global space through depictions of projects like the new obelisk in Piazza Navona and books on themes related to translation and occult forces like those of the Jesuit Athanasius Kircher.[5] Ogilby's project also drew upon the sense of more totalizing imperial spectacles gleaned from the illustrated travel diaries of those like Sir Thomas Roe and Peter Mundy, who had seen among other things the great Mughal works of Akbar and his successors, and most directly from extensive Dutch travel literature and printing projects. The image created by Ogilby and others in London should not, however, be seen as a Leviathan generating a single centralizing focus and imperial agenda. By the 1670s, the abstract image of monarchical power radiating outward from the metropole would also be a tool through which people in Asia, Africa, the Americas, and the British Isles themselves could influence politics in London, pulling the city and the crown more deeply into the conflicts and cultural productions occurring at the frontiers of its networks. Moreover, these shifts and changes, especially as the Manchu conquest of China disrupted the global silver trade, made London as an emporium immensely successful, giving it both the global reach and the financial power to buttress the image of an increasingly powerful monarch.

As part of the coronation festivities Ogilby orchestrated, the East India Company sponsored its own pageant to show support for such image making. In the years immediately before the Civil War, the increasingly wealthy and overlapping memberships of the Levant and East India Companies in London dominated City politics, using the court of aldermen, the customs farm, and the East India Company itself as sources of authority.[6] Private traders had disrupted that authority in the Commonwealth years. As part of his broader efforts Cromwell recognized a new joint-stock in 1657, and Charles II quickly gave the Company a new charter in 1660. When the king passed the East India House, one of the young sons of the Company's director Sir Richard Ford appeared dressed in Mughal garb with two black slaves saying, "Stay, Royal Sir, here comes an Indian, Who brings along a full fraught Caravan Of perfect Loyalty and Thanks, to pay; As Your due Tribute, on this Glorious day." Ford's other son, also in a Mughal vest, arrived on a camel also led by two black slaves and other attendants, scattering jewels, spices, and silk to the spectators. The Mughals were ostensibly the tributary here, a fiction that was in fact the opposite of the actual situation at the Company's presidency in Surat. As a way of promoting this new image for Londoners, in the pageant the Company itself paid tribute directly to Charles and distributed largess to the city itself as if it too was the obsequious child bowing to parental authority.[7]

What the Company portrayed for the eyes of Charles II and the court was a vision of empire built of Asian tribute and supported by African slavery. This symbiotic relationship with the old East India Company and a proposed new Guinea company for trading in African slaves, to be run by the future James II (James, Duke of York), would in combination give the new king, according to East India's writers, an "Imperial Title . . . the same in Deed, which Spain's proud Crown vaunts but in Name," so that divinely "Which Heavens grant! . . . we never see the Sun set on Your Crown, or Dignity."[8] Unlike the more modest late sixteenth- or early seventeenth-century attempts to place London into global trading networks, the Company here encouraged Charles II to see himself as the successor to the *imperium* of the Spanish Hapsburgs in the sixteenth century, a power to be wielded against "encroaching Holland's Rival Force." The languages here were mixed, an old one of corporate bodies paying tribute and translating between sovereignties coupled with a new almost mechanical world of interaction and "rival force." As with Hobbes's *Leviathan*, the image of absolutism concealed such tensions.

A coherent image also became important to the East India Company as a way of challenging the private competition to its trade that had arisen during the 1640s and 1650s as well as the increasingly diverse demands being placed on it by Asian states and merchant intermediaries. Silk and cotton in the second half of the seventeenth century represented a hope for making Asian trade profitable again after its dissolution between 1654 and 1657, but the Company now needed to make more complex deals in what historians call the forward system, which involved prepayment of silver for inland production of cotton cloth in India.[9] Cloth did replace pepper and spices by the 1680s as the most profitable commodity, but making this switch required fundamental changes, which included exporting of larger amounts of silver to the Indian Ocean and Southeast Asia, shipping Indian textiles to Southeast Asia, and moving copper and gold from East Asia to India in what was called the "country trade." The Restoration court wanted a relatively closed and tightly regulated circuit in the Atlantic in order to generate taxation revenue from a new sugar and slave economy, an "absolutist" and classically "mercantilist" trade based on new revisions of the 1651 Navigation Act. But the East India Company needed monopoly protection and a unified economic policy in London simultaneously coupled with a far more flexible approach to gathering coin and bullion in Europe and to developing trade circuits in Asia.[10]

Rather than a pure form of mercantilism, the Navigation Acts of 1651,

1660, and 1663 negotiated between these mutually dependent visions of a tightly reorganized British Atlantic world and a highly mediated and multipolar Asian trade. The first Navigation Act of 1651 tried to avoid any conflict by explicitly stating "that [neither] this Act nor anything therein contained, extend not, nor be meant to restrain the importing of" commodities from the Mediterranean (the Levant Company) and that beyond Cape of Good Hope (the East India Company).[11] After Charles II arrived in London in May 1660, he quickly revised the first Navigation Act that September to limit the amount of bullion that could be exported by the East India Company to £70,000 and to require that all goods coming into England from Asia, Africa, and America arrive on English ships with crews consisting of 75% Englishmen. Even with such "mercantilist" gestures, the 1660 act still kept the language of not restraining the Levant and East India trades.[12]

When the coronation pageant occurred the following April, the East India Company wanted to show the king that it, like the African slave trade, could be a more integral part of his strategies, while at the same time hoping for an easing of the new regulations since its export of bullion and cooperative efforts with Asian merchants and sailors more strongly challenged mercantilist designs. In December 1660, Andrew Riccard, governor of the East India Company, had written a short history of the Asian trading system for the Council of Trade to suggest the layered and competitive history of a trade that "was managed of old by the Chinese, then by the Moores of Cambaya or Guzeratts" and later the Portuguese.[13] The Dutch and English had succeeded on the basis of invitations to trade, "patents," and the capital stock generated by such patents rather than building fortresses like the Portuguese. According to Riccard, the Dutch and English strategies emulated the older dominance of Chinese and Gujarati trading networks and were thus more historically and legally appropriate. For the Company's strategy to be effective, however, it needed to extend its trading networks deeper into East Asia. In 1661 the London merchant and Company factor at Banten, Quarles Browne, presented Sir Thomas Chamberlain, the Company's deputy-general, with an account of the trade of Cambodia, Siam, China, and Japan in which he argued that the company should expand into East Asia relying upon the interdependence of commercial networks in the South China Sea just as the Selden Map depicted.[14] This was a big selling point for the revised 1663 Navigation Act, which dropped all language about the Levant or East Indies, making all claims of royal sovereignty over ocean traffic essentially Atlantic ones. This act also added strikingly new language about "free liberty for exporting" money and gold or silver bullion, after registering such exports with the customs house.[15] Armed with a permanent joint-stock and a radically freed bullion and currency market in London, the Company's plans to expand into East Asia could now be realized.

To please the court, the new act implied a royal monopoly on the gold trade with Africa, perceived as a way of making the inner circle wealthy enough to exert significant political power. James, Duke of York, hatched up this scheme in the fall of 1660 with a group of courtiers led by Prince Rupert, who during naval exploits for the Royalists in 1652 had retreated to the Gambia River. They planned to locate and seize West African mines of gold like the Spanish had done with silver in the Americas.[16] Protected by forts captured from the Dutch and maintained by state funds, this gold would then go on East India Company ships to India and Southeast Asia, where in addition to silver, gold was also in demand. The East India Company and the new Africa enterprise put together agreements to this effect in October 1662. The latter received a new charter as the "Company for the Royal Adventurers into Africa" in January 1663. At the same time preparations were made to send a large standing army to Bombay, a gift negotiated as part of Catherine of Braganza's dowry, with the idea of making it a secure bullion trading center in India akin to what the Dutch had at Batavia, the Spanish at Manila, and the Japanese at Nagasaki.[17] The design for the new gold coin, the five guinea first struck in 1663, echoed the East India Company's coronation pageant iconography—two African slaves surrounding an elephant (the castle was added in 1674).

Only a tiny proportion of the East India Company's currency exports, however, would end up being gold brought by the Royal Adventurers.[18] As early as 1663, after a series of disastrous voyages that reaped mostly small loads of elephant ivory, the Royal Adventurers decided that the slave trade would be more lucrative, especially with trade to the North American and Caribbean colonies under crown regulation. Coastal West African kingdoms, having long experiences trading with Europeans and North African Muslims, were reluctant to part with gold, but they could use imports of weapons and gunpowder to capture slaves.[19] England went from taking 23,000 enslaved people across the Atlantic between 1601 and 1650 to 115,200 between 1651 and 1675, most of those after 1663. At the same time, the Portuguese and their Dutch investors, responsible for forcibly taking 439,500 and 41,000 people across the Atlantic between 1601 and 1650, had a combined share of the slave trade approximately equivalent to the English in the period between 1651 and 1675.[20] The African Company profited from the slave trade itself, and the crown generated customs revenue from importing Virginia tobacco and Barbados sugar.

The 1663 Navigation Act, also known as the Staple Act, did not simply set up the basis for a mercantilist Atlantic economy using colonial reexports through London after the disappointment of the gold trade; it also enabled a profound financial revolution in London in relation to Asia. Since the beginning of the century and blossoming in the 1650s, private traders had been bringing back diamonds from India as a valued commodity that was also easy to move privately both in Asia and in Europe.[21] In 1664, the directors of the East India Company officially recognized this trade, simply requiring a 2% commission on imports from Company stockholders and 4% from nonstockholders in return for complete legitimacy.[22] Some of this trade involved newly arrived Portuguese Jews in London, linked into the coral, emerald, gold, and silver trades in Europe, Africa, and Portuguese India (Goa), through Livorno and other cities. In the absence of a direct source of African gold, the Company increasingly turned to these same London Jewish merchant bankers—notably Jerónimo Fernandes de Miranda and Gomez Rodriguez (d. 1678) and his son Alphonso—among other merchants and brokers in Amsterdam to supply gold from a wide variety of sources to ship to India.[23] A large-scale money market, with its roots the sixteenth-century maneuverings of Gresham and the bullion agreement in the Cottington Treaty with Spain (1630), developed in London around the East India Company's need for silver and gold. Bullion came from a variety of bankers; perhaps the most important up until 1676 was Edward Backwell, who was also one of the principal suppliers for the mint and government credit.[24] All of this occurred in a highly decentralized manner, without institutions like the Amsterdam bank (1609) or the future Bank of England (1694) and through the mechanisms and extended networks of urban London.

Opening London's bullion trade with Asia produced striking results. Between 1601 and 1650, the Dutch VOC exported approximately 425 metric tons of "silver equivalent" to Asia. During the same years the London East India Company sent out 250 metric tons. From 1651 to 1700 the Dutch sent out 775 metric tons, but the figure for the East India Company was 1,050 metric tons—four times the bullion exports of the Company in the first half of the century and 35% more than the Dutch in the same period.[25] Instruments of credit and banking expanded rapidly in London, including mortgage loans issued by scriveners, commercial loans issued by brokers, foreign exchange facilitated by brokers, and finally the issuing by the Goldsmiths of "accountable notes" in exchange for deposits that could circulate as the equivalent of paper money.[26] In *The Key of Wealth or a new way of improving trade* (1650), William Potter had argued that increased trade would increase paper obligations, creating in the process new wealth. Gen-

erating assets based on the development of London's trade and exchange
with Asia had the effect of diversifying and securing such assets through de-
centralization along merchant, banking, and other networks so they could
not be easily seized or forcibly borrowed upon by the monarchy like the
notorious "pepper loan" of 1640. This did not produce a robust court as
much as a robust set of constituencies in Parliament increasingly tied to the
financial systems and exchange networks of London.

Why had this dramatic shift in exchange not happened before? In part
it was an unintended consequence of the legal opening of London to free
trade in bullion, but it also responded to a profound shift in global exchange
patterns that occurred between 1660 and 1690, which disrupted the "silver
cycle" of ca. 1540–1660. The Qing Empire dramatically reduced its importation of silver as a result of war and economic depression, which had ripple
effects on currencies, bullion, and commodities trading throughout Asia
and in Europe, Africa, and the Americas as well. As the Selden Map suggests, the "late Ming" (1580s to 1650s) commercial expansion on the south
China coast and in the Yangtze Delta had been supplied with silver by intricate exchange trees of Chinese merchants reaching out to Japan, Manila, the
Malukus, Banten, and Batavia as well as a host of other emporia. Presumably this expansion would have continued even after the Qing conquest in
1644, and it did in the late 1640s and 1650s despite some disruption due to
warfare. But in 1655, things fell apart in an attempt to rein in the Japanese-
born Fujianese merchant, pirate, and Ming loyalist Zheng Chenggong. After
negotiations for Zheng's surrender collapsed in 1655, he began systematically attacking private shipping and attempting to invade Fujian from his
island bases. Qing attempts to blockade his trading networks failed, networks that provided key transshipments of silver from both Nagasaki and
Manila. Blockade turned into a ban on foreign trade and coastal clearances
in an effort aimed not to regulate European ships, but those of Chinese
merchant networks. The result was a thirty-year economic depression (ca.
1660–90), punctuated by the "Three Feudatories" civil war (1673–81) and
characterized by a return to deflation problems not seen since the 1580s,
including a dramatic erosion of agricultural incomes due to overproduction
of rice, cotton, and other agrarian commodities and a general contraction of
commerce.[27] The islands and territorial kingdoms of Southeast Asia linked
by Chinese, Japanese, and European merchant networks began to develop
new exchange patterns to compensate. The Mughals and the sultanates of
Mattaram and Golconda in India and the Safavids in Persia, as well as an
array of territorial kingdoms in Southeast Asia, all sought to expand their
own emporia by exporting cottons and silks ordered in advance with ready

silver and gold. The London East India Company and private traders associated with it, unlike the Spanish tied to Manila or the Dutch to Batavia, became uniquely positioned to move different currencies and bullion as well
as various types of cloth around economic circuits fragmented by the Qing
economic depression to help rebalance exchange.

As the global economy drove shifts in London's economy, bringing larger
bullion flows through the city, Charles II tried and largely failed to capitalize on these changes by extending absolutist policies to Asia and using the
new colony of Bombay to directly challenge the Dutch. In February 1662,
he put together a standing army to ship to Bombay under the command of
Sir Abraham Shipman as governor and general of the city.[28] Sir George Oxinden, president of the Surat Factory from 1662, advocated creating a standing
army funded by the crown and using force to expand the English presence
in India.[29] The soldiers arrived in the fall of 1662, but the Portuguese refused
to let them land, forcing them to spend two years on the small island of
Anjadip off Goa. Most fell sick and died, including Shipman himself in October 1664. When the survivors finally did land in Bombay, the long period
of deprivation encouraged a kind of lawlessness, including a 1665 incident
in which soldiers from the Bombay garrison seized and looted Indian ships.

Oxinden, whose interest in Charles II's absolutist strategies in Bombay
seems to mostly have been about defraying Company expenses, then managed the crisis. A consummate broker, he knew Gujarati and other Indian
languages as well as Portuguese after a stay in India from 1632–39. He
had failed at private ventures to China and elsewhere in late 1650s, giving
him good reasons to stay away from angry London investors, as lawsuits
mounted and printed broadsheets appeared demanding restitution in London during the 1660s and 1670s.[30] In this particular crisis, he got Charles
to transfer authority for Bombay to the East India Company and turned the
survivors of the Anjadip ordeal into the East India Company's "1st European Regiment."[31] Oxinden was the kind of broker that an English presence
required in India, working for crown and Company but comfortable acting independently and able to raise capital and sell diamonds privately in
London. He died in 1669, and his massive mausoleum in Surat, which also
contains the body of his brother Christopher (d. 1659), was in many ways a
statement of the kind of independence of translation and transaction that
sustained London's interests in India. The London broadsheets and lawsuit
testified to his and his family's continuing and conflicted ties to the city.

Instead of building up a military presence in Asia, which both the Dutch
and later in the 1680s the French would do, the Company increasingly pursued military partnerships in which it brokered diplomatic and arms deals

between the English crown and sovereigns in Asia to allow the latter to more effectively wage their own wars. The shift in the relationship with Banten during the Restoration is a good example. The new sultan Abdul Fatah (aka Ageng Tirtayasa, r. 1651–72 and jointly until 1683), after having taken on the VOC in a war from 1656 to 1659, saw the East India Company along with French, Danish, and Fujianese merchant organizations as useful in pursuing a growth strategy during the Qing depression by increasing pepper production. Chinese and English merchants helped him build a navy in order to take over diamond fields on Borneo's Landak River in 1661 and solidify control of southern Sumatra. He commissioned the *Sejarah Banten* (ca. 1660), a historical chronicle of Muslim pilgrims converting Java, to serve as a kind of countercolonization narrative to the Dutch. In December 1664, Sultan Abdul wrote to Charles II thanking him for a shipment of "Great Ordnance muskets" (cannons) and asking for nine more as well as gunpowder and bullets to stop "the Imperious Hollanders . . . supported by their Jappan, China, Tunckeene [Tonkin, or Vietnam] and Amboyna Trade."[32] With the transformation of the Bombay project and the redeveloped relationship with Banten, London's fiscal and trading relationships with Asia were left open-ended, encouraging the integration of brokered relations, financial exchanges, translations of sovereignty, and a global outlook more deeply into the life of the London itself (figure 30).

ASIA AND THE PROBLEM OF RESTORED SOVEREIGNTY

Responding to the dynamics that increasingly made London not only an emporium in Europe but also a key node in the development of two rather distinct economies—slavery in the Atlantic world and bullion exchange in Asia—desires emerged for a more comprehensive image of both London and England's global role. This division of economic zones, which gave a certain degree of stability to Charles II's emerging state over the next quarter century, rested on two false ideas. The first was that the institution of slavery could remove the necessity of translation in the Americas and Africa as well as keep such issues of translation out of London itself. The simultaneous eruption of King Philip's War and Bacon's Rebellion in 1675 and the subsequent difficulty in comprehending and managing them, as well as the increasing colonial dependence on the crown for authority and military support over the next century, were good examples of this failure. Cotton Mather blamed King Philip's War on too much translation on the frontier (everything from commerce to long hair) and too little effort being made towards conversion, while the lack of a coherent colonial policy on slavery

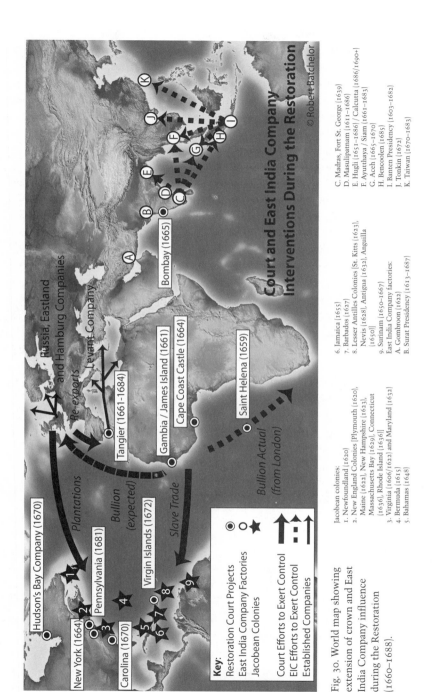

Fig. 30. World map showing extension of crown and East India Company influence during the Restoration (1660–1688).

© Robert Batchelor

Court and East India Company Interventions During the Restoration

Hudson's Bay Company (1670)

New York (1664)

Pennsylvania (1681)

Carolina (1670)

Virgin Islands (1672)

Russia, Eastland and Hamburg Companies

Levant Company

Re-exports

Tangier (1661–1684)

Gambia / James Island (1661)

Cape Coast Castle (1664)

Saint Helena (1659)

Bombay (1665)

Plantations

Bullion (expected)

Slave Trade

Bullion Actual (from London)

Key:
◉ Restoration Court Projects
○ East India Company Factories
★ Jacobean Colonies

Court Efforts to Exert Control
EIC Efforts to Exert Control
Established Companies

Jacobean colonies:
1. Newfoundland (1620)
2. New England Colonies [Plymouth (1620), Maine (1622), New Hampshire (1623), Massachusetts Bay (1629), Connecticut (1636), Rhode Island (1636)]
3. Virginia (1606/1622) and Maryland (1632)
4. Bermuda (1615)
5. Bahamas (1648)

6. Jamaica (1655)
7. Barbados (1627)
8. Lesser Antilles Colonies [St. Kitts (1623), Nevis (1628), Antigua (1632), Anguilla (1650)]
9. Surinam (1650–1667)
East India Company factories:
A. Gombroon (1622)
B. Surat Presidency (1613–1687)

C. Madras, Fort St. George (1639)
D. Masulipatnam (1611–1686)
E. Hugli (1651–1686) / Calcutta (1686/1690+)
F. Ayutthaya / Siam (1661–1683)
G. Aceh (1665–1670)
H. Bencoolen (1685)
I. Banten Presidency (1603–1682)
J. Tonkin (1672)
K. Taiwan (1670–1683)

before the war was replaced in New England by both Grotius's (and Spanish) "just war" theory of capture and an expanded Native American slave trade.[33] The second false idea stemmed from a basic contradiction between the image of a closed system of sovereignty led by a paternalistic monarch, which in many ways tried to revive medieval cosmography, and the need for dynamic approaches to exchange and translation like those of merchants largely free of state control moving currencies between Asian ports to support and supply the mechanisms of state power. Because of this, the 1663 settlement of the third Navigation Act actually opened up fundamental questions in London about the nature of government, questions intensified by fire, plague, and Dutch invasion in the mid-1660s. All of this increased the distance between the image of monarchal stability under Charles II and the realities of London's emerging practices of global exchange.

The desire for a world picture was not new and can be seen in the renewed popularity during the Restoration of the clearly dated texts of the early seventeenth-century compiler Peter Heylyn, who might be seen as the cosmological counterpart of the patriarchal authority described by Robert Filmer in the 1630s. In Heylyn's earliest work, *Microcosmus or A Little Description of the Great World: A Treatise Historicall, Geographicall, Political, Theological* (1621), which came out his Oxford lectures on historical geography, he wrote that world pictures appealed to kings because "the hearts of Princes are in a manner boundless, one world is not sufficient to terminate their desires." Against atomists and Epicureans, Heylyn wanted to demonstrate the dynamic power of history and geography to remove the sense that the world was "randome," "unstable," or conversely "like a dead carkasse [that] hath neither life nor motion at all."[34] Commonwealth officials seized Heylyn's library, but he nevertheless managed through a compilation of pieces often directly lifted from Hakluyt, Purchas, Botero, and Pierre d'Avity to produce his most famous work, the four-volume *Cosmographie* issued by Henry Seile in 1652. The revised 1657 edition was then further expanded after the deaths of both Heylyn and Seile by his entrepreneurial widow Anne Seile and the publisher Philip Chetwind in new editions in 1665–66, 1669, 1670 (with a 1667 appendix), 1674, 1677, and 1682. In 1682 and 1683, two competing biographies of the author came out, testifying to the enduring popularity of the *Cosmographie*.[35] These editions suggested to the extent to which formerly royalist visions like those of Heylyn's and Filmer took on a new life as Restoration clichés about the derivation of political authority from natural sources.

At the same time as such clichés began to take hold, even to the exclusion of more subtle approaches like that of Hobbes, the dynamics of Asian trade increasingly created complex problems for sovereignty that the informal settlement between Company, crown, and Parliament could not address effectively. This was particularly true of the landmark 1668 legal case *Thomas Skinner v. East India Company*.[36] In 1654, the old East India Company faced bankruptcy and sold off its shipyards, houses, forts, and goods, abandoning its Banten factory between 1656 and 1658 and making a public notice of the closure of its books on the Royal Exchange in London. Skinner, who like a number of other Londoners wanted to get involved in East Asian trade, established a trading station within the principality of Jambi on Sumatra, and the Pangeran Ratu granted him a small 2.5-kilometer-wide island called "Bhrala" (Berhala) that was well-placed at the entrance of the Straits of Malacca. There he planted cardamom and contracted with several Chinese to cultivate a pepper garden, but in May 1659, he narrowly escaped death when a surgeon from the newly rechartered East India Company and six of his servants attacked him in the streets of Jambi with knives. Company factors looted Skinner's house, seized the goods in his warehouse at Quale along with those of his brother Frederick Skinner, and forcibly boarded his ship the *Thomas* with a small army of one-hundred hired fighters armed with *keris*. Company factors also detained the local diamond merchant and banker Peter Denis de Brier, who had 15,000 *reales* worth of Skinner's diamonds. Thomas appealed to the pangeran, who ordered everything returned and repaired, but the Company's servants absconded to Banten, claiming rights conveyed by Cromwell to seize the property of anyone trading against the October 1657 charter. Left destitute, Skinner had no choice but to return to London in 1661 for compensation, filing a claim that November as a citizen of the Corporation of London.

The case paralyzed government decision making. The Privy Council did not know what to do; any decision could give the East India Company either too much or too little power in relation to London citizens. The Company denied as much of Skinner's argument as it could, questioning among other things Skinner's citizenship, with claims he was part Dutch and had backing from Genoese capital. They also implied that Skinner was an actual traitor, having somehow helped the Dutch attack the Company's factory with his own army of 40 Europeans and 150 local mercenaries, using guns and swords rather than the *keris* and clubs wielded against Skinner and his people. After six years, including plague and fire, Charles II sent the matter to the House of Lords in 1667. The Lords awarded damages to Skinner, but the Company then got the House of Commons to claim the Lords had exceeded their jurisdiction. The Lords and the king's lawyers countered that the king should superintend conduct of all lesser authorities using courts and writs. The stalemate worsened through 1668–69, as orders to imprison both Skinner and the East India Company's governor came from the respective houses. Finally, in February 1670 Charles II asked for resolution of the case, and when the two houses refused him, he ordered all proceedings erased from the record, using absolutism as a kind of last resort.[37] The deep fractures in sovereignty revealed by the case needed to be hidden. The East India Company won by default, and in doing so had made almost as much of a mockery of the institutions of English government as the Dutch had when De Ruyter towed away the English flagship the *Royal Charles* from the Medway in 1667.

With the merchant networks connected to London's exchange relations having outgrown the capacity of Westminster to effectively manage them, the validity of the chartered joint-stock company itself came into question. Stockholder lists for the new Royal African Company of 1672 and the East India Company in 1675 show holdings predominantly by people who would a decade later be Tories. These stockholders increasingly acted according to political rather than purely mercantile interests.[38] In the early 1670s, a broadsheet appeared in the streets of London arguing that under Queen Elizabeth joint-stocks had been seen as "but one scruple better than monopolies" and that such expedients for fostering trade had outlived their usefulness (figure 31). London merchants in Asia or Africa, as the broadsheet pointed out, were no longer perceived as radically foreign—as either "devils" or "gods." Letters patent, including the "phyrmand or chop," were now globally acceptable instruments of government for recognizing rights of merchants to trade and contract freely. Knowing how to translate themselves into different religious and linguistic settings, London merchants no longer needed to maintain expensive forts as the absolutist backers of the joint-stock suggested. "Now we are so well acquainted with the Customs both of the *Persians* and *Indians*, after almost one hundred years Converse and Traffick with them," read the broadsheet, "that instead of seeking to them for their Commodites, we are courted for our Gold and silver."

The Royal Society of London, chartered by the king in 1663 and never visited by him, was a more direct attempt by its members to formulate a science or *scientia* that through its rationalization of language and images would centralize and manage translation. Frustrated by his own inability to make this happen in Ireland in the 1650s, Robert Boyle took on the task of making a coherent image of the world through communities of witnesses drawn in part from Oxford scientific and translation circles from the 1640s and 50s.[39] The clearest effort to manage language along these lines came with the publication by John Wilkins, the secretary of the Royal Society,

The Eaſt-India and Guinny-**TRADE**,
As now under a **JOYNT-STOCK**,
Conſidered, and proved prejudicial to the
KINGDOMS INTEREST.

THE plauſible Pretence of maintaining of Forts, is the grand Argument uſed by the Favourers of the *Guinny* and *Eaſt-India* Companies, why that Trade ought to be managed in a Joint-Stock.

At the firſt Settling of Factories in thoſe Countries, it might require ſome extraordinary Charge; but now we are ſo well acquainted with the Cuſtoms both of the *Perſians* and *Indians*, after almoſt one hundred years Converſe and Traffick with them, that inſtead of ſeeking to them for their Commodities, we are courted for our Gold and Silver.

The *Turky* Company, though not in Forts, yet their Expences in maintaining of an Embaſſador at *Conſtantinople*, Conſuls at *Smyrna*, *Aleppo*, *Cyprus* and *Tripoli*, together with the Allowance of Factorage, doth equal if not exceed the *Eaſt-India* Company's Charge.

'Tis about fifteen years ſince the *Eaſt-India* Company gave liberty for Subſcriptions; and they did not admit of, neither do they now employ, one half of the money that was ſubſcribed; but rather chooſe to ſupply their Neceſſities by taking it up at Five *per Cent.* Intereſt.

The whole Stock doth not really amount to above Four hundred thouſand pounds ſterling. One man hath by degrees obtained one two and thirtieth Part. Three or four have amongſt them an eighth Share of the Stock. Several Widows and other perſons are conſtrained to ſell, by reaſon Dividends are deferred and unduly made for that very purpoſe: So in few years, it will become a Monopoly.

They have ten or fifteen Sail of Ships annually in their Imployment, and in *Perſia*, the Great *Mogul*'s Territories, *China*, *Japan*, Eaſt-ſide of *Africa*, the Eaſt and South Sea Iſlands, they have not above twenty Trading Ports where Factories are eſtabliſhed.

The Trade of the *Eaſt-Indies* may admit a Stock of two Millions ſterling, employ eighty Sail of Ships of like Burthen, eſtabliſh ſixty Factories; and though their gains may amount to but Ten *per Cent.* and the Joint Stock that now is, gain Fifty *per Cent.* yet it is equal to the Kingdoms Intereſt. Beſides it will, by a modeſt Calculation, advance His Majeſties Cuſtoms Seventy thouſand pound *per Annum*, increaſe Mariners and Navigation, and infallibly ruin the *Dutch*, whoſe great Prop, next to their Inviſible Bank, is their *Eaſt-India* Joint Stock.

The Kings Cuſtoms and Additional Duties amounts to about Seven hundred thouſand pounds ſterling *per Annum*: So by computation, the annual Trade of this Kingdom Inward and Outward, is Fourteen Millions *per Annum*: Now the *Guinny* and *Eaſt-India* Companies have the Grant from *Cape Verde* to *Cape Bona Eſperanza*, the *Red Sea*, Eaſt-ſide of *Africa*, *Perſia*, *Mogul's* Country, Coaſt of *Cormandal*, Bay of *Bengal*, *China*, *Japan*, the Iſles in the Eaſt Ocean and in the South Sea. There is not above Eighty legitimate Merchants concerned in theſe Joint Stocks, that have ſo great Privileges, and ſo large a Scope to trade in: Now in His Majeſties Dominions there are upwards of One hundred thouſand Traders to Sea; and all the Places and Kingdoms, that the *Turky*, *Hamburgh*, *Eaſt-land*, *Ruſſia*, and *Greenland* Companies, *Italian*, *Spaniſh*, *Barbary* Merchants, the Adventurers to *New England*, *Virginia*, and *Caribbee* Iſlands, are not as large again as what the *Guinny* and *Eaſt-India* Companies challenge by Grant.

UPON all theſe Conſiderations it is humbly propoſed and deſired, that the *Eaſt-India* and *Guinny* Trade be reduced to the ſame Regulation and Method that the *Levant*, *Hamburgh*, and other Patent-Companies enjoy, who had greater Reaſons for a Joint Stock than can now be pretended; and though their Priviledges were granted them in the Infancy of Trade, when we were ſo great Strangers in remote parts, that in ſome Countrys we were taken for Devils, and Navigation for Conjuration, and in other parts adored as Gods, and thoſe Mariners that paſt *Cape Verde*, and entred the Streights of *Gibralter*, were regiſtred in *Hackluit's* Voyages; yet notwithſtanding all theſe Difficulties, the Wiſdom of Queen *Elizabeth's* Senators was ſo great, that they would never admit of Joint Stocks, eſteeming them but one Scruple better than Monopolies.

The Conſtitution of the *Hamburgh*, *Eaſtland*, *Ruſſia*, and *Levant* or *Turky* Company is, That thoſe who have ſerved Seven years to a Freeman of the City and a Member of the Company, may challenge their Freedom. Any man who will leave off Shop and Ware-houſe-keeping, for twenty five or fifty pounds may require it. All Noble-men and Gentle-men may demand it. When they are admitted, they are ſworn to obey ſuch Orders, as the major part of the Company ſhall Enact; which formerly were only to order Leviations and Conſolage upon Goods, for defraying Embaſſadors, Agents and Conſuls Expences, maintaining our Capitulations and defending our Priviledges.

Fig. 31. "The East-India and Guinny-Trade as now under a Joynt-Stock Considered, and proved prejudicial to the Kingdoms Interest." The National Archives, CO 77/15 Eliz-Charles II, f. 138, ca. 1672–1675, photo by Robert Batchelor.

of his *Essay towards a Real Character* in 1668. Wilkins was adept at working London patronage networks, rewarded as a prebendary of St. Paul's and Bishop of Chester in 1668 for siding with Matthew Hale and the moderates in an effort to reconcile dissenters to the new Restoration order. Recognizing the power of an imperial language along the lines of Latin, Wilkins had noted that Chinese allowed communication across a large kingdom but was far too complex to be effective as a philosophical language and that artificial commercial languages like Malay offered more potential.[40] In an almost neo-Filmerian gesture, Wilkins returned to the genealogical family and gave the systematic and structural frame of the language itself as much weight in generating linguistic authority as the words themselves, a kind of image of the split between dynastic structure and the dynamic world of exchange and translation.

In a repost to Wilkins's call for an artificial generative grammar, John Webb in his 1669 *Historical Essay* suggested that through an eternal and divine order handed down through Noah, the "primitive language" of Adam had survived the ravages of time in China. An old collaborator of Inigo Jones, Webb found himself in retirement in Somerset in 1669 after having been passed over for the promised post of Royal Surveyor in favor of Christopher Wren. Webb nevertheless dedicated his volume to Charles II as an indirect rebuke and an idealized model of coherence and purity, which "was out of one body and one Offspring peopled so at length grew into one body and form of Empire."[41] Webb's suggestion that Chinese was somehow divine contrasted sharply to Wilkins's more mundane critique of it as a "general Character." Contemporaries were split. The linguist John Beale wrote to John Evelyn that he thought Chinese might work better than what Wilkins proposed, while others looked more directly to Malay.[42] Webb and Wilkins can be read as trying in very different ways to buttress an older form of world picture. But both Webb's and Wilkins's scholarship also pointed towards a more contemporary problem—the need to find a new language to address gaps in Restoration sovereignty. This was a growing problem because devolution of power had created a range of actors around the world looking to London for signs of authority.

ABSOLUTISM AND JOHN OGILBY'S WORLD PICTURE

John Ogilby was the key figure who put forward an imagistic strategy distinct from the linguistic ones of Wilkins and Webb. His ties to Scotland and Ireland as well as relationships with Royal Society circles and Dutch publishing gave him substantial resources and unique perspectives.[43] In contrast to Boyle's failure to practice natural history in southwest Ireland during the Interregnum, Ogilby made his name and money in Dublin as a theater manager in the late 1630's, suffered through the Civil War and returned to London during the Commonwealth to become a translator of antiquity—Virgil's *Works* (London: 1649, 1654, 1658, and Dublin: 1666), Aesop's *Fables* (London: 1651, 1665), and Homer's *Iliad* (London: 1660, dedicated to Charles II). With the Restoration, he moved to the Strand in London and developed innovative sales techniques in relation to print, including newspaper advertisements and lotteries for large expensive editions with copperplate engravings. But his translation of the *Odyssey*, the "most Ancient and Best Piece of Moral and Political Learning," was ill-timed, coming out on May 25, 1665, as the plague spread in the docks to the east of London, and then his entire stock of classical translations burned in the Great Fire. As a result, Ogilby completely changed direction in his publishing and embraced the activities of Dutch presses and their global engagement in an effort to map London's place in the world. His timing was right, coming at a period of great uncertainty about the emerging economies of the Restoration, the nature of Charles II's sovereignty, and London and the East India Company's role in relation to Asian trade. He placed London within a framework translated out of Dutch efforts that emphasized both its global reach and its Qing-like imperial dominance over England and the British Isles. The monarchy of Charles II could be seen as the ideal synthesis of commercial power and imperial authority.

Ogilby's first publication along these lines appeared in 1669, a translation of Johan Nieuhof's account of the Dutch VOC embassy (1655–58) to the new Qing Empire, printed in 1665 in Amsterdam by Jacob van Meurs.[44] Copyright for Ogilby's translation had been granted one month after the passing of the Rebuilding Act on March 20, 1667, just before Wilkins and Webb published their own perspectives on the Chinese language.[45] The English translation appeared late, after successful editions in French (1665), German (1666), and Latin (1668) already made the question of Dutch and Qing relations a European one. Nieuhof synthesized a number of narratives with his own, borrowing from early Jesuit accounts by Trigault and Ricci and more recent Qing-era books from the 1650s by Semedo and Martini.[46] His picture of the Qing was in many ways optimistic, despite depicting damage from the transition and drawing attention to Dutch failures in establishing permanent trade relations with the Qing and maintaining their base in Taiwan after 1661.[47]

The printer van Meurs had commissioned 150 copperplate engravings based on Nieuhof's sketches, suggesting that China and its urbanized economy could be effectively translated in ways other than directly through the language itself. Nieuhof worked from a range of Chinese printed and visual sources, and Amsterdam engravers added further fanciful details to fill out the blank areas of the sketches to allow for a kind of double reading, a comfortable exoticism "repackaged for European consumption" that disguised a degree of openness to translation.[48] This use of so many detailed copperplate engravings contributed to the delay in Ogilby's publication, since such expertise in copperplate engraving in London was relatively new, including examples in Ogilby's on Charles II's coronation as well as Robert Hooke's *Micrographia* (1665).[49] In this sense, Ogilby translated what had previously been doubly foreign images—the superior technical ability in Holland to produce images as well as elements from the extensive commercial culture of Chinese woodcuts.

To Nieuhof's material Ogilby added a translation of a letter from the Jesuit "John Adams" as well as selections from the Rome-based Jesuit Athanasius Kircher's 1667 *China Illustrata*, which linked the publication to the image-making efforts of Rome in this period as well.[50] "Adams" was Johann Adam Schall von Bell of Cologne, the first Roman-trained and from April 1664 a papally sanctioned "mandarin," who as the head of the Beijing observatory had been the great "antagonist" of the Dutch in negotiations there. Both works were problematic because of their strong Catholic connections, but the inclusion of the Jesuits counterbalanced Dutch republicanism. Schall in fact informed the Shunzhi emperor not only that the Dutch threw off their monarch but also that "if these people ever get footing, upon pretense of Commerce in any place, immediately they raise a Fortress, and Plant Guns (wherein they are most expert) and so appropriate a Title to their possessions."[51] Kircher's project, on the other hand, was explicitly about Jesuit practices of translation, including in the original a reengraving, transliteration, and translation of the Nestorian Stele found in 1625. Like Webb, he wanted to place China in a classical and Judeo-Christian history linked to both hieroglyphics and Hebrew, and his "illustrations" were meant as authentic religious emblems.[52] Even though Ogilby appeared to be merely copying, reengraving, and resetting already complex assemblies of various Qing, Dutch, and Roman cultural materials about the nature of empire, he also reframed them by comparing accounts. In the process, he demonstrated to Londoners how complex global images were already being produced in Amsterdam, Rome, and Beijing.[53]

Success with the *Embassy* encouraged Ogilby to take on a bigger and more comprehensive atlas project that would surpass the efforts of Heylyn and George Humble from the 1620s.[54] In May 1669, he printed a "Proposal" announcing his official move from the "Greek and Latin Paper Kingdoms" he had previously published to "a New and Accurate Description of [the World's] four Quarters . . . teaching them English." Africa would be the first volume because "though not remotest, [it is] yet furthest from our Acquaintance."[55] By November he had obtained a warrant from Charles II (attached to the dedicatory material in *Africa*) that extended printing rights from earlier warrants issued in 1665 and 1667 for a term of fifteen years. In June 1670, the king also waived the customs on 20,000 reams of high-quality folio paper that could be obtained only from France, something he had done for the Nieuhof volume as well.[56] The atlas appeared in seven volumes between 1670 and 1676—*Africa* (1670), *Atlas Japannensis* (1670), *America* (1671), *Atlas Chinensis* (1671), *Asia* (1673), *Embassy to China* (rev. 1673, some editions bound with the *Atlas Chinensis*), and *Britannia* (separate editions in 1675 and 1676), with a planned eighth volume on Europe remaining unfinished at his death. He described the project as a "Reducement of the whole World" and in his dedication to Charles II for *Africa* "A New Model of the Universe, an English Atlas . . . in our Native Dress, and Modern Language." Macrocosm in microcosm, English modernity linked to Stuart glory, these were the tropes of absolutist image making, and enough Londoners found them compelling that, like Samuel Pepys, they entered lotteries for the chance to own such an expensive production. Into the 1680s, the East India Company still recommended reading the Japan and China volumes to captains and factors going to Siam and potentially onward, and copies could be found in the library at Madras.[57]

For the atlas project, Ogilby used exclusively Dutch sources, notably Arnold Montanus's volume on America and a series of volumes by Olfert Dapper, all lavishly illustrated and produced by the printer van Meurs based on the success of the Dutch edition of Nieuhof.[58] Ogilby had formed a partnership with van Meurs, probably from the Nieuhof volume itself, in order to drive out his main London competitor, Richard Blome, and ensure rights to future volumes.[59] The turnaround time was shorter than that of the Nieuhof volume, with Ogilby's *Africa* appearing in 1670, two years after its publication in Dutch. The new Dutch volumes were mostly dedicated to the Grand Pensionary of the United Provinces, Johan de Witt (1653–72), and they all appeared after de Witt had passed the Perpetual Edict (1667), permanently abolishing the office of Stadtholder in the province of Holland and excluding Charles II's nephew William of Orange (b. 1650) from the office generally. As such, the volumes attempted to show the superior power of urban and republican commerce to monarchy, politically part of the early stages of the "Radical Enlightenment."[60] The last volume used by Ogilby came out in 1672, the year that in a series of critical events de Witt resigned and was subsequently killed, William of Orange became Stadtholder, and after the failure of William's efforts at an alliance, France and England declared war on the Dutch.

Although using the same material as the Dutch publications, Ogilby's agenda was different— an ideal future vision for a globally oriented and ex-plicitly monarchical Britain centered on London and Charles II that would be tied to trade with Qing China and Japan, the slave trade in West Africa, and slavery in the Americas. In 1670, neither the East India Company nor the defunct version of the old Africa Company had established solid trading relations in these regions. In the 1660s, Charles had acquired two fortified positions in Africa. In his dedication to the king, Ogilby highlighted Tangier in North Africa, which protected the Mediterranean trade of the Levant Company and had come with the dowry of Catherine of Braganza in 1661, as "Your own Bright Star . . . Your Metropolis, Your Royal City Tangier, which Seated on the Skirts of the Atlantick, keeps the Keys both of the Ocean and In-land Sea." It had been declared a free city in 1668 with its own corpora-tion and an expensive standing army maintained by the crown. Ogilby made no mention of the other English possession, the Cape Coast Castle, which had been captured from the Danish West India Company along with a series of Dutch forts along the Gold Coast by Admiral Robert Holmes in 1664, be-ginning the second Anglo-Dutch War. The Royal Adventurers to Africa had farmed it out in 1668 to a group called the Gambia Adventurers for a period of seven years. For Ogilby and Dapper, the principal locus of the European presence in "Negroland or the Countrey of the Blacks" was still the Dutch castle of São Jorge da Mina (Elmina Castle) in Ghana, which the Dutch had captured from the Portuguese in 1637 and was now the center of their slave trade.[61] But Charles II's new charter of the Royal Africa Company in 1672 would authorize the building of more forts in West Africa and the keep-ing of standing armies there on the model of Tangier. In 1670 both *Africa* and the "atlas" of Japan, which was really a collection of Dutch embassy and exploration accounts, indicated the stronger presence of the Dutch in West Africa, China, and Japan and the failures of English ambitions in the 1660s. Ogilby appeared like his contemporary Sir William Temple to be pushing the crown to compete actively with Dutch commercial republican-ism, building on similar sentiments in London at large.[62]

The next volume, *America* (1671), came on the heels of Shaftesbury and Locke's *Fundamental Constitutions of Carolina* (ca. 1669). The constitu-tion replaced two earlier efforts (1663, 1665) at defining the new colony of Carolina, which like New York (acquired from the Dutch in 1664 and named after James, Duke of York) was meant to be a pillar of absolutist strategy in the Americas. Carolina, chartered in 1663, would be a desti-nation for slaves following the models of Jamaica, Barbados, and especially Virginia, which passed a Slave Code in 1661, and Locke and Shaftesbury had tried to moderate this with guidelines about religious toleration and conversion of slaves and Native Americans.[63] In Ogilby's vision Carolina, through the importation of slave labor, would also be a potential replacement for England's lack of an effective trade with Qing China and Tokugawa Japan. He made a series of claims to show this, including that a kind of tea grew in Carolinas (probably the "black drink," mildly hallucinatory and strongly purgative). He noted that Nanjing was same latitude as Carolina, and that like the Yangtze Delta, Carolina was a fit place for planting "Silk, Ginger, Indigo, Tobacco, Cotton, and other Commodities fit to send abroad and furnish foreign Markets."[64]

Along with this imperial vision of an Atlantic consumer economy based largely on goods that could replace Asian imports came a different emphasis on the Chinese economy itself in the *Atlas Chinensis* (1671, misattributed by Ogilby to Montanus). This synthetic text, culled from VOC archives, could be read as critical of or at least anxious about the effeminacy and luxury of absolutist monarchy.[65] The image of a boating party, exemplary of many in the book, with Mandarins eating off of porcelain and drinking tea, cruising pleasurably in sight of villas and commercial ships, suggested both commerce and luxury and paralleling Dutch anxieties about their own consumption and "embarrassment of riches."[66] But for Ogilby the atlas also revealed how the circulation of people, objects, and media could produce a sense of unity and how religion, commodified through images, became a source of cultural stability and even commercial potential rather than political conflict. The roads and waterways described by the Pieter van Hoorn embassy (1666–68) were both mechanisms of transportation for military, state, or commercial purposes but also drivers for expanding a consumer economy.

The printed image could also be a medium of translation rather than simply exotic collection. To demonstrate religious beliefs in Qing China, the *Atlas Chinensis* included reengravings of four popular Chinese woodblock prints showing various images of Bodhisattvas, Laozi, and Kong Fuzi. One depicted an actual tourist pilgrim destination, the *Changshu Taoyuan jian* (Changshu Peach Springs Stream 常熟桃源澗) near Suzhou and Lake T'ai (figure 32). There on Mount Yu was the tomb of Yan Yan (506–443 BCE), Kong Fuzi's only student from the south of China, as well as a tourist site meant to evoke the Peach Blossom Springs, which the writer Tao Yuanming (365–467) had immortalized as a kind of "otherworldly" Confucian and Daoist paradise (*shiwai taoyuan* 世外桃源).[67] In the Buddhist emphasis of this particular print, values of retreat and pilgrimage carried over into a path of meditative shrines in the natural landscape. The Dutch and then English reengraving of the woodcut in copperplate retained the parallel perspective of the original and the sense of movement through interstitial spaces. The

Fig. 32. John Ogilby, *Atlas Chinensis* (London: 1671), before 571; from Olfert Dapper,
Beschryving des Keizerryks van Taising of Sina (Amsterdam: Jacob van Meurs, 1670),
facing 109. The title reads 常熟桃源澗 [*Changshu Taoyuan jian* "Changshu Peach Springs
Stream"]. The Getty Research Institute, Los Angeles (84-B22302).

accompanying description labeled it as an image of the Guanyin bodhisattva
("Pussa" for *pusa* 菩薩), whom both Dapper and Kircher thought was the
same as the Greek Cybele and Egyptian Isis. The similarly fantastic descrip-
tion suggested deluded pilgrims who "must go through several by-ways and
Chambers, and along a steep bridge, which at the bottom is Guarded by a
Man sitting on a Tyger: At the Door of every Apartment, a Priest of this
Goddess keeps Guard, which will first be Brib'd, before they will permit the
Pilgrims to enter."[68] The intimations of corrupt luxury and idolatrous mix-
ing of images echoed certain Chinese elite discourses concerned about pop-
ularization of culture, but the engraving also suggested a coherent and complex space with its own logic and techniques. Taking a lead from Kircher
and the Jesuits, who may have been relying on the same archive of images,
this logic was to a certain extent understood as translatable through the
medium of print, and the growing Restoration fashion for such images on
porcelains saw this aesthetic spreading through the domestic spaces and
new coffee houses of London.

Ogilby explicitly wanted his atlas to be "modern" in the way that Dapper, Kircher, and the image makers of the Qing were. The possibility of

translating spatial and visual techniques from China to Europe on the basis of their modernity rather than their antiquity made the volumes on China, and in particular Ogilby's reinterpretation of them, unique. The clearest indication of this is the connection between Ogilby's reprinting of the *Atlas Chinensis* and his own subsequent road atlas project in Britain, a project that would form the basis of the informational, transportation, and postal system that the eighteenth-century fiscal-military state was built upon. In the *Atlas Chinensis*, Ogilby included a table from Dapper that had been previously translated from a Chinese source entitled "The Roads, and Distances of the great Cities of China one from another." It stressed the "exceeding curious" practice of measuring distances between cities in China. Dapper had quoted directly from Chinese sources: "a Scale of Furlongs written by the Chineses themselves, though not from all Places, as they have it, but onely from the Metropolis of each Province to that of another in the following manner: Peking from Kiangning is 2425 Chinese furlongs [*li*, 里, ca. 645 meters]." Thus began a table of fourteen distances, followed by other tables of routes not involving the capital at Beijing.[69] The data came both out of popular merchant manuals like the Wanli era (1573–1619) merchant handbook *Shangcheng yilan* (商程一覽 "Commercial Journeys, An Overview"), which contained a map and distance tables for the "Thirteen Routes of the Two Capitals" (*Liang jing shisan zhu* 兩京十三者) as well as the work of late Ming and early Qing scholars who were interested in developing a comprehensive list and table of distances as a way of both unifying commerce and abstracting empire from dynasty during a time of transition. Some of the figures match or come close to those in cited in Gu Zuyu's (顧祖禹) *Dushi fangyu jiyao* (讀史方輿紀要 "Essence of Historical Geography"), which was written between the 1630s and 1660s and circulated in manuscript.[70] The most identifiable institution using these roads was the Qing postal system, a Ming inheritance that created an idealized space of transparent communications "which speedily convey all Letters, the Emperor's and Governor's Edicts from one place to another, by which means, nothing that is strange or News happens, but in a few days, it is spread though the whole empire."[71] Both Dutch embassies had moved efficiently along the water and road systems of China, and the opening map of Nieuhof's volume, juxtaposed to the ground plan of the Forbidden City, showed the inland journey along the trunk route for travel from the south to the north of China.

The tables of distances in Dapper and accounts of the road and postal system in the Qing Empire proved to be an inspiration at the level of technique for Ogilby's projects of mapping London and visualizing the British road system in the 1670s. Ogilby had become adept at using London as the hub of the new postal system, and in February 1672, the year following the release of the *Atlas Chinensis*, he gave a list of people in various cities in England who could procure copies of his English atlas, "for the ease of such of the Gentry as are far remote from the Author's dwelling."[72] That same month, he announced his new project for mapping Britain as no mere map-view but an "Ichnography," literally "track" (*ikhnos*) writing. The new diagrammatic method would leave behind his more representational competitors like Richard Blome's 1673 *Britannia* well as new editions of Speed's classic *Theatre of the Empire of Great Britaine*, first published in 1612, derived in part from William Camden's *Britannia* and more generally from methods pioneered by Ortelius. In doing this, Ogilby received the direct encouragement of London's aldermen, who also commissioned a full map of rebuilt London to be vetted by a raft of former and future lord mayors as well as Robert Hooke as the city surveyor. As the map of London dragged on, with delays and lack of funding, Ogilby focused on the first volume, "England and Wales," of his planned six-volume *Britannia*, in particular an unprecedented mapping out of the roads that fed into London in a series of scroll-like strips, which were then compiled into a full map of England (figure 33). Offering an image of London's relations with England on the model of the Qing road system proved easier that showing the complex and changing city itself.

Part of the reason for this was that road mapping could be tied to one of Charles II's more absolutist projects at home: the development of a postal system. In the dedication to *Britannia*, Ogilby echoed the themes of the coronation pageant and credited Charles with having "laid open to us all those Maritin [*sic*: Maritime] Itineraries, Whereby We Trade and Traffique to the several parts and ports of the world."[73] Domestically, Charles's postal system act of 1660 had revised a previous postal system created under the Commonwealth in 1657, establishing the General Letter Office as a royal monopoly and giving it to James as Duke of York to manage.[74] By 1667 it had become a significant source of crown revenue, although most of this revenue was not being directed back into the improvement of roads and services.[75] In a subtle critique of such neglect, in which London as an emporium was better connected to the world than to the kingdom, *Britannia* was designed to "improve our commerce and correspondency at Home" by registering the king's "high-ways . . . from this Great Emporium and Prime Center of the Kingdom, your Royal Metropolis." Ogilby made his appeal not only directly to Charles and James to modernize the roads in the name of empire and emporium but also to a more popular audience, since revenue for the road atlas would come primarily from sales rather than subsidy,

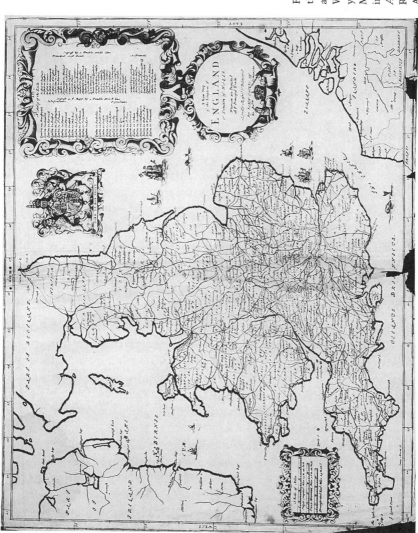

Fig. 33. "A New Map of the Kingdom of England and Dominion of Wales Wheron are Projected All ye Principal Roads Actually Measured and Delineated." in John Ogilby, *Itinerarium Angliae* (1675). The Getty Research Institute, Los Angeles (94-B20405).

especially with cheap portable editions.[76] To naturalize this and give it classical authority, Ogilby followed the legitimizing strategy of Kircher and referred the whole project to ancient "Persian Princes" who had "Stationary Distances through their vast extended territories exactly Registered and Enumerated," and to Alexander and Julius Caesar who had followed their example. Charles as sovereign and James as his postmaster would revive Britain to the imperial glories of antiquity rather than directly emulating contemporary China. This philosophy of the monarchical state as responsible for rationalizing space in turn helped justify James's authority to seize the private penny post after 1682, which was seen as a potentially subversive institution for transmitting nascent newspapers and pamphlets. With the establishment of post offices in market towns in 1683, the number of postal employees began to grow rapidly as part of a more tangible shift towards absolutist strategies and a fiscal bureaucracy.[77]

Ogilby's work offered Londoners an image of both a coherent world and a coherent Britain centered on London, as well as new strategies for making absolutism linked with commercial empire more of a reality than an image. It was the closest that London had come to having its own pictorial narrative of global processes, its own interface with the world. Still, even with the road maps, Ogilby's partner William Morgan had to push hard to achieve sales to make up for the vast expenses that the various projects entailed. They nevertheless represented efforts by Londoners to stress the commercially and globally organized underpinnings of the image of English absolutism under Charles II and James II.

BROKERING THE ABSOLUTIST IMAGE: INTERVENTIONS FROM BOMBAY AND TAIWAN

If Qing China in works like those of Webb and especially Ogilby came increasingly to be seen as a model in relation to absolutist strategies of the state, the economic depression in the Qing and military disruptions meant that shifting patterns of commerce in the Indian Ocean and Southeast Asia began to play even more prominent roles in shaping London's commercial relationships during the Restoration. Francis Lamb's 1676 "A New Map of East India" marginalized China while highlighting the imperial possibilities in India as well as Taiwan (newly depicted as a single island as on the Selden Map) in a world centered on Southeast Asia and the river systems feeding into the Indian Ocean. Charles II's and the East India Company's new projects in Asia, initially envisioned as imperial extensions of absolutism, were themselves transformed by shifts in the silver cycle that opened opportunities for smaller polities and merchant networks across South and Southeast Asia to develop more robust emporial strategies. The failure to make Bombay into a royal outpost with a standing army in the 1660s encouraged the East India Company in cooperation with Surat brokers to reconfigure its trade in Asia dramatically. By the 1670s new EIC relationships in Bombay and Taiwan produced the surprising result of texts being sent in Persian and Chinese from these areas to help London merchants and the English crown develop more complex understandings of sovereignty, exchange, and religion in both the Indian Ocean and South China Sea. Such dispersed knowledges created longer-term commitments in London to merchant networks in these regions.[78]

With the exception of Bombay, all of the Company's factories in Asia in the 1660s remained in tributary relation to Muslim political authorities, notably the two presidencies (Surat and Banten). The other key location, aside from the Mughal port at Hugli in Bengal, was Fort St. George on the Tamil- and Telugu-speaking Coromandel Coast, the Company's only fortified factory until the acquisition of Bombay. The land had been purchased in 1639 indirectly from the Nait or governor of the Carnatic before the collapse of the Vijayanagara Empire. The fort was almost abandoned in 1661, but by 1664 the sultanate of Golkonda had finally consolidated its hold on the region. By an agreement of 1658, the Company paid 380 pagodas in tribute each year to Golkonda, a sum raised to 1,200 pagodas in 1672, which lasted until 1687. Heavy reliance in each of these cases on Muslim empires, which for London more generally also included the Ottomans and the Moroccans in the Mediterranean, caused nervousness in London for both financial and religious reasons. Anglican conversion efforts in relation to Muslims consistently foundered from Morocco to Java.[79] Charles II's dowry gifts of Tangier and Bombay in theory helped allay such concerns by creating fortified Anglican enclaves in the Mediterranean and Indian Oceans, while plays like Dryden's strongly monarchist and last heroic tragedy, *Aureng-Zebe, or the Great Mogul* (performed 1675, printed 1676) helped legitimize broader tributary relationships in London.[80] These were issues of public discussion in London. Three months before *Aureng-Zebe* appeared on stage at the Theatre Royal in Drury Lane, an elephant was sent from Surat and auctioned at Garraway's coffee house across town in Exchange Alley. Thomas Garraway had been the first to sell Chinese tea in London at his old coffee house, the Sultaness Head in nearby Sweeting's Rents, from 1658 until it burned in 1666. Public viewings of the elephant at the center of London cost three shillings.[81]

Certainly some of this public discussion was a mix of cosmopolitanism and exoticism that had its roots in both the panoply of imported commodities as well as the fashionable and fantastic lives like those of the versatile Sherley brothers (Thomas, Robert, and Anthony), whose exploits in the Safavid and Ottoman Empires in the early sixteenth century were legendary. But it also suggested how the networks of an emporia economy fractured traditional humanist conceptions of language, including uncertainty about whether languages like Persian belonged in the canon. Edmund Castell and Edward Pococke's financially disastrous *Lexicon Heptaglotton*, produced with the printer Thomas Roycroft in London in 1669, a companion volume to the Polyglot Bible that included a Persian-Latin dictionary left among Jacob Golius's papers (d. 1667), was a kind of swan song in this regard. More explicitly scientific texts in Arabic and especially Persian began to garner attention, such as the copies of Ulugh Beg's catalog of stars used both by John Greaves in the 1640s and early 1650s and Thomas Hyde in 1665 for publishing parallel translations in Latin.[82]

In particular, Arabic began to be treated less as part of the array of humanist sacred languages, and more as the key element for comprehending a distinct technical, legal, political, and philosophical culture with its own cosmological underpinnings that exhibited the universal power of a broader "human reason" (*ratio humana*). Pococke, once a chaplain for the Levant Company, published with his son in 1671 a Latin translation of Ibn Tufail's twelfth-century *Hayy ibn Yaqzan* (literally "Alive Son of Awakening," translated as *Philosophus Autodidactus*).[83] Announced in the 1674 *Philosophical Transactions*, it was a popularized mixture of Ibn Sina's reading of Aristotle and Plato mixed with Sufi concepts of wisdom and intuition. It dovetailed with François Bernier's admiration of the Sufis among the Mughals as translated by the Royal Society's secretary, Henry Oldenburg, and it could be used to support the idea of natural religion. The text tells the story of Hayy ibn Yaqzan raising himself from infancy on an island in the Indian Ocean, achieving greater wisdom than can be gleaned from the books of prophets. The allegory could be read as making individualistic claims about the possibility of reason emerging inductively through pure observations of nature by an individual outside of the Christian or Islamic tradition, an argument that paralleled the cultural claims made by Dapper, Ogilby, and Bernier about non-Christian religions, but it also suggested that texts could have a kind of autonomous power in themselves, becoming distinct by translating and juxtaposing a variety of traditions. At the same time he published the *Philosophus Autodidactus*, Pococke also put out an Arabic translation of the Anglican catechism and liturgy.[84] Although the catechism had little effect in the Arabic-speaking world and it failed to spread autonomously in the manner of the *Hayy ibn Yaqzan*, the choice of translation suggested that the ability to define clearly the institutional claims of the state church had become more important as part of a process of mutual understanding than any broader Christian humanist or even Protestant campaign for religious conversion. Such failed conversion and the appeal of sectarianism was precisely the story of Hayy, who himself could not convince the population of Salaman's Island of the value of his ideas.

This growing recognition of sectarianism and its relationship to language drove translation in this period, and it was accompanied by an awareness of the ways that conventional assumptions about language and religion had been overthrown.[85] Robert Boyle, luminary of the Royal Society as well as a director of the East India Company, in particular became increasingly concerned that the focus on languages used in biblical and classical hermeneutics, including Arabic and Persian, was too limiting. As the new head of the Corporation for the Propagation of the Gospel in New England (1649, rechartered 1662), he helped encourage the "Moheecan Bible" in Massachusett (*Wôpanâak*), which was printed at Cambridge in the Massachusetts Bay Colony by John Eliot from 1661 to 1663 for use in training Massachusett pastors at Harvard. But its failure as a conversion tool was demonstrated spectacularly in King Philip's War (1675–78), where the killing of the Harvard graduate John Sassamon (Wassausmon) and advisor to Metacomet (King Philip) in early 1675 became the central catalyst for the conflict, and in the frontier conflicts of Bacon's Rebellion in Virginia (1675–76).[86]

Garnering more attention were languages coming from the East India Company's mercantile relations. As problems erupted in North America in 1675, Boyle saw a new opportunity, when Sultan Abdul Fatah of Banten switched from writing letters to Charles II in Arabic to writing in Malay using Jawi script.[87] Boyle seemed to think, perhaps encouraged by Pococke and Wilkins but certainly by the Bodleian's librarian Thomas Hyde, that Malay would be an ideal alternative to Arabic in Southeast Asia. The sultan used both at Banten, and the Bodleian held texts in Malay. The Jawi script used in these texts derived from Persian versions of Arabic, which had been brought along with translated Sanskrit texts to Sumatra in the fourteenth century. Hyde wanted Christ Church, Oxford, to become the great English center for training Asian missionaries, developing programs in Arabic, Persian, mathematics, and Malay through the sponsorship of the East India Company.[88] Boyle paid for five hundred copies of the 1677 *Jang Ampat Evangelia* put together by Hyde, Thomas Marshall, and Bishop John Fell.[89] But a letter from the Company's factor at Madras, Streynsham Master, sent to Samuel Masters, a divine and fellow of Exeter College, Oxford, in December 1678, listed numerous reasons why such a translation project would not succeed. The fact that nobody spoke Malay in India, the indifference of the English clergy, the lack of churches, and the subtle doctrinal differences between conformists and nonconformists all made Anglican missionary work impractical.[90]

So despite the appearance of a push outward for translation coming first from Oxford and then from the Royal Society, efforts to dominate language from centralizing institutions gave way starting in the 1670s to relatively autonomous efforts from places as diverse as Surat, Banten, and Taiwan. These increasingly set the agenda for translation in London. In northwest India, from about 1674, mutual efforts by the Anglo-Irish Gerald Aungier, the Company's president at Surat and the governor of Bombay from 1672 to 1677, and merchant groups in Surat led to a series of copies of Avestan and Persian texts coming to London, many from the Parsi Zoroastrian community at Surat and neighboring Navsari, the religious center of Zoroastrianism in Gujarat.[91] The Parsis were important Persian diaspora merchants, interesting to Hyde and others in London and Oxford because they had a pre-Islamic religion. At Surat, scribes knew both Avestan and Persian. Most of the books Hyde obtained were made by low-level priests (*herbad*) working in Navsari, and like the popular and synthetic sixteenth-century *Saddar*, they came out of the Parsi experience in Mughal India. Some of this textual production was probably connected with the charitable work of Rustom Maneck (ca. 1635–1719) of Surat, a prominent Parsi merchant and subsequently a broker (*dalal*) for the Company, an example of the kind of trader having great success in the late seventeenth century. More broadly, however, it came out of closer relations between priests and wealthy laymen in *panchayat* assemblies adopted by the Navsari Parsis from 1642, a more general function of the growing wealth of Parsi brokers and merchants.[92] Over the next twenty years, Hyde himself used such texts to complete his magnum opus *Historia Religionis Veterum Persarum* (1700). He also used his copy of the *Khordeh Avesta*, produced by the herbad scribe Hormuzyar in 1673, to make an Avestan font he described as "Persian."[93] This was a clear shift in terms of translation, from a reliance on Dutch institutions as in Ogilby and an emphasis on biblical scholarship with humanist goals towards a direct engagement with merchant networks connected to London in order to construct alternative histories that could challenge current models of scholarship.

Hyde also tried to collect texts appealing to a pan-Persian identity. He obtained through Aungier a prose abridgement of Abolqasem Ferdowsi's famous *Shahnama*, the first in England. By the time it reached Oxford, it had the English title, "Shahnamah Nussier or A Chronicle of all the Kings of the Persees." It was so rare in Surat that another "learned Herbud" scribe, Khwurshid son of Isfandiyar (the latter name like Rustam being an important character in the text), was as Aungier wrote in the copy, "very loath I should part with it before he had taken a copy of it, but it could not be done, our ships being soe near their departure."[94] The herbad's frustrated desire to hold onto the text reflected not only rarity and cultural valences like personal names but also the subversive ambiguity in such storytelling. Despite the fact that the *Shahnama* is explicitly about kings, its broader theme is that of restless subjects coping with dynastic instability.

The willingness of Parsis in Surat to make copies of these books and to sell them to English buyers had much to do with restless subjects. Under the Mughal Emperor Akbar (r. 1556–1605), the courtly language and education in madrasas had shifted away from Turkish (the language of Babur and Humayun) towards Persian. Akbar's famed religious pluralism, which most importantly abolished the poll tax or *jizya* for non-Muslims, had been one of the reasons that the Parsis of Surat had thrived along with Hindus and other non-Muslim groups under the umbrella of Persian.[95] While Aurangzeb still supported the expansion of Persian, he also between 1669 and 1679 introduced measures, including temple taxes and *jizya*, to cull more revenue from Hindus and other religious groups like the Parsis who had risen in status either through the bureaucracy or trade.[96] In Surat, such measures caused particular hardship because the city was sacked twice, in 1664 and 1670, by the emerging polity of the Marathas to the south of Bombay in their war with the Mughals.

Intertwined with these grand political shifts of empire were multidimensional popular moves towards religious reformation, that might be described as a sectarian hardening driven by trade prosperity and increased textual production of manuscripts. The Marathas, for example, seem to have been driven by a revival of Shiva worship (Shaivism) that more broadly involved copying and disseminating Sanskrit texts. Likewise, both the Mughals and Bijapur were in the throes of a Sufi ascetic revival, missionary *faqir* ("Fakirs") playing prominent roles in court and spreading texts like those of Sultan Bahu (1631–91) from the Punjab. To this should be added the blossoming of Armenian Christianity, Judaism, and Parsi Zoroastrianism in the region, all of which were connected to broader trade diasporas and all of which used new wealth to build libraries of texts. Christian influences were complex; a variety of Protestants, Catholics, Armenians, and others coming from Europe and the Ottoman Empire, and even internally as Koli ("Cooly") Christians from India and Luso-Indian Catholic communities saw a kind of revival through autonomy. In the 1670s, Hindus and Parsis in particular were ready to shift allegiances and began to open their libraries to help the English understand their complex needs; Armenians, Jews, and other groups would follow their example.

The regulations for Bombay in 1670 compiled by the Aungier explicitly used Medici Livorno's relation to the Ottomans as a model for a viable center of money-lending and brokerage. To abet this, Aungier requested free trade for inhabitants of Bombay not only in Indian ports but also for Banten, where Gujarati merchants would briefly reinitiate a trade in 1674, as well as more generally "ports and islands of the South Seas." The caveat was that such trade would not be carried out by English freemen trying to compete directly with the Company.[97] He even floated the idea that Bombay should be a completely free port for five years. The stakes were high, demand for silver in Persia, gold in India, and copper in both along with the need to have ready money for advance cloth payments meant that London was in a better position than the Dutch at Batavia to take advantage of the economic shift away from the Qing and toward the Indian Ocean.[98] The new Bombay was also supposed to be a safe place to store commodity precious metals like Dutch Batavia. In addition to a rebuilt fortress, it would have its own mint and the ability to collect money sent by London and in turn to use that to pay for cotton cloth production in advance, a requirement for the developing cotton trade. John Fryer, who visited between 1673 and 1675, had a very precise phrase for this, explaining, "In this Exigency on either side, the Martial as well as Civil Affairs, are wholly *devolved* on the Merchants."[99] The Company's first army designed to replace the one sent by Charles II was a mixed group of about 300 English soldiers, 400 "Topazes" (creole Luso-Indians), and 500 or so Marathi-speaking Bhandari militia. The Bhandari, whose small principalities on the surrounding coast had been supported by the Portuguese, had formed a kind of special guard with privileges of blowing a horn, and they now composed over one third of this rather complex military entity. The company still had a chance to create the bullion emporium envisioned first by Charles II and James, Duke of York, although in doing so they had already had to significantly transform the concept of a standing army in ways that would prefigure the kinds of colonial warfare common in the eighteenth century.

Aungier could use the interest of people like Hyde in Oxford in commissioning copies of Zoroastrian and secular Persian texts as a way of convincing Parsi elites that the English would not follow the new Mughal policy of religious discrimination.[100] Because Parsi brokers, like Rustom Maneck, had money in different locations, they now also had greater status as community

leaders. They could dole out charity when taxes or warfare devastated families, and they could also make larger deals in relation to migration. Aungier
himself agreed to give the Parsis land for a Tower of Silence (*dakhma*) near
Bombay's Malabar Hill. He made similar deals with individual Hindus.
Nima Parakh, a Bania (the Hindu merchant caste) whose extended merchant
family worked out of Diu, made moving to Bombay contingent upon land
for a house and warehouse as well as freedom of religion and religious rights
(no animal killing, no Christians or Muslims in their compound), rights to
free trade, to due process of law, to bear swords and daggers, and not simply
to tolerance but to civility and respect. Aungier explained that the East India
Company did not force conversions, about which "the whole world will vin-
dicate us," and generally acceded to all Parakh's terms.[101] Similar deals were
subsequently made with Jewish, Armenian, and Sidi (descendants of East
African slaves, with their own coastal kingdom) merchant families. Like the
North American colonies, Bombay was an emerging space of religious liberty for practical reasons of recruiting immigrants. But the rights demanded
and extended in Bombay were in some ways more radical than the religious
debates taking shape in London and Carolina around the unratified *Fundamental Constitutions* (1669, revised 1682) that so occupied Shaftesbury
and Locke. It set the stage and became the model for deeper commitments
and the longer-term urban relationships between London and Mumbai, with
Madras (Chennai) after its incorporation in 1687, as well as the new trading
center at Calcutta (Kolkata) in Bengal from 1690.

This mutual translation between the Company and merchant networks
could also be seen in the 1670s in East Asia, where Fujianese merchants at
Banten helped the Company develop a relationship with last Ming loyalist
regime on Taiwan under Zheng Jing (d. 1681). The Zheng family took the
Fujianese merchant networks shown on the Selden Map for granted, and
because they were at odds with the Qing, these shipping routes took on
an even greater importance. The quasi-apocryphal biographies of Li Dan,
Zheng Zhilong, and Zheng Chenggong (supposedly born in Hirado in 1624)
all testify to the complex histories of movement and acquisition of skills
in seventeenth-century maritime East Asia that had become de rigueur for
powerful figures. While Zheng Chenggong and Zheng Jing certainly had ambitions on the mainland, they also looked to build a powerful maritime presence by borrowing skills and taking substantial territory from the Spanish
on Luzon and of course the Dutch in Taiwan. All of this suggests why Zheng
Jing might have thought it necessary to continue printing from Taiwan the
old Ming calendar of the Yongli pretender (1646–62, the "Southern Ming")
during the 1670s and to send copies to London between 1671 and 1677. This

was a way of pulling a diverse space of merchant networks together through a uniform system of time that was an alternative to the expansion of Qing territorial empire and its Beijing-centered scientific projects.

Taiwan in the seventeenth century was a complex frontier zone, as the Selden Map suggested, increasingly important for building linkages along critical silver supply routes from Japan and Manila both in the boom years of the early seventeenth century and during the midcentury bust. From the late sixteenth century, merchants and settlers moved onto Hainan Island, Penghu, and Taiwan itself in larger numbers along these trading corridors, changing the nature of piracy and trade in the region.[102] The Ming had begun to pay closer attention to Taiwan because of pirate and smuggling havens there in the late sixteenth century, launching an anti-pirate exploratory expedition of the island in 1603. In addition to small settlements of Fujianese and Japanese traders, they found a complex texture of languages, warring village "micropolities," and an economy largely based on deer hunting and fishing. The late Ming ethnographic mission to Taiwan had emphasized the lack of a calendar, mathematics, and a bureaucracy as a sign of barbarism; as Chen Di explained in 1603, "Here there are still people who do not have a calendar, who do not have officials and superiors, who go about naked, and who use a knotted string for calculations."[103] Calendars were in this sense seen not only a sign of political legitimacy but also a tool like the compass for civilizing barbarians and establishing rhythms across space and time.

The Zhengs tried to use shipping networks similar to those of Li Dan and the Selden Map to lay claim to the legacy of the Ming against both the Dutch and the Qing and to make Taiwan central to silver and gold bullion transfers in East Asia. The lead in doing this had previously been taken by the Dutch, and this explains why the Nieuhof volume was such a lavish production in the 1665, just three years after the loss Fort Zeelandia on Taiwan to Zheng Chengong (Koxinga) had resulted in a dramatic reversal of their Asian strategies.[104] As the English were leaving Japan, Li Dan helped handle the negotiations between the Ming and the Dutch on Penghu (simply 彭 Peng on the Selden Map) in 1623. After Li Dan's death in 1625, his old employee and troublesome protégée Zheng Zhilong also worked with the Dutch and by 1628 had taken control of his master's fleet. The Dutch, as they fortified Zeelandia between 1624 and 1634, initially benefited from large migrations of people fleeing droughts, famines, and exceptionally cold winters in south China beginning around 1624. At the same time, Zheng Zhilong sponsored migration, creating a peonage system in exchange for transport, land, and small amounts of capital to encourage sugar and rice production. With the collapse of the Ming, Zheng Zhilong had remained an

ally of the Longwu emperor until his defeat by the Qing in 1646, but his son
Zheng Chenggong remained loyal to the Southern Ming under the Yongli
emperor in Yunnan. Zheng Chenggong had needed Taiwan as a defensible
base for his fleet when the Southern Ming finally fled into Burma in 1661, a
possibility that had largely been prepared by events in the 1620s.

Like the loss of Forts Amsterdam and Orange on the Hudson River,
which led to the establishment of New York and New Jersey (1664–65),
the loss of Fort Zeelandia and Taiwan was an important setback for the
Dutch in the 1660s, despite their continuing strength against the English
in Europe and their acquisition of the nutmeg-producing Pulo Run in the
Bandas. The Dutch tried, through an informal naval alliance with the Qing,
to drive out the Zhengs in 1664, resulting in a second defeat by Zheng Jing.
Chinese merchant networks at Banten, probably working in tandem with
the Zhengs, opened up negotiations with the English. Like the factory established at Tonkin in 1672 (another Zheng base of operations), Zheng Jing's
Taiwan seemed to offer possible markets for cloth and drugs from Surat, the
Coromandel Coast, and Bengal that would help develop those trades. Chinese silk and other artisanal manufactures like porcelain and lacquer could
be brought back to Europe, but more importantly gold, silver, and copper
from Japan and silver from Manila could be sent to India to further fund
the production of cloth there.[105] The first two East India Company ships in
May 1670 employed a Batavia merchant named "Sooko" (or "Succo") and
brought eight other Chinese on board to help navigate, translate, and broker
negotiations. Initially the only commodities at Taiwan to be found in trade
were deerskins and sugar, sent to both Japan and China, but Zheng ships
also brought in Japanese copper and gold (koban coins, 小判 called "co-
pangs") as well as small amounts of Manila silver. Indian cottons (calicoes)
turned out to be useless in any trade with mainland China—the cotton industry in the Yangtze Delta supplied strong and cheap cloth in "great quantities"—but they could be transshipped to Manila and Nagasaki and sold
there along with English woolens.[106]

As part of negotiations for a formal treaty (signed October 1672), the
Zhengs wanted to convince the East India Company of their own legitimacy
as representatives of the Southern Ming Yongli emperor, who had been dead
for almost a decade. The Company's Captain William Limbrey wrote dis-
missively of Taiwan, which appeared to him a rather large military camp of
around 70,000 soldiers, "The King is the only marchant and hath monopo-
lized the goods of the country, viz. skins & sugar, with which & some few
Chyna goods as silke &ca he drives a considerable trade to Japon, which
furnisheth him with moneys to maintaine his army & is the best prop of

his kingdom, sending 12 or 14 juncks yearely thither."[107] To feed this large army, junks filled with rice came from Siam by way of Penghu in exchange for commodity metals. This situation became urgent when the Japanese shut down silver (although not gold or copper) exports in 1668, which had important effects on both the Dutch and Zheng operations as well as merchant networks across East Asia.[108] Printing the calendars and even sending them to England was part of a larger effort to stabilize and redefine the temporal and commercial system for the benefit of Chinese merchants who had complex credit relations across East Asia from Nagasaki to Banten. Calendar publishing in Taiwan was a business question as much as a religious or diplomatic one, involving popular legitimacy, especially among the networks of Hokkien merchant families who sustained Zheng power across East Asia and among former tributaries to the Ming from Japan to Siam. The Zhengs needed to create credit for themselves, and the calendar, like the dot offshore Quanzhou on the Selden Map, marked a point of technical equilibrium. The Company's presidency at Banten and by implication London represented important constituencies for attracting silver, merchants, and weapons, so as part of a gift to the Company the Zheng regime included fifty printed calendars in January 1671 and another fifty in November 1672, a practice that continued to some degree at least until 1677.[109] The East India Company records indicate complex accounts of debts and credits among Chinese merchant families, with interest and due dates calculated according to both the English and the Chinese calendars.[110] Archives with consistent dates allowed both Chinese merchant networks like Feng's and the East India Company to create durable relationships across multiple markets despite rapid change.

Despite this, it still seems surprising that scarce resources were devoted to printing calendars in the name of the dead Yongli emperor from 1670 as part of a regional and indeed global propaganda effort to convince the world of the legitimacy of Zheng Jing's regime against the Qing. But Zheng Jing also tried to take advantage of a period of uncertainty and infighting at Beijing, when, especially during the Three Feudatories Rebellion (1673–81), the question of legitimacy and restoring the Ming (fan Qing fu Ming 反清 復明 "reverse the Qing, restore the Ming") was particularly strong among the southern provinces ruled as vassal states (fan 藩) under princes (wang 王). Moreover, from 1645 into the nineteenth century in Japan and then in Korea, there was substantial resistance to Qing calendar reform, while for the Nguyen in Vietnam the Ming Datong calendar adopted in 1644 remained in use until 1812. In maintaining those linkages, the Yongli calendar translated better. The Qing Shunzhi emperor himself tried to downplay the "Western learning" (*xixue* 西學) aspects of the newly adopted Qing calendar that had initially been emphasized by Adam Schall, ordering it to be called in 1654 *Shixian li* (時憲曆) or "Temporal Model Calendar." Between 1664 and 1670 the Jesuits lost their influence over calendrical composition, but under orders from the young Kangxi (r. 1661, coronation 1667), the Jesuit Ferdinand Verbiest would oversee it from 1678, after publishing a 2000-year calendar called the *Yongnian lifa* (永年曆法) or "Eternal Calendar."[111] Still, at first reading, the Zheng calendars seem to be a rather desperate form of resistance to these combined efforts of the Qing and Jesuits. In Beijing, even Mei Wending, who argued against joint Jesuit and Ming calendar reform efforts in his 1662 *Lixue Pianzhi* (力學偏執 "Superfluousness of Calendar Learning"), had nevertheless rejected the previous Ming calendar as obsolete.

The oddest aspect of the Taiwan calendars from the 1670s is that they marked time using the reign era of a dead pretender, giving them a strange timeless quality. The Yongli emperor (1646–62) Zhu Youlang, the grandson of the Wanli emperor, had been strangled to death by a Ming general after having been returned to Yunnan from Burma in 1662.[112] The calendars themselves use blue indigo ink for mourning and appear quite conservative in their use of the old Ming calendar. The literal meaning of the claimant emperor's reign name Yongli—"perpetual calendar" (永曆), itself an allusion to the ten-thousand years of the name "Wanli" (萬曆)— only added to the sense that the Zheng calendar was a repetition of earlier Southern Ming attempts to both repeat and reframe the problem of sovereignty as something divorced from the rituals and instruments of the court itself. This is echoed in other phrases on the cover like *huangli* (皇曆 "imperial calendar") and the last words in the title *tongli* (統曆 "governing calendar"). Mirroring official calendars, Zheng Jing's included a "Diagram of the Position of Spirits for the Year" (*Nianshen fangwei zhitu* 年神方位之圖) that mapped out yearly spirits in relation to a *luoshu* magic-square of nine colors and numerically coded "palaces" (*gong* 宮) or "flying stars" (*feixing* 飛星) surrounded by the twenty-four compass points from the *luoshan* (figure 34).[113]

The concept of "restoration" (中興 *zhongxing*), a politicized phrase attached to the reign name from the late 1640s reflecting ideas about a military reconquest led by Yongli, appears on all of the almanacs, and it defined a kind of parallel between the situation of Zheng Jing and Charles II.[114] At one level, the supporters of Charles II wanted to put forward an image of royal domination of time and space, inaugurated with the founding of the Royal Greenwich Observatory in 1675 under John Flamsteed. This echoed

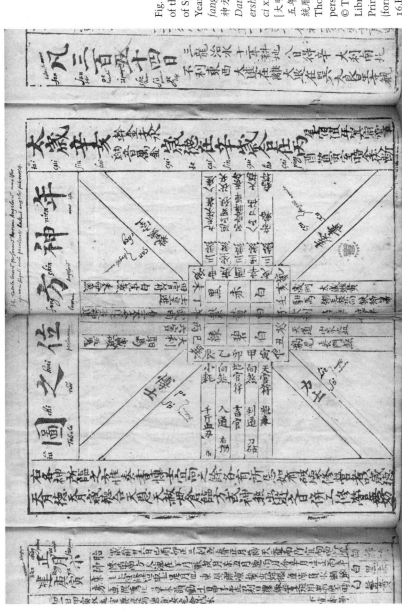

not only the Observatoire de Paris (1671) but also the older Beijing observatory (1442, with new instruments in 1673 by Ferdinand Verbiest) and that of Ulugh Beg in Samarqand (1420). But the acceptance of the calendars and sending them on to London marked an acceptance by the East India Company at its Banten presidency that neither time and space nor religion could be imposed in East Asia. In 1671, while preparing for a Japan voyage that would launch from Taiwan, the Banten presidency also ordered the cross of St. George removed from its flags, flying a secular flag with red and white stripes. The captain was to explain to the Japanese that the Company's servants were "not Christians as the Spaniards or Portugueses for that wee abominate their priests and toyes and are of the same religion as the Hollanders, worshipping the ever-living God." In distinction from the Dutch, the factors were to describe "Our government monarchical, as that of Japon," and that the "horrid murther of our King . . . was perpetrated by a parcel of rebels in a time of rebellion and is noe maxim in our government."[115] This was a world of monarchies that needed to define cosmologies for legitimacy and stability while at the same time to remain open to radically different cosmologies.

The general assumption in both making and collecting these calendars was that time, having lost certain aspects of its sacred authority, now required technical interventions to maintain continuity across space. Time had to be translated and devolved from the direct control of both church and king. Henry Coley, who gave Elias Ashmole his copy of the calendar, was a self-described "Philomath" responsible for taking astrology out of the realm of Renaissance secrets and making it wildly popular during the Restoration. He published a yearly astrological almanac or "starry messenger" from 1672, which carried on after his death in the mid-1690s, as well as a mathematics textbook linked to astrology and a *Clavis astrologiae* (1669, 1676). The repetitive and mathematical aspects of astrology, in both England and Taiwan, became a privatized ritual substitute for the loss of a cosmological center, whether Rome or Beijing, and Ashmole had actually bound his copy together with the popular mathematics treatise, Cheng Dawei's *Xin bian zhi zhi suanfa tongzong* [新編直指算法統宗 "Newly Arranged Straightforward and Reliable Systematic Treatise on Mathematics"] (1592).[116]

Hyde himself was especially interested in the concept of the table or "tabula" (圖, *tu* "diagram") in this regard as an image of a translatable and portable cosmographic framework, subsequently a theme in his book on Asian games. His copy of the calendar was the only one in which the text is transliterated beyond the title page, including extensive notes on the *Nianshen fangwei zhitu* (年神方位之圖) or "Diagram of the Position of Spirits for the

Year." This perhaps stemmed from an interest in *tianshen* (天神), Matteo
Ricci's attempt to translate the Christian concept of "angels" and Aristotelian cosmology more generally. In October 1683, as the Qing were consolidating their hold on Taiwan, Ferdinand Verbiest presented his compilation
of translations the *Qiong Lixue* (窮理学 "Fathoming the Study of Principle")
to Kangxi, writing of how Aristotle's methods of syllogism and inference
were the foundations of knowledge. While angelic mediation in the form of
fangwei (方位 "compass direction") was being emphasized in London as a
way of understanding the Zheng calendars, Verbiest was explicitly removing
all references to both God and Christian angels (*tianshen*) from this work
on Aristotelian cosmology in a failed effort to surreptitiously introduce it
into the exams system.[117] Hyde understood the table in the calendar as mapping out the mediation of angelic forces as "tabula locorum personarum
horum Angelorem anno 1st nam singui anni peculiores habent angelos pa-
dronos."[118] He read this in relation to a broader Chinese tradition and saw
such tabular books in the tradition of Fuxi's ("Fo hi") invention of the *Yijing*
trigrams and the *taijitu* or yin-yang symbol, examples of which he had in
his own personal collection of Chinese books (figure 35) as well as on the
Chinese compass that accompanied the Selden Map.[119] Such technical and
linguistic devices hovered between ancient tradition and the artifice of the
game, mediating elements for translating time and space. Echoes of this can
be seen in Chickamatsu Monzaemon's play about Zheng Chenggong, "The
Battles of Coxinga" (国姓爺合戦 *Kokusen'ya Kassen*, 1715); towards the end
of the play, the history of the period becomes a transcendent game of Go.

However, the game of the transcendent image of time was mirrored by
real consequences of the complex entanglements of credit and debt generated by these engagements in Taiwan. Towards the end of Zheng rule, debt
rooted in paper instruments marked by the dual calendars as well as the uncertain legitimacy of the regime drew the Company more directly into military affairs. As a yearly condition of trade with Taiwan, the English were to
send with every ship to Taiwan, among other items, two hundred barrels
of gunpowder and two hundred matchlock guns.[120] Once the invasion of
coastal China and the islands offshore Fujian Province began in 1674 as part
of the Rebellion of the Three Feudatories (1673–81), even more direct help
from the Company was desired. Larger shipments of guns and powder (five
hundred muskets and three hundred barrels) began in October 1674, and a
mortar was sent that proved effective in the mainland battles. Such shipments were fueled by hopes in Banten and London that the Zheng alliance
with Wu Sangui and the rebels on the mainland would result in an exclu-
sive English relationship with a an imagined post-Qing China.[121] Military

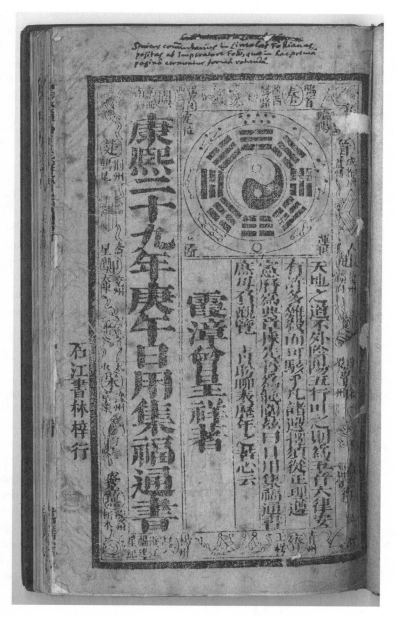

Fig. 35. *Kangxi ershiju niangeng wu riyong ji fu tongshu* [康熙二十九年庚午日用集福通書] with manuscript annotations by Thomas Hyde reading, "Sinicus commentarius in Lineolas Fokianas, positas ab Imperatore Foki, quae in hac prima pagina cernuntur forma rotunda." © The British Library Board, Print 15298 a.32 (formerly Royal 16 B X).

linkages and increasing involvement in debt relations encouraged further
capital investment, tying the Company more deeply into Chinese exchange
networks. By late 1675, the East India Company was expanding all of its
operations in East Asia in anticipation of Zheng getting them "a free trade
for China upon good terms"—stationing six factors at Banten, five factors
and four writers at Taiwan, and two factors in Tonkin and building a new
260-ton ship, the *Formosa*, "to remain in the country."[122] The Surat presidency was enthusiastic about the project, especially to help buoy struggling
Bombay, where Japanese copper could be used in the mint and Japanese gold
would give returns of 40%.[123]

Such hopes in London, Surat, Bombay, and Banten put the Taiwan factors in an awkward position. Zheng asked for six to eight gunners to help
train his forces in August 1675, holding up payments of large promissory
notes (one for three hundred chests of Japanese copper) during the negotiations. Despite worries of a loss of neutrality that would pull the Company
into the kinds of politicized relationships that the Portuguese and Dutch
had, the factors nevertheless persuaded four gunners to aid Zheng even as
they were explaining to Chen Yonghua that "it was not customary in any
place where the Honourable Company has a factory to require theire servants to goe to the wars for another prince."[124] As an immediate reward
for this shift in principles, Chen allowed the Company to set up a second
factory on Amoy (Xiamen) in 1676 off Quanzhou, essentially a direct trade
with mainland China. As things began to collapse, the trade at Amoy turned
almost entirely to guns and bullion—a kind of immediate realization of
the mediated global circuits of silver, gold, and copper kept in the paper
accounts. Zheng Jing himself retreated to Amoy from his former base in Fujian by the summer of 1677, something the factor Edward Barwell referred
to as "this sudden mutation of government."[125] London still sent *reales*,
guns, and powder as late as August 1681, and Josiah Child wrote to Zheng
as "King of Amoy," requesting a "royall phyrmand or chop" for free trade,
not limited to particular buyers or sellers, and to guarantee "the common
justice of all nations." He mentioned precedents—the "King of Indostan"
(the Mughal emperor), the "Emperour of Persia," the "King of Gulcondagh,"
and the "King of Siam."[126] But Amoy and Taiwan by this point were increasingly speculative sovereignties—jointly wished for and constructed in
writing and print by the Company in London, the Zhengs, and the range of
networks that connected them.

At the point when this bubble of speculative sovereignty in print—
whether Ogilby's *Atlas* or Zheng's almanacs—reached its greatest heights,
such translation strategies were also being revealed as extremely fragile and

fragmented. Despite the increasing amounts of money that the Company was putting into expansion of trade in East Asia—pulling London deeper into complex global merchant, currency, and sovereignty networks and thus changing the way that translation took place in London and Oxford—it was not producing direct profits for the Company itself.[127] Concerns about the growing role of interlopers and brokers led the Company to look to the crown to enforce the Company's rights to trade, but such efforts had little or no effect. Even though the East India Company tried to operate during the Restoration as a monopoly with the support of the crown, its exchanges had linked London more tightly with the fluid and dynamic world of Asian trade. London had become a hub for a variety of often centrifugal forces that were stretching its networks and translations beyond recognition and even beyond the capabilities of someone as talented as Ogilby or Flamsteed to represent them as a coherent and sovereign cosmological image. A clear need had emerged for a new system of understanding the world, one that better comprehended the decentralized effects of forces in a world of divergent concepts of space and time.

CHAPTER FIVE

The System of the World

Newtonian space, time and matter are no intuitions. They are receipts
from culture and language. That is where Newton got them.
—Benjamin Whorf, "The Relation of Habitual Thought and Behavior
in Language," *Language, Culture, and Personality*, ed. Leslie Spier
(Menasha: Sapir Memorial Publication Fund, 1941), 88

The city transports the mind.
—Sir Josiah Child, *Discourse of the Nature, Use and Advantages of
Trade* (London: Randal Taylor, 1694), 3

1687: GLOBAL REVOLUTIONS

Big historical claims have been made about the revolutions in physical
science and Parliamentary sovereignty in England for the years 1687 to
1689—a social, political, economic, and epistemological "birth" of modernity. But the final years of the 1680s saw dramatic changes on a global scale
as well. Emphasizing the European or regional dimensions of these changes,
Steve Pincus has described the odd coalitions and changes produced in England by "Catholic modernity" and the French conception of the absolutist
state.[1] On a global scale the picture becomes more complex. England and
Siam would experience interrelated "revolutions" in 1688, each replacing
its king because of suspicions of foreign (i.e., French) domination. Some
historians have even argued that the Indian Ocean region as a whole went
through a revolutionary period in which conceptions about the sources of
sovereignty changed.[2] In East Asia, Qing China finally subdued the last

of the Ming loyalists, and the silver cycle that had stalled in the second half of the seventeenth century picked up speed once again. The Treaty of Nerchinsk (1689) settled the Qing Dynasty's claim to the Amur River basin against Russia, setting the stage for a century of Qing expansion far beyond the old frontiers of the Ming and the Great Wall as well as helping define Peter the Great's shift towards Europe and frontier expansion to the northeast. In many ways these events outpaced the ability of people and institutions in London to comprehend them. They also provoked a broader reflection on the process of translating knowledge and institutions as well as the nature of sovereignty, giving both the Glorious and Newtonian revolutions a global character as systems for comprehending the world.

By the 1680s many Londoners could see the successful results of centralizing absolutist trends around the world, both in terms of politics and of science. This included the use of large standing armies to occupy territory and the management of confessionalism by monarchs or emperors. Corresponding failures of English absolutist efforts in Tangiers, Bombay, and elsewhere were well known, and in England Charles II's and James II's efforts to exert more control domestically over urban corporations met with substantial resistance. Meanwhile, success stories came in from across Europe—France under Louis XIV; Sweden under Charles XI; Leopold I's defeat of the Ottomans in 1683; Frederick III's "Danish Law" of 1683. As worrisome as European efforts could be, expansionist and centralizing policies by more distant empires also had begun to directly impact London's interests—Moulay Ismail ibn Sharif in Morocco, the Mughal emperor Aurangzeb in Golconda and Bengal, Somdet Phra Narai in Siam, and Kangxi's Qing all shut down trading stations and routes. The Qing in particular seemed to some a mere barbaric conquest but to others an almost quasi-republican phenomenon. As Gabriel de Magalhães described the Manchu conquest in a book quickly translated from Portuguese into French once it had arrived in Paris and then almost immediately into English in 1688, "All those that revolt pretend that it is by the decree of Heaven, that sent them to ease the People opprest by the Tyranny of their Governours," which according to Magalhães resonated with a deeply held cultural belief, "insomuch that there is hardly one among them that does not hope to be an Emperour at one time or other." So he argued, a series of groups from peasants to Buddhist sects "make a Profession of creating new Kings, and establishing a new Government in the Empire."[3] This was not absolutism based on dynastic divine right but a vision of the Chinese emperor as a Machiavellian prince enthroned by faction and religious enthusiasm. Not only Dutch and

French versions of modernity but also Moroccan, Mughal, and Qing ones had started to show better results in dominating territory and trade during the 1680s than strategies coming out of London.

Contemporaneous with this rise of absolutism, a multifaceted crisis emerged between 1682 and 1688 in London's exchange strategies in Asia. In 1682, the Dutch removed the English from their presidency at Banten, just as a celebrated embassy had arrived in London that summer to increase the sale of arms and solidify a relationship with the English crown. That same year in December, the Company's factory at Ayutthaya burned, supposedly at the hands of interlopers. In 1683, the Taiwan factory closed, and Bombay tried to break away from both Company control and English sovereignty more generally. As tensions with both the Dutch and French grew, outright war also began against both the Mughals and Siam in 1686 and 1687.[4] East of India, only Tonkin in northern Vietnam and the struggling new factory of Bengkulu on western Sumatra remained.[5] During the crisis, the Company's director Josiah Child, a strong supporter of both Charles II and James II, wanted the crown to adopt Dutch and French methods and to make direct military interventions with both Royal Navy ships and standing armies of crown-supported soldiers. Fears that London's exchange relations in Asia, Europe (the Baltic and Mediterranean), and the Americas might collapse became intertwined with anxieties that James II was building a Catholic-style empire.[6] The incorporation of Madras (Fort St. George) as an English town in 1687, its population of over 100,000 people second only to London in sheer size, was symptomatic of the uncertain changes underway. Its governor Elihu Yale was ordered to make a census by gender, occupation, and religion (English, Portuguese, "Moors, and "Gentoos"—gentiles or Hindus), so that Company and crown would have some idea of what they were getting into.[7]

In 1683, in response to the collapse of Banten, the East India Company under Child formed a "secret committee" that quickly became involved in putting out fires.[8] Its first task would be to suppress a rebellion that began in December of 1683 in Bombay, "to give such necessary orders for reducing the said island unto the obedience of His Majesty's authority."[9] Paradoxically, the rebellion against Company rule emerged through the leadership of Richard Keigwin, a military commander and devout royalist. Still under the control of the presidency at the Mughal-controlled port of Surat, Bombay had since the mid-1670s suffered from being in the midst of intermittent warfare between the emerging Maratha Empire in the Deccan and the Mughals. Aurangzeb ultimately moved his court, capital, and a massive army to the Deccan at Aurangabad, and he forced the English to let him use

Bombay as the winter quarters for the fleets of his Sidi allies and his own ships. The Mughals attacked or occupied Bombay Harbour several times using the mercenary Sidi admiral Yakut Khan, putting great stress on the city and forcing tributary concessions in 1672, 1673, and 1680, as part of a broader assault on the Marathas. By October 1679, Keigwin was helping the Mughal navy against the Marathas, in the early stages of a war between the Marathas and the Mughals that would last twenty-seven years. Looking at the bottom line, Child and the East India Company directors decided to cut military expenditures and recall Keigwin. What some at the time called Keigwin's "revolution" was framed simultaneously as anti-Company, anti-Mughal, and pro-English. He declared the city a free-trade zone, imprisoning the Company's governor and issuing passes for trade in the name of Charles II to all of India, Persia, and Arabia. He wanted Bombay to be an independent political corporation under the loose sovereignty of the Union Jack—the flag and crown as symbols rather than substantial presences. Sir Thomas Grantham, who in the late 1670s suppressed Bacon's Rebellion in Virginia and had been sent out in July 1683 to settle a dispute with the Safavids over Gombroon and to regarrison Bombay, retook the city, pardoned Keigwin in November 1684, and convinced John Child, the Company's governor and president at Surat, to excuse the would-be revolutionaries.[10] This attempt at revolution had been an example of both crown and Company losing control over devolution, resulting from the understanding that neither absolutism nor Company rule had been particularly effective in responding to the changing political situation in India.

While Keigwin had tried to open up a free trade zone in the western Indian Ocean, the Company was trying to use the courts in London to stop efforts by interlopers in the eastern Indian Ocean at the Coromandel Coast and Aceh. As with Bombay and the Marathas in the west, there was a core conflict between the Mughals and the Qutb Shahi sultanate of Golconda, which had around 1683 stopped transferring some of its tributary taxation payments to the Mughals. Because Golconda was the center of the gold trade, this had implications for regional exchange, involving most immediately Bengal, Pegu, Siam, and Aceh as well as the Dutch in Sri Lanka and Java. The closely watched test case, *East India Company v. Thomas Sandys*, lasted from 1683 to 1685.[11] Sandys had his ships seized for trading with Aceh as well as Masulipatam and Porto Novo on the Coromandel Coast. Unlike the 1668 *Skinner* case discussed in chapter four that involved a series of different elements of the English government, the Sandys case was handled purely by Chief Justice Jeffreys in a sign of growing crown control over disputes in London. Not surprisingly, Jeffreys came down firmly on the side of royal prerogative over foreign trade, which also meant a certain kind of victory for the East India Company itself. His was perhaps the only clear statement of a totalizing policy of English mercantilism ever made.

Yet even though it set the stage for dramatic changes in Company policy in 1686, including war against the Mughals and Siam, the court's decision in *Sandys* also revealed how deeply the legitimacy of exchanges, contracts, and translations between London merchants and various Asian princes and merchant networks made without the approval of the English king had permeated London. Sir John Holt, arguing for the East India Company as plaintiff, tried to use both Grotius and a series of legal opinions rooted in medieval opinions about trade with Islamic regions to argue that kings as defenders of the faith had a right to limit trade with infidels and pagans on the basis that they do not have valid laws. Sandys's defense council Sir George Treby found this "absurd, monkish, phantastical and phanatical," noting that Holt had radically misconstrued Grotius. A more practical argument suggested that trade in Asia from London involved considerable investment up front and significant competition from both "Indians" and Europeans.[12] But the case, as Judge Jeffreys noted, really turned on the question of the law of nations as it related to the sovereignty of princes. Holt had contended that all trade is founded on the covenants of princes for religious reasons, "lest the manners and morals of the people should be corrupted by the example of foreign nations," no doubt a worrisome argument for those concerned about a Catholic and French-leaning king. Sandys's lawyers followed Selden's description of complex legal recognitions in various areas having historically established legal and legitimate practices of dominion over the sea, and they thus responded that trade was free according to the covenants of common law made independently by merchants from cities.[13] Using logic similar to that which Keigwin had used in Bombay, Sandys's lawyers had cleverly made sovereignty into an issue about the equal rights of London's citizens to engage in contracts; by implication any merchant or prince who also claimed legal authority to trade could do the same.

The basic problem faced by Judge Jeffreys in resolving this impasse was that the crown's actual sovereignty and naval capacity did not extend as far as the merchant networks of London. The Restoration image of a monarch with global sovereign reach was largely that, an image. Conversely, in terms of the common law or the law of cities, as the Attorney General Robert Sawyer explained, "The principal part of foreign trade is transacted beyond sea, where the common law can have no cognizance, but is confined within the compass of the four seas," nor "can foreigners be presumed to have cognizance of the municipal laws of this kingdom." The core problem of sovereignty was whether it translated from the narrow seas of Britain to the broader ocean of exchanges at any level of "cognizance." Judge Jeffreys argued that certain trades needed "companies and societies," primarily because of capital expenses (discovery, forts, factories, making leagues and treaties), but that "societies to trade" are not monopolies since they receive patents for the "public" good rather than acting as private persons.[14] The gesture towards the public rather than state sovereignty is remarkable here especially coming from Jeffries because it called attention to a gap in sovereign power. As Miles Ogborn has noted, the case was also a landmark in terms of East India Company printed propaganda, which shifted from trying to convince the court in private through the "secret committee" toward the goal of influencing public opinion and especially that of Parliament.[15] Princes did not define the nature of trade by their agreements; instead, trade came out of contractual relationships that the state then could recognize in the interests of the public. Here was also a reason why Parliament should play a more prominent if not sovereign role. The publicness of the Company was a solution to the crown's inability to extend sovereignty and to the increasing volume of commerce by a wide variety of Asian and European actors in both the Indian Ocean and East Asia. No wonder that as secretary to the Admiralty, Pepys took very careful notes.

Indeed, much of this publicness had already emerged during the summer of 1682, when a Bantenese embassy had been toured around London and even invited to the wedding reception of Sir Josiah Child's daughter (figures 36 and 37). The Exclusion Crisis had seen the emergence of competing political newspapers, which the crown had largely shut down in 1680. These began to reemerge in 1681 and 1682 in papers like Nathaniel Thompson's unlicensed but Tory *Loyal Protestant* (March 1681–March 1683) and the short-lived radical Whig newspaper the *London Mercury*, published by Thomas Vile and Richard Baldwin. Political fractures also encouraged competition between the once closely linked Levant and East India Companies over whether the former could establish a second East India Company, a subject covered extensively in the new papers.[16] The *London Mercury* lasted almost concurrently with the Bantenese embassy from April to October 1682, when the government shut it and several other papers down during the controversial elections for lord mayor and the trial over London's corporate status. To radical Whig printer like Vile, it was important to demonstrate and indeed exaggerate connections between the crown, the East India Company, and foreign powers to show the roots of excessive or absolutist authority coming out of global exchange practices. Vile in this way echoed the conspiratorial narratives of the Popish Plot, when in one of his many

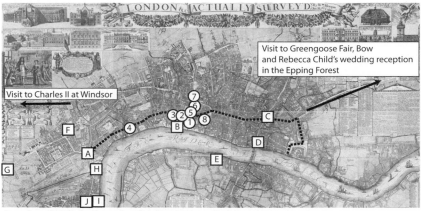

London 1682: As Seen by the Bantenese Embassy

Procession route of the Bantenese embassy   ▪▪▪▪▪
Printers and booksellers associated with the embassy   ○
Places visited by the embassy   □

Fig. 36. Map of printers and engravers associated with and locations visited by the Bantenese embassy (1682) overlaid on William Morgan, "London, &c. actually Surveyed" (London: 1682). Reprinted London Topographical Society, 1904. Image courtesy of the Library of Congress, photo by Robert Batchelor.

1. Henry Hills, Royal Printer in Blackfriars Publishing House
2. Henry Rhodes, Fleet Street near Bride Lane
3. Robert White, Fleet Street near Inner Temple Gate
4. John Lloyd, Stand print dealer
5. John Oliver, Old Bailey near Ludgate Hill
6. Walter Davis, Amen Corner, former site of Royal College of Physicians

7. John Overton, shop outside Newgate
8. Robert Prick. St. Paul's Churchyard

Places visited by the embassy:
A. Embassy stays Charing Cross (Lionel Emps, Charles II's bow maker's house)
B. Duke of York's Theater (Edward Luttrell sketches from life)
C. East India Company Offices, Leadenhall Street

D. Monument to the Great Fire
E. Bear and Steer Fight, Beargarden, Southwark
F. Major Foubert's Riding and Mathematics Academy, Rupert Street
G. Funeral and Burial of the Embassy's Cook, St. James Park
H. Whitehall
I. Houses of Parliament
J. Westminster Abbey

stories Titus Oates argued it had been a Jesuit-Muslim conspiracy by the former Barbary slave Adam Elliot, a tale refuted by a member of the Moroccan embassy in January 1682.[17] The procession through London, the tourist itinerary, and the profusion of printed images of the Bantense embassy in the summer of 1682, some of which seem to have been directly supported by the East India Company, could be either read as a positive sign of London's extensive commercial relations and success in translating languages of exchange or as negative signs of the growing absolutist tendencies of the crown and its close relationship with Child and the East India Company. News of the loss of Banten and Taiwan in 1682–83 and Keigwin's rebellion in 1683 were very public blows to the Company's image, which neither the resolutions of the Sandys case nor affairs in Bombay could remedy.

So when Child took the East India Company to war against both the

Fig. 37. The principal ambassador Kiai Ngabehi Naya Wipraya from Banten with
Javanese (Carakan, top) and Malay (Jawi, bottom) inscriptions as well as an oval in
Italian by Jacob Collins and Nicholas Yeates after Henry Peart, "Keay Nabee Naia-wi-
praia principall Ambassador from Sultan Abdulcahar Abusnazar King of Suro-soan,
formerly called Bantam," line engraving by Robert White (London: Walter Davis, 1682).
© Trustees of the British Museum, 1848,0911.727.

Mughals and the Siamese, this was to be a war by a "public" company for the "nation." In a much-cited letter from June 9, 1686, to Fort St. George, Child wrote about creating "formidable martial government in India," "a politie of Civil and Military power," and a "well-grounded sure English do-minion in India for all time to come." This is often interpreted as a bid for empire or corporate sovereignty, but the historical interaction is more com-plex.[18] The immediate cause of this war had been the further tightening up of Mughal sovereignty in the 1680s, which had already had a substantial influence on the development of Bombay in the 1670s. The Mughal war against the Marathas spread south and east, leading to the conquest of the Bijapur sultanate in 1686 and Golconda in 1687. To fund this war, the Mu-ghals instituted a customs duty of 3.5% at Surat in 1680, and also began to enforce it in Bengal at Hugli in 1684. Job Charnock, the company's factor at Hugli, refused to pay even when put on trial by the Mughals, resulting in the arrest of several of the Company's soldiers. In a series of reprisals, the governor of Bengal (Aurangzeb's uncle) finally sent a large cavalry force in December 1686 to evict the English from Hugli entirely, forcing them to move to Calcutta. In February 1687, the Company began to attack Mughal shipping, forts, and cities, culminating in failed naval assault on the eve of the Glorious Revolution. The war did not go well, and John Child had to surrender Bombay to the Mughals in 1689 after an extensive siege and the sack of the Company's new fort at Mazagon. After paying large fines, the Company made a treaty with Aurangzeb in February 1690, securing the re-turn of Bombay and restoring rights to trade at Calcutta, but only as mere traders. Such changes demanded radical reworkings of both the East India Company and the English state, especially in the aftermath of 1689. Bom-bay, Madras, and Calcutta became pillars of the English presence in India and global cities in their own right not because of the success of British im-perialism but in many ways because of its initial failure.

Perhaps the greatest changes came at Madras, which Child during the war in the late 1680s reenvisioned as an emporium and garrison that would literally be part of England. When the process of incorporating Madras began in 1687, it was first defined as a regency. A heated debate took place in the Privy Council over whether incorporation should be under the crown's Great Seal or the Company's Broad Seal, whether India be an extension of the crown through corporate charters or an extension of the governor and Company of Merchants of London trading into the East Indies. Ultimately both were used, although the document was clearly a Company product.[19] In September 1688, however, the Mughals defeated Golconda, and Madras not only had to pay tribute to the Mughals but aid them in future wars

against the Marathas. The slippage in terms of growing sovereign respon-
sibilities as distinct from sovereign claims had begun in earnest in India.

The extension of sovereignty through urban corporations had also been
an attempt to secure points of contact with non-Company merchants and
in so doing reduce the potential space for interloping and outright rebellion. With the loss of Banten, the Company also hoped to bring Chinese
merchant networks back to India, which would ultimately short-circuit the
possibility and need for country trade in Southeast Asia let alone direct
trade with China.

> Mr Hoath tells us that there are some Syamers and Chineses that do in-
> habit at Atchoone [Aceh] possible Mr. [Elihu] Yale may find means when
> he is there to persuade some of the latter to trade and reside with you
> at fort St. George of whome if you had any considerable number they
> would in probability greatly encrease the trade of Maddras as tis known
> they have been the making of Batavia. They are excellently well ac-
> quainted with all the Crooks and small Places in the Island of Sumatra
> and the South Seas, and if they were settled with you would in your
> Sloops and other small Crafts find means to procure great quantities of
> Pepper and the best sort of Cassia Lignum if the Dutch had twice as
> many forts and factories as they have in those parts.[20]

The idea of encouraging the Fujianese to revive the old trade routes dating back to Zheng He (and depicted on the Selden Map) was not as fanciful
as it might seem. Two ships from Madras that arrived in Amoy (Xiamen) in
late 1687 returned with an embassy in February 1688 from the new Qing
governor Shi Lang.[21] But Chinese merchants never followed, and direct trade
from Bombay and Madras to Amoy remained risky and intermittent. Again
consolidation of territorial empires in Asia took the lead in shaping trade
relations, and while the East India Company reorganized its affairs in the
aftermath of the Glorious Revolution, the Qing legitimation of European
trade at Guangzhou (Canton) ultimately made such efforts unnecessary
after 1699.[22]

If reorganization in India seemed at best uncertain by 1688, the Company then also saw its efforts falling behind the Dutch and the French at
that time in terms of trade with the Qing. The Qing had in 1684 opened
the Amoy (Xiamen) maritime customs office and four mainland locations
(Macao, Zhangzhou, Ningbo, and Zhenjiang) to deal with trade in the coastal
provinces, which slowly devolved into over sixty offices.[23] The Dutch under
Nikolaas de Graaf traded at Guangzhou in the winter of 1684–85, enter-

tained with banquets and fireworks after negotiating with local officials.[24] They also sent an embassy to Beijing under Vincent Paets from 1685–87, although like the previous two it failed. Nevertheless, the opening of new customs offices led to a rapid expansion of the Chinese diaspora trade, enabling the Dutch to easily obtain goods at Batavia.[25] Because of this, they actually stopped sending ships to trade directly with the Qing in 1690. Moreover, Dutch attacks on East India Company factories in Masulipatam in Coromandel and in Bateng-Kapas in west Sumatra kept the Company of the opinion that, "By all which it manifestly appears that the Netherlands East India Company are still pursuing their design of engrossing ye whole East India Trade by violence, injustice and oppression, if timely remedy be not provided to prevent their continuall encroachments to which we humbly submitt to your Majesty's great wisdom."[26]

The French also in the 1680s appeared to have gained a significant amount of initiative by co-opting the Jesuit mission that the Belgian Ferdinand Verbiest had revived at Beijing in 1670. Through his work on calendar reform, world and star maps, and the observatory, as well as translations in the 1670s and 1680s, Verbiest had become an intimate of the emperor Kangxi, convincing him to restore the Jesuits at court after a highly politicized regency during the early years of his reign (1662–69) had resulted in their exile and imprisonment.[27] Louis XIV's confessor Père François de La Chaise sympathized with Verbiest and the accommodationists, encouraging them to come to Paris and Versailles. One of those exiled and then brought back to Beijing was another Belgian Jesuit, Philippe Couplet, who was required to remain in Europe from 1683 to 1692 because of disputes over accommodation strategies in Rome with Innocent XI. In the late 1680s, Couplet helped Paris became the great publishing center for Jesuit translations of Chinese historical chronology and the four "Confucian" books. As the exiled Huguenot and former secretary to Louis XIV Henri Justel wrote to the Royal Society in 1685, Couplet's acquisition of 124 Chinese books for the Bibliothèque Royale in Paris meant that the Bodleian no longer had a larger collection.[28] The title page to the translation of the Confucian four books suggested that it offered access to a vast and unknown library. This flurry of publications in Paris, while meeting with a sympathetic response from James II and his followers, also raised the specter of strange hybrids of priestcraft in the guise of "Catholic modernity." The concluding warning of Hobbes's *Leviathan*, against both Catholics and dissenters corrupting the vision of a rational and absolute state, would have resonated in this regard: "But who knows what this Spirit of Rome, now gone out, and walking by

Missions through the dry places of China, Japan, and the Indies, that yeeld him little fruit, may not return, or rather an Assembly of Spirits worse than he, enter, and inhabit this clean swept house, and make the End thereof worse than the Beginning?"[29] The fact that James II hung Godfrey Kneller's picture of Shen Fuzong, the "Chinese convert" and protégée of Couplet who visited London and Oxford in 1687, in the room next to his bedchamber was a testament to how willing the crown had become to follow the lead of French absolutism in trying to resolve contradictions emerging from global exchange and sovereign limitations.

The sense that the Dutch and French global translation enterprises and knowledge networks were leaving behind those of London began to play out in the pages of the *Philosophical Transactions* in 1686 and 1687, when it came to light that Jesuit cooperation with Kangxi and the Dutch with Peter I in Russia had erupted into a war between the two expanding empires in the Amur River basin. The first gleanings of this in London appeared in a 1686 account of Kangxi's 1682 and 1683 expeditions to the Songhua River and to "Western Tartary" after the suppression of the Three Feudatories Rebellion and Taiwan. The article depicted the Qing as a consolidated absolutist power. Kangxi had a massive standing army of 70,000 soldiers and courtiers that he took with him on a northern tour in order to keep them trained and active, making him "the greatest and most powerful monarch of Asia, having so many vast Estates under him, without being any where interrupted by the Territory of any Forrein Prince, and he alone being as the Soule which gives Motion to all the Members of so vast a Body." The Jesuits were clearly helping Kangxi to develop his absolutist project although they were by no means in control of it. The letters as a whole showed Kangxi as the enlightened emperor, merging Chinese and European systems by sitting with Verbiest and a pocket version of his star map, "inquiring the Hour of the Night, by the Stars in the Meridian: Pleasing himself to shew to all the Knowledge he had acquired in these sciences" and reciting the names in both "Chinese and European languages," to supplement Kangxi's Manchu. Verbiest had gone along to survey elevations, star positions, magnetic declinations, the height of mountains, and distances between locations, portraying his own cartographic work as updating Martini's *Atlas* and the work of Ming cartographers like Luo Hongxian.[30] The Jesuit strategy was reportedly to "insinuate into the minds of those Princes, by Means of the Mathematics," and thus challenge the influence of Lama Buddhism on the Manchus with a systemic approach to mathematics, cartography, and cosmology. They could then use the efficient postal system of the Qing Empire and the reestablished Jesuit networks to spread such ideas to the south. To demonstrate how this plan would work, Verbiest's table of distances across the empire was included in the articles.

The principal point of the articles in the *Philosophical Transactions* was that the English did not really understand what was happening in terms of the growth of empires in Asia and had largely been left out of the process. "It may seem wonderful that the Author of these Letters makes mention in his former of a kind of Warr between the *Oriental Tartars* and the *Moscovites*, notwithstanding the extream distance, these People appear to be from one another in our Geographical Charts; but those who know how much the *Moscovites* have extended the Bounds of the *Empire* along the *Tartarian* Sea, will judge the thing less difficult, besides those who have seen these Countrys, have made Discoveries much differing from those which our Geographers have informed us of hitherto."[31] Russia and the Qing seemed to be developing their own absolutist strategies for conquering Central Asia with the Dutch and French playing supporting roles, and as a result information about this process seemed to be rapidly arriving in both Paris and Amsterdam while London lagged behind. The French were getting maps and letters both from Moscow and from their contacts in Beijing, while the Royal Society had to content itself with these translated letters from Verbiest and a stray letter from Moscow that explained that troops were being raised there "to go to War with the *Chinese*." In 1687, the former ambassador to England and burgomaster of Amsterdam Nicolaas Witsen came out with what Robert Southwell, the president of the Royal Society, referred to in the 1687 edition *Philosophical Transactions* as a "Columbus like" map of Tartary (figure 38). Witsen bragged of his constant correspondence with Moscow and Constantinople and yearly letters from Beijing, which allowed him to gather information from a global network.[32]

Perhaps the strongest example of a growing sense in London that the translating processes involved in commerce and knowledge needed to be rethought was not Banten, Taiwan, or Bombay but Siam, which lay at the trade and diplomatic intersection of the Mughals, the Safavids, and the Qing. The war with the Mughals in 1686 quickly became a war declared against Siam as well. After losing a factory in Cambodia in 1659, the East India Company had from 1661 considered expanding trade at Siam, but they and the Siamese suffered from Dutch harassment. The Siamese king Somdet Phra Narai (r. 1656–88) offered the Company Patani on the eastern side of the Malay Peninsula in 1678, an important center of Chinese trade at the end of the trunk route on the Selden Map, if the Company could offset the influence of the VOC. But Narai was also courting the French and the

Fig. 38. Carel Allard, "Tartaria Sive magni Chami Imperium" (Amsterdam: ca. 1690), a reduced single-sheet version of the map by Nicolaas Witsen in six sheets (Amsterdam: 1687). Biblioteca Nacional Portugal.

Safavids, and by 1682 Siam developed a bad reputation in London because of interlopers and Narai's talented and multilingual minister Constantine Phaulkon (born Konstantinos Gerakes/Gerakis; aka "Constance"), a Greek sailor and until 1678 East India Company employee who was suspected along with the interloping brothers George and Samuel "Siamese" White of having a hand in the burning down of the EIC factory in 1682. In November 1685, responding to earlier Siamese embassies to Isfahan in 1669, 1680, and 1684, a Safavid embassy arrived in the western Siamese port of Tenasserim, where as in Ayutthaya and Mergui there was a substantial Persian merchant community.[33] Phaulkon and the Whites had encouraged Siam to engage in a naval war with the dynastic allies and trading partners of the Safavids, the Nawab of Masulipatam and the sultanate of Golconda in 1684 and 1685, which both helped the Mughals and resulted in broad persecution of Muslim traders and purges of Muslim officials in Siam.[34] A number of Islamic merchants resident in Siam from Makassar, Java, India, Persia, and other areas fled to the Coromandel Coast, including the English fort at Madras, which was a tributary to Golconda.

Phaulkon and the Whites seem to have miscalculated, for in 1684, they also had encouraged Narai to send an embassy to France on an East India Company ship by way of London and Calais. The French, having had their fleet sunk by a combined Dutch and Golconda naval force in 1674, were natural allies for Narai, but their primary successes in Siam during the 1660s and '70s had been through the Société des Missions Étrangères rather than the Compagnie des Indes. After the French had managed to set up a trading factory in 1680, Narai sent an embassy to Paris on the first ship out, which foundered off Mauritius. Sending the second embassy through London seemed to be an effort to play upon rumors of the kind of increasingly tight Anglo-French relations that had resulted in the Exclusion Crisis. If it were not for the divide between the Company and interlopers in Siam, the embassy might have had more of an impact during its stopover in London. Phaulkon and the White brothers, hoping to be legitimized by Charles II, sent a pair of hanging scrolls having maps of the heavens and earth with explanatory texts in Chinese.[35] The emissaries also brought a copy of Ignatio à Costa and Prospero Intorcetta's *Sapientia sinica*, which contained a short biography of Confucius as well as parallel-text and transliterated Latin translations of fourteen pages from the *Daxue* (大學 "Great Learning") and the first five sections of the *Lunyu* (論語 the "Analects") printed in China, which ended up in the hands of the Cambridge scholar and fellow of the Royal Society, Nathaniel Vincent.[36] In 1682, Vincent was trying to convince Pepys to help him publish his book the *Conjectura Nautica* and support the use of his invisible ink cipher he called the *Cryptocoiranicon*, so the idea of translating the Latin translation of the *Daxue* into English possibly seemed a way into the London intellectual circles that he, being in Cambridge, felt marginalized from. In 1685, Vincent annotated an old sermon he had preached before Charles II, and praised Confucius and the idea of a meritocratic bureaucracy as the proper path to follow, including two paragraphs of a sample translation.[37]

The scrolls White sent appear to be reprints from Ming woodblocks, and they were designed to show the related orders of terrestrial empire and celestial pattern. One entitled 皇輿地圖考 (*Huang yuditu kao*, "Imperial World Map Verified") shows the fifteen provinces of the Ming with data on their administrative structure from the census (figure 39). At the top of the map, Beijing is described as having 8 magistrate or prefectural seats (*fu* 府), 18 submagistrates (*zhou* 州), and 115 counties (*xian* 縣). At the bottom of the map the eight prefectures are listed along with the counties under each prefecture. In the center is a reference map showing the empire itself, one that is strikingly similar to the map of the Ming on the Selden Map and most likely related to the same encyclopedia map. In some ways, this scroll on its own seems, like the Zheng calendars, to exhibit a kind of nostalgia for the traditions and tributary model of sovereignty defined by the Ming, which should have appealed to London given the efforts to reengage the Fujianese trade diaspora in places like Sumatra.

The second scroll was celestial rather than terrestrial, entitled 通華經緯 圖考 (*Tonghua jingwei tu kao*, "Verified Complete Chinese 'Warp-Weft' Diagrams") (figure 40). *Jingwei* (warp-weft) refers to the patterns of astronomical forces. In 1673, as part of his new set of instruments for the Beijing observatory, Verbiest had made an ecliptic armillary sphere called *Huangdao jingwei yi* [黃道經緯儀 "Ecliptic Warp-Weft Apparatus"], which measured the longitude and latitude of celestial bodies. As with many Verbiest projects, the sphere was an attempt to change the meaning of warp-weft, which instead of longitude and latitude previously meant a weaving together of the cosmos from a central axis. In Chinese geography, *jingwei* was the basis of the city and the graticule on the map—the *jing* of nine north-south streets crossed by the *wei* of nine east-west ones in the *Zhouli* to create the ideal capital city.[38] But Beijing on the terrestrial scroll map is not centrally located. Instead, the celestial map, which in Beijing would have implied a connection between the patterns of the imperial capital, the empire, and the stars, now suggested that stars, and in particular stellar coordinates as ways of defining space, were as important as imperial centers. This made the map an invitation to collaboration with the new royal observatories in London

Fig. 39. Terrestrial scroll map of China brought to London in 1684 with the Siamese embassy, *Huangyu ditu kao* [皇輿地圖考, in seal script]. © Bodleian Library, University of Oxford, 2009, Sinica 123/2.

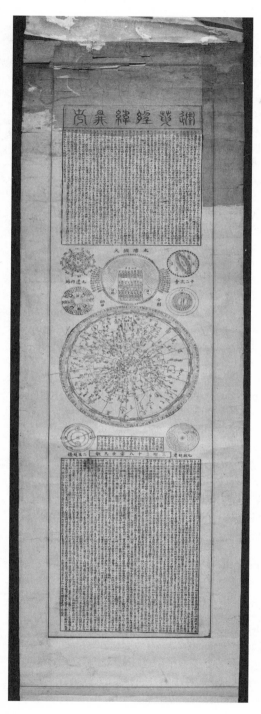

Fig. 40. Celestial scroll map in Chinese brought to London in 1684 with the Siamese embassy, *Tong hua jingwei tu kao* [通華經緯圖考, in seal script]. © Bodleian Library, University of Oxford, 2012, Bodleian Library Sinica 123/1.

and Paris, an argument that a relationship with Siam with its own linkages to old Ming cosmological techniques could in fact become more productive scientifically than the Jesuit one with Beijing and the Qing. Yet the East India Company, now fearing that the Whites might be another Keigwin, did nothing to encourage such relationships.

In Paris and Versailles, the city and Louis XIV's court feted the Siamese embassy even more than Londoners had with the Bantenese in 1682, with articles appearing in the *Mercure galant* and *Gazette de France.*[39] Louis XIV decided to send two prominent but unpredictable envoys—the Chevalier Alexandre de Chaumont, a distinguished naval officer and recent convert from Calvinism, and the Abbé de Choisy, who in his youth had liked to cross-dress but now worked with the Missions Étrangères—along with six Jesuit scientists. The French brought several large telescopes from twelve to eighty feet long as well as pendulum clocks, microscopes, thermometers and barometers, an air pump, and a Romer-style orrery illustrating the Copernican system, as well as reference materials. The Siamese returned to Paris with a third embassy in September 1686, which did not stop in London. The reception was carefully researched in an effort to translate Siamese protocol, and Louis XIV, who had seemed to the first embassy a bit too casual and informal, tightened up the image of French royal absolutism.[40] The 1686 publications of the two emissaries and one of the Jesuit scientists presented a powerful image of what was happening in Siam, setting the precedent for the publication of the French editions of Confucian texts the next year. From London, it appeared that France had rapidly used both Confucian knowledge and systemic astronomy to set themselves up to have strong alliances and intertwined governance strategies with both Siam and the Qing.[41] While James II may have admired the new global reach and international style of Louis XIV's court, he was far from achieving anything similar.

In the meantime, London's own global trade routes were rapidly collapsing. By June 1686, Child had decided that given these conditions only state intervention could salvage the Company in Asia, that India should be England's Siam, and that the English interlopers in Siam would have to be suppressed. He had the Privy Council recall all English subjects from the service of foreign princes in July 1686, an order targeted especially at Samuel White, so the Company could go to war with both the Mughals and Siam.[42] In 1685, Golconda had launched a formal complaint with the East India Company at Fort St. George over White's capturing ships under a cover of war with Siam. Fort St. George responded in April 1687 with a declaration of war against Siam, after a Siamese ship with military supplies had already been captured in Bengal.[43] With London's approval, in June 1687, an expedition was sent from Madras for Mergui, where White was the Shahbandar. All the English interlopers were informed that as loyal subjects of James II they were now at war with Siam. Siamese forces retaliated on the night of July 13, killing between 50 and 60 people in the compound surrounding Samuel White's house and forcing a general retreat. The French embassy arrived in September 1687 with six warships, and 1400 men, including 600 soldiers under General Desfarges, 300 artisans, and 15 Jesuits.[44] Responding to perceived threats from the Dutch and English, Narai allowed the French to establish fortified positions in Bangkok and Mergui and a garrison at the capital, all of which was reported back to London. So in September 1688, when the Glorious Revolution began, France appeared on the verge of becoming the new Dutch in East Asia, having introduced a syncretic or accommodationist version of "Catholic modernity" in Siam and possibly in the Qing Empire as well. A book appeared in Paris by the priest Nicholas Gervaise celebrating both the progress of Christianity and the importance of the 1686 emissaries to Louis XIV, who claimed that Narai saw him as a "model."[45] When the Glorious Revolution began in England in September 1688, nobody in Europe knew that Narai had been arrested and Phaulkon executed in May and June 1688, and that the new Siamese monarch Phetracha had forced the French to withdraw with heavy casualties from their fortress at Mergui and retreat from the fortress at Bangkok after a four-month siege in November 1688 in what the French would call the Siamese Revolution. Londoners, however, were aware as they invited a Dutch monarch to take the British throne that Child's two wars against the Mughals and Siam and his faux-absolutist strategy in Asia had in fact strengthened the hands of the French and Mughals and left London further from regular trade with the Qing than they had been even in 1684.

THE SEARCH FOR NEW TRANSLATION METHODS

The relatively sudden ability of the French court to use absolutism as a vehicle for translating not simply politics and commerce but religion and science on a global scale has to be understood in relation to the desire of courts like the Qing or Siam to engage in and indeed foster that process. The Qing in fact drove the process in an effort to achieve the kinds of successes demonstrated by the triumph of commerce in the late Ming, reopening in the 1680s the silver cycle that had largely shut down since the collapse of the Ming. The rapid revival of trade in East Asia after Qing successes in consolidating their empire and the desire by Siam to achieve similar results led to the spectacular but largely opportunistic and temporary successes for the French in the middle of the 1680s, culminating in both the Confucian translations and the massive military embassy to Siam in 1687. This resurgence of Qing, Siamese, French, and indeed Mughal state power fundamentally undermined efforts towards contractual commercialism coming out of London under phony absolutism and even the more militant commercial imperialism of the Dutch.

New efforts in London related to translation, especially the translation of Mandarin, suggest how transformative these changes coming out of Asia were. Nathaniel Vincent's initial effort to embrace Confucian meritocracy in 1685 in the aftermath of the Siamese embassy had been profoundly conservative. He translated the first sentence of the *Daxue* as "The intent of great Men in Knowledge and Instruction, does consist in the enlightening of our Spiritual Power conveyed to us from heaven, by the Virtues." Actually quite short in Mandarin, 大學之道在明明德 (*da xue zhi dao zai ming ming de*, "The great learning is the way into understanding brilliant virtue" or "illustrating illustrious virtue"), that sentence nevertheless contained some very resonant words including the core concepts of *xue* ("learning") and *de* ("virtue") as well as *dao* ("the way") and *ming* (twice, meaning "understanding" or "bright" and the name chosen for the previous dynasty). Vincent interpreted the Latin of the *Sapientia Sinica* quite broadly, a passage originally translated in Aristotelian terms of "the soul" and "animal appetites." He instead focused on the process of enlightening ("illuminando") as a translation of *ming*.[46] Vincent's translation reversed the role of *dao* and *de*, way and virtue, so that virtues translated and conveyed power from heaven (*coelo* as an interpolation of *dao*) through "great men" rather than learning as a more general path towards virtue. In some ways, this conformed to a kind of Protestant vision of enlightening, as expressed by Milton in the prophecy to Adam, "Reveal to Adam what shall come in future days, As I shall thee enlighten," in which enlightening remained a top-down process from heaven. The rest of Vincent's translation tried to domesticate such authority in order to buttress a conservative Protestant symbolic order, a conservative Enlightenment in which Confucius like Filmer advocated the "rightly ordering" of the family in tandem with "enlargement of the understanding" and "the perfection of knowledge" in order to "strengthen the mind."[47] Yet Vincent's rather crude effort to naturalize and domesticate Confucian thought as a familiar system that universalized Protestant ethics, something Confucius supposedly achieved introspectively by "penetrating that most perfect Harmony of Nature in Humane reason," also suggested a much broader shift in translation practices in London that began in 1685.

Aside from Vincent's, four new approaches to translation emerged in

London between 1685 and 1687 in relation to the changes occurring in Asia, helping to alter fundamentally both conceptions of knowledge production and ideas about the role of the state in London. Perhaps the clearest and most direct approach was formulated in a 1685 essay by Sir William Temple, who had retired from active London life to his estate at Moor Park in Surrey. His essay, "Upon the Gardens of Epicurus: or of Gardening in the Year 1685," included the famous translation of *sharawadgi*:

Their [the Chinese] greatest reach of imagination, is employed in contriving figures, where the beauty shall be great, and strike the eye, but without any order or disposition of parts, that shall be commonly or easily observed. And though we have hardly any notion of this sort of beauty, yet they have a particular word to express it; and where they find it hit their eye at first sight, they say the sharawadgi is fine or is admirable or any such expression of esteem. And whoever observes the work upon the best Indian gowns, or the painting upon their best screens or purcellans, will find their beauty is all of this kind, (that is) without order. But I should hardly advise any of these attempts in the figure of gardens among us; they are adventures of too hard achievement for any common hands; and though there may be more honour if they succeed well, yet there is more dishonour if they fail, and 'tis twenty to one they will; whereas in regular figures, 'tis hard to make any great and remarkable faults.[48]

This passage, cited throughout the next century in relation to the landscape garden movement, was in 1685 and again in 1690 when it first appeared in print a skeptical critique of ambitions to translate, recommending against attempts by "common hands." Whether or not *sharawadgi* actually translated a Chinese or as some have proposed a Japanese phrase, Temple's take on language here tried to mirror the action of Chinese and Japanese aesthetics as seen in porcelains and screens, a kind of striking ("hit their eye at first sight") sense of beauty without order that was, to a geometrically trained European, simultaneously radically different from any familiar "notion."[49] The approach to translation articulated in this essay was the opposite of that of contemporaries like Dryden or Vincent. Dryden wanted Virgil to "speak English," and Vincent seems to have had similar goals with Confucius. Here Temple, the great advocate of "ancient virtue," introduced a radically foreign and artificial concept in an effort to return a sense of coherence. *Sharawadgi* remained in a state of constant translation, quasi-intelligible or semi-translated as Emily Apter might suggest, against what the French Jesuits along with Vincent were trying to do in making "Confucius" a familiar and stable figure, who could justify newly constituted absolutist and patriarchal models across East Asia from Beijing to Lopburi. "Confucius" like *sharawadgi* was an invented neologism, but the former made the patriarchal state universalizable rather than, as in Temple's formulation, keeping open the possibility of linguistic exchange.[50]

Temple sought to avoid absolutism or "Catholic modernity" and at the same time to use translation to preserve traditional authority relationships. His "Essay on Heroic Virtue" agreed substantially with the kinds of sentiments Vincent put forward about Confucius: "It is that Empire where all Nobility is from Worth and Knowledge, where none are born great but those of the Royal Family, where Men are honoured and advanced then only when they deserve to be so," where the good state is a space for the production of such virtue.[51] But Temple's praise of Confucian values, which was also placed in a universal context that included lawgivers like Mango Copac in Peru, made a relatively conventional if uniquely global statement about the virtue of top-down ancient political models. In fact, unlike his contemporary Isaac Vossius, the former secretary of Grotius who probably inspired much of Temple's writing about China, Temple's interest in the apparently "modern" and universalizable aspects of China, like the compass, printing, or even the extent of urbanization, was comparatively muted.[52] *Sharawadgi* put forward the possibility that in fact much of what might seem natural about such translations was merely an elaborately crafted appearance, and the word *sharawadgi* itself suggested that certain concepts were so radically different they could not even be grasped. Here Temple displayed far more skepticism than Locke, whose constructivist vision contended that ideas or words presenting challenges for the translator were merely cultural shorthands for "complex" assemblages rooted in the historicity of language due to changes in "custom or opinions."[53] Temple did not have such confidence that language could be broken down so easily into simple ideas, and thus it was best, as Voltaire would later say, to cultivate one's own garden. He did, however, suggest that at certain moments a kind of striking translation was necessary (*sharawadgi*) to deflate the confidence of those who thought translation could work absolutely, like the syncretic cult of "Confucius" being pushed by the Jesuits in Paris, Beijing, and Ayutthaya.

Following on the heels of Vincent's patriarchy and Temple's radical difference, Robert Hooke offered a third alternative in the *Philosophical Transactions*. Hooke placed more confidence in artifice itself.[54] Aside from Thomas Hyde at Oxford, Hooke and his friend Francis Lodwick were part of the handful of people in the London area during the 1680s actively trying to learn Chinese.[55] Hooke worked with Lodwick and Wilkins on this question for many years, claiming that like Wilkins he had tried to translate the Lord's Prayer out of a Chinese manuscript owned by Lodwick but had lost it in 1666 during the Great Fire of London. The process convinced Hooke that Chinese was not of a "literal" (i.e., phonetic) character. He subsequently obtained a Chinese dictionary with pictures and one of the Taiwanese calendars. He used these to have direct experience in translating Chinese numbers, which in 1686 he engraved for the reader along with a Chinese abacus. Hooke and Lodwick would finally meet with three visiting Chinese in 1693, but even with Lodwick's help Hooke found the similarity of pronunciation for different characters difficult to master.[56] Such challenges pushed Hooke towards making structural and systemic comments on translation rather than focusing on individual words and concepts like Temple.

In 1686, Hooke distanced himself from older efforts like that of Webb to claim that Chinese was a kind of natural or original language, shifting the emphasis to the artificial—"Whether or not there ever were any Language Natural, I dispute not: But that there have been, are and may be artificial Languages tis not difficult to prove." Following the kinds of aristocracy of merit assumptions that Temple put forward, Hooke argued that Mandarin was an "invention" imposed by "thinking and Studious men" like himself and similar to mathematics in this regard. The Chinese simply employed a "differing method," and it was possible to see such "differing thoughts of Invention" at work in every language. In fact, Hooke seemed most interested in demonstrating not only divergences between languages but also the possibility of using a variety of different languages, each with their own method and inventive possibilities, for the pursuit of science.[57] Chinese characters were compounds of strokes, the true "letters, elements or particles out of which more compounded characters are constructed or contexted," and Hooke noticed radicals, that characters could be composed of other characters, "crowded together" into a square. Hooke saw such visible "compounding" as the "great singularity" of the Chinese language, which made it different from all others in the world. This also made it appear to be a "real" or "natural" language because the reasons for the compounding have been forgotten, which meant that the language must now simply be memorized "by roat." He included meaningless practice exercises from a Chinese educational primer in his engraving, 年五人大先仁八之 (nian wu ren da xian ren ba zhi), to show examples of cursive and seal script in characters traditionally copied and assessed in the teaching of writing.[58] He linked differences in language with differentials in media and its translation, such as the invention of printing in China and in Germany. Following

Temple's skepticism, Hooke agreed that a direct translation of technique was impossible because the different compositional methods worked better for each language, but differences in language "gave a hint to the Inventors" of printing and that "such an intimation was enough to an Ingenious Artist to improve the first Contrivance, and make it more accommodate to the literal way of writing with us." Invention, as the Chinese have shown with their language, was a kind of "compounding" that produced "singularities."[59] Any assumption of naturalness along the lines of Webb or Vincent, implying that some kind of moral language or content in terms of values could translate universally, was a mistaken understanding of the differential and artificial character of language itself.

If Temple's and Hooke's efforts were skeptical commentaries about the proposed translation projects of Vincent, Vossius, and Lodwick as well as the French Jesuits, those of Thomas Hyde at Oxford responded to the critique that artificiality could not translate. The interest in China in the late 1680s as well as the arrival of Shen Fuzong in 1687 triggered a massive translation project by Hyde, involving not only cataloging books but also working with Chinese maps and measurement systems. The only printed products of this effort aside from the Bodleian library catalogue itself appeared in Hyde's two books on Asian games in 1689 and 1694 and an appendix to Edward Bernard's book on measurement.[60] Interest like that expressed by Hooke and Lodwick in trying to derive systemic aspects of language from Chinese had led to some interest in Chinese games, and Henri Justel had written to Edmond Halley about Chinese chess in 1686.[61] Hyde the librarian liked figures and tables, a penchant he shared with the older artificial language, cryptographic and mathematical circles at Oxford that had been influential in forming the Royal Society.[62] But Hyde's approach in cooperation with Shen was different. When Hooke examined the Taiwanese almanacs, he focused on numbers, easy to recognize and prominent on the title page and elsewhere in the book, closely examining the *"tabula"* or *tu* (圖), a word that described tables, pictures, charts, and maps.[63] Hyde amassed a broader collection of tabular writing samples, often with calendric terms as well as weights and measures—in Chinese, Sinhala (Sri Lanka), Manchu, Telugu (Coromandel), Malayalam (Malabar), Siamese, and Sanskrit as well as more well-known Near Eastern languages.[64] In pursuing this research, he was preparing in Oxford both the institutional and scholarly grounds for a broad and global comparative linguistics on systemic grounds. When James II came to the Bodleian and mentioned the translation project of the French Jesuits, Hyde gestured to the shelves of books and manuscripts collected by Selden and Laud.

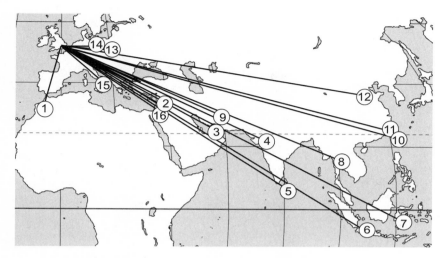

Fig. 41. Probable origin of books and manuscripts from Asia in Thomas Hyde's private library in the 1680s (with British Library call numbers).

1. Morocco (?) [Arrows of Lottery, Asian Reg.16.B.27–28]
2. Aleppo [Asian Reg.16.B.21]
3. Ormuz [Asian, Reg.16.B.3]
4. Surat [Asian Reg.16.B.1–2, B. 5–8, B.12–15; B.19, B.23]
5. Sri Lanka [Asian Reg.16.B.3, B.20]
6. Banten [Print OR.70.BB.9]
7. Giolo * [Thomas Hyde, An Account of the Famous Prince Gliolo, Son of the King of Gilolo (London: 1692) Print 10825 b.47]

8. Siam [Asian Reg.16.B.4]
9. Persia [Asian Reg.16.B.21]
10. Taiwan [Print 15298 a.30]
11. Fujian Province [Print 15298 a.32]
12. Beijing [Verbiest books, Asian Reg.16.B.3; Print OR.74.b.6]
13. Poland ["On the Slavonian and German Languages" (1672) Asian Reg.16.B.22]

14. Berlin [Andreas Muller, Disquisitio Geographia & Historica, de Chataja (Berlin: Rungianis, 1671), Print 10055.ee.32]
15. Rome [Giovanni Molino, Dittionario della Lingua Italiana Turchesca (Roma: Antonio Maria Gioiosi, 1641), Asian Reg.16.B.24]
16. Nablus [Asian Reg.16.B.3]
*Printed in London

Hyde wanted to take a middle road between the ancients and the moderns. Like Hooke, he was concerned to challenge the ideas of Andreas Müller that there could be a simple key to unlock the secrets of the ancient Chinese language. He had little confidence in the artificial language schemes of Wilkins, Dalgarno, or Lodwick, and he did not agree with Hooke that Chinese had once been a purely artificial language. Unlike them, Hyde had English-Chinese dictionaries in his personal library that he relied upon in addition to getting direct translations from Shen.[65] In addition to having access to the Bodleian, Hyde most certainly had the best personal library of Asian books in England in the 1680s (figure 41). He also kept quite current about technical debates involving cartography. When Shen Fuzong went in the summer of 1687 to the Bodleian to help Hyde catalog the Chinese books there, attempting to develop a cartographic practice that would translate into Qing imperial interests in the north as well as the East India Company in the south was high on the list of tasks. Hyde pulled out the Bodleian's collection of Chinese maps, and they went to work on translating the Chinese

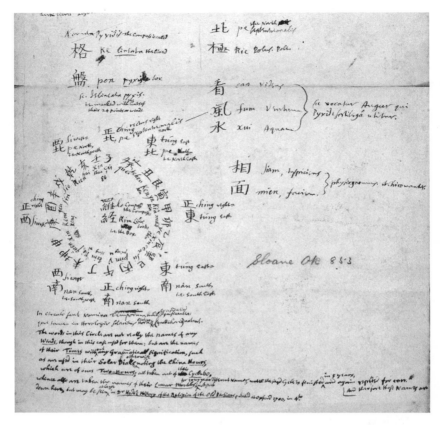

Fig. 42. Part of Hyde's and Shen's notes on the compass at the top of the Selden Map.
© The British Library Board, Asian Sloane 853.

compass rose on the Selden Map (figure 42). They also tried to sort out the
depictions and names of areas to the north of the Great Wall, made confus-
ing by a mix of transliterations and possible errors in writing characters in a
cartouche in the upper left of the map.[66] In a similar way they examined the
terrestrial scroll sent by White from Siam with Shen in 1687, transliterating
provinces, marking Japan and Hainan Island, and taking separate notes on
the areas north of the Great Wall in particular.[67] Coupled with this kind of
translation that would appeal to Royal Society audiences, Hyde also wanted
to show that no theorist in Europe really had a firm grasp on translating
Asian texts. To do this Hyde had Shen go through Müller's *Disquisitio Geographia* and correct the characters that had been provided by Leiden's Jacob
Golius, while Hyde himself corrected the Persian taken from Ulugh Beg.[68]

The arrival of Couplet's protégé Shen presented an opportunity to subtly challenge two beliefs about the closure of Chinese culture—Couplet's claim to Chinese as an original and pure language kept so by constraining commerce and Hooke's idea that languages could be invented in isolation as systems. Shen's father was a Nanjing physician who had converted to Catholicism probably in the 1670s, a stroke of luck since most of the Bodleian's collection consisted of medical texts. Shen arrived with Couplet in Holland on a VOC ship in late 1683, traveling to meet Louis XIV in 1684 and Innocent XI in 1685. They found themselves in the midst of the dispute between king and Pope over the Jesuits, the French church, and accommodation in China. In Paris, Shen aided Couplet and Melchisédech Thévenot in putting together another "clavis sinica," and he then left Couplet in Paris, going to London with the Jesuit Father Spinola in March 1687 in the hope that James II might be sympathetic to the Jesuit mission. Shen was at Oxford in June and July. In a little over a month, he and Hyde accomplished a great deal of translation work, including cataloging and annotating over sixty Chinese books and maps, and their correspondence continued when Shen returned to London.[69] Neither Spinola and nor Couplet himself, who came to London in December 1687, ever left London or met with Hyde. They departed for Lisbon with Shen in April 1688 so Shen could finish his Jesuit training and prepare to return to China.[70]

Shen's visit to London and Oxford revealed that he was not altogether convinced of Couplet's Confucian project and James II's involvement in it. James II, who visited the Bodleian in September, asked whether Hyde had a copy of Couplet's Confucius and "whether the Chinese had any divinity," to which Hyde replied that they were all heathens with idols representing an ancient but denigrated familiarity with Christianity. Boyle, whom Shen went to see in late July, merely asked about how many characters Chinese had, to which Shen replied that he himself knew between ten and twelve thousand, and he asked how different Mandarin was from the language of the common people and how widespread it was apart from the "Mandarins," "magistrates," or "literati."[71] Hyde for his part was eager to suggest to Boyle the falseness of Couplet's project, how Shen "doth not praise" the philosophy of Confucius, and explaining as well the incorrect impression given by the name "Confucius," "Cung-fu-çu, where *Cung* [孔 *kong*] is the family name, signifying *foramen*, nothing to the purpose: but *fu-çu* [夫子 *fuzi*] signified master."[72] For Hyde, the Jesuit approach to translation merely produced new and false idols like "Confucius" out of bad translations, and if Hyde's version is to be believed, Shen the good Christian convert shared some of these suspicions.

Hyde's focus on games was an effort to shift translation efforts away from Mandarin and towards an Indo-Persian center. Shen's supposed contempt for "Confucius" was useful in this regard, as was the fact that among the few things he had brought to London was an actual printed Chinese boardgame. Hyde's basic argument was that chess, the great systemic game of strategy and interaction, had been transmitted to both Europe and to China out of India. It had reached Europe through Persian, then Arabic, and finally Jewish mediators, notably Abraham Ibn Ezra, Ibn Yehia, Jehudah Halevi, and Maimonides. Because the Exchequer was modeled after this game of kings or *Shahs*, chess had been at the core of the success of the English fiscal state since the Norman Conquest, and Hyde's knowledge of Persian, Arabic, and Hebrew allowed him to demonstrate this.[73] For his part, Shen could show in book two of Hyde's work how Chinese chess was derivative of Indo-Persian forms and how games had been corrupted in China into an emphasis on dice ("Nerdi") and fortune rather than strategy. As Hyde suggested in book one, the beauty of chess, nobly organized according to the form (*idea*) of a battle, was that it did not have its basis in chance, luck, or fortune ("cause, sorti aut Fortune) nor a fictional astrology ("ficti, ut Planetarum et Astroum"), but ancient rather than modern truth ("veri").[74]

Translation and comparative work in archives was key to proving all of this. Hyde derived word chess from Persian "Shah" and used his copy of the Surat *Shahnama* to show that it came from India, but he also cited ancient Irish chronicles sent from Dublin by Narcissus Marsh as well as Caxton's *Game and Play of the Chesse* (1474), the second book printed in English, to show how interwoven it was with British history.[75] Chess was no mere metaphor for translation or good government but a model for both. Within chess were "compositions" or "problems" (Ar. and Per. *mansub* بمنسو), literally derivative questions of order or "predicates," and Hyde focused in particular on detailing chains of moves or turns, which implied infinite possibilities. This geometric, tabular, and strategic logic was a kind of diffused universal, varying like chessmen according to different nations and craft traditions, but nevertheless having a core "authentic" form. The reciprocal model of the game also paralleled Selden's arguments about the necessary external recognition of laws and Hyde's own translation relationship with Shen. As Hyde explained, the two principals in the Exchequer were also like players of a game, mutually ensuring that laws were followed and entries double-checked. It was no accident that Hyde dedicated the first book to Sidney Godolphin, secretary of the Treasury.

Chapter eight of the *Shahludi*, on Chinese chess (象棋 *xiangqi*), and the more extensive sections on Chinese games in the 1694 *Nerdiludi*, both composed with Shen, were designed to show that Chinese games revealed a degenerate rather than virtuous system of government. Chinese chess was not actually centered on the king but on the general of a standing army, the kind of shift to bureaucratic absolutism detested in James II. All the pieces were military and specifically designed for the Chinese system of government, very different from Indian chess although Hyde noted that it was a "model." Rather than a culture untouched by commerce and bounded by walls, Chinese chess revealed the long integration of China with Asia as a whole, staged as a battle across the dividing river or *jie he* (界河; for Hyde "Kia ho"), which Hyde thought was the Yellow River or Huang He (Hyde, "Whang Ho") because as on Chinese maps it divided the Han areas of China from Tartary, Tibet, and India. Paralleling his emphasis on making chess Indian, Hyde made a long digression about the Chinese invention of gunpowder, noting that Tavernier had instead attributed it to India and that Boyle described eight-hundred-year-old cannons in Pegu (Burma). Ultimately like chess, gunpowder was an Indian invention that the Chinese appropriated.[76]

Book two, finished after the Glorious Revolution, went further along these lines, but it too seems to have been hatched out of Hyde and Shen's discussions about chess. It concerns other eastern games grouped under the collective title of *nard* ("nerd" for Hyde), the Arabic precursor to backgammon, in addition to "Trunculorum" (*Latrunculorum* or "robbers"), alluding to dice games. These games of chance were the opposite of chess. Hyde dedicated this treatise to the Whig radical John Hampden Jr., who had been arrested with Sydney as part of the Rye House Plot and in late 1689 coined the phrase "Glorious Revolution."[77] In the aftermath of the revolution, Hampden fought the expansion of the excise, which he claimed that in tandem with "an army of 100,000 men, Englishmen will be as enslaved as Asia."[78] The sense that Shen too had suspicions of bureaucracy seems to be indicated not only by his choice of the missionary life but also the paper gameboard he brought with him across the world, an emblem of cynicism about the Chinese meritocratic bureaucracy and standing army entitled the "Promotion of Mandarins" or *Sheng guan tu* (陞官圖, lit. "promotion official diagram"). Based on dice, in this game courtly advancement derived purely from luck rather than learned achievement on the exams.[79]

Shen, who left China in his early twenties even though he knew a significant number of characters and was the son of an upwardly mobile doctor, had almost certainly failed to get a place in the bureaucracy. The game was an odd souvenir of his former life in Nanjing, usually played by young men waiting to take their exams in large almost city-like testing centers.[80]

The metaphor of Chinese chess had also been an important code among those with torn loyalties in the transition to the Manchus. The notorious loyalist-turned-collaborator Qian Qianyi (d. 1664), who had served the remants of the Ming at Nanjing in 1644–45 before surrendering, explicitly described in his poetry the war between Zheng Chenggong and the Qing as "haiyu qi" (海宇棋, "chess in sea space").[81] A game like *Sheng guan tu* had an air of cynicism about it, reconciliation with the Qing after the decisive defeat of the Yongli and Zheng concept of "restoration," mixed with a kind of coded loyalty to Ming values. Although Hyde's interest in games preceded Shen, dating back at least to his efforts to collect Persian texts from Surat in the 1670s, Shen clearly came to the efforts at translation with a set of metaphoric associations that linked the tabular sciences of cartography with the strategic and classical mappings of the game.

Hyde explained that the large and intricate tabular format of "Promotion of Mandarins" gave the impression not of a rational structure but an "uncharted forest or labyrinth" (*inexploratae Sylvae aut Labyrinthi*) from which even a masterful player could not escape (*unde Lusor tyrocinium agens se extricare nequiverit*) (figure 43). Because of this, he had the engraver add explanatory Roman numerals. He called the game a "revolving scene" (*ambientium scena*), which had "no place for skill or experience" (*nec arti aut Peritiae locus est*) and into which the subject is literally "thrown" (*jacere*).[82] A long classical tradition of cynicism (or sour grapes) about positions in Chinese bureaucracy was first expressed in relation to this game by Tang dynasty author Fang Qianli, writing in a vein similar to Hyde about how under the ancients the worthy were promoted but later the worthy demoted and the base promoted.[83] *Sheng guan tu* was a disposable game, printed on paper rather than a solid board like chess or backgammon (雙六 *shuang liu*), and clearly debased in relation to the highly esteemed and frequently depicted strategy game of Go (圍棋 *weiqi*).[84]

As Hyde printed it, the game became a damning critique of the learned meritocracy of the Chinese government emblematized in Couplet's translation and the Confucian morals advocated by Vincent as well as a spectacular emblem of the kind of translation work he had done with Shen. The player (the pieces on the upper left) advanced through the bureaucratic hierarchy of empire. Roman numerals XXXI–XXXIII, representing governorships *zhi* (知, transliterated as "chi") of the minor, major, and great administrative units, and Hyde translated as "civitatem" what were actually counties (*xian* 縣), subprefectures (*zhou* 州), and prefectures (*fu* 府). Despite its snakes-and-ladders appearance, the game was not linear. The dice could throw someone into a wide variety of positions, enabling craps-like betting on rolls, which

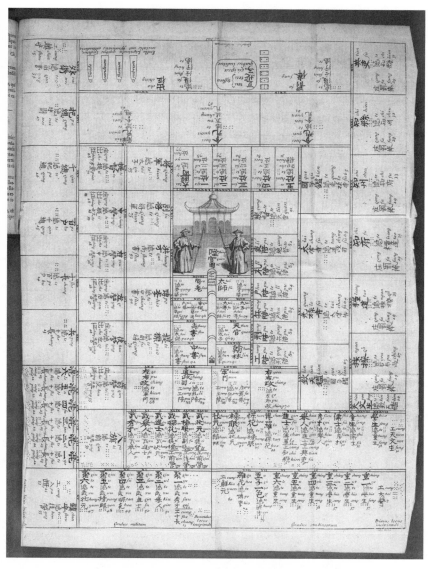

Fig. 43. Thomas Hyde and Shen Fuzong, "De Ludo Promotionis Mandarinorum," in *Historia Nerdiludii* (Oxford: Sheldon Theater, 1694), between pp. 70 and 71. This item is reproduced by permission of the Huntington Library, San Marino, CA.

was really the substance of the game. Rolling doubles—like the double fours depicted in several places on the diagram—had a particular significance in depicting virtue ("te" or *de* 德) as well as attributes like ability (*cai* 才). Hyde gave a series of instructions for dice throws, but the game itself seems unplayable from the instructions. It is unlikely that Shen and Hyde commiserated about their respectively marginalized positions over a game of promotions, but in Hyde and Shen's translation it became a figure of the shared ideal of translation—no doubt under a good king and a universal religion—and the failure of that ideal because of the corruptions of courtly life, ideology, inherited systems, and chance. This fits with the overall tone of Hyde's *Nerdiludi*, with its telling dedication to Hampden, conveying a spirit of pessimism about the events of 1688–89 in which theoretically deductive systems of authority like the university or the court devolved from an ideal of rational order into seemingly random inductively generated hierarchies.

Hyde's concept of the game as a matrix of translation and of politics—language games and game theory—is more fully developed than the striking translations of Temple, the artificial systems of Hooke, or the universal virtue of Vincent. At a basic level it was dialogic, involving at least two players and in the case of Hyde and Shen an actual cross-linguistic partnership resisting the dominant ideological trends of their larger patronage and social networks. Yet, unlike Temple's and Hooke's skepticism about translation, Hyde's approach left little room for unpredictable outcomes and was in many ways reactionary. Hyde and Shen's isolation in the archive and use of the metaphor of the game barred what Amartya Sen calls "social realizations," in which the partnership could in turn leverage other processes.[85] Hyde's game model only charted the decay into modern corruption from ancient virtue, ultimately leaving an aestheticization of everyday life in the endless repetitive loops of the game, the shared frustrations of an Oxford librarian and a young scholar convert from Nanjing.

Hyde's books were not published until after the Glorious Revolution, by which time this message of courtly corruption was largely irrelevant due to the rise of Parliament and party politics. In 1700, when he finally published his book on Persian religion, a substantial work of scholarship focusing on the Zoroastrian religion, the interest in studying Persian and Arabic in England was quickly vanishing. The book famously ruined Hyde financially. Hyde also worked on a project for printing Malay with Boyle during the years leading up to the Glorious Revolution, but the newly cast type was never used. Finally, Hyde also became involved along with the Oxford chaplain and collector John Pointer in the circus-like promotion of Prince Giolo in London, who had been brought back from the Malukus (the island of Halmahera or Giolo) by William Dampier only to be exhibited for his tattoos in London, subsequently dying of smallpox in Oxford in 1692. Hyde found a Dutch merchant familiar with Sulawesi to translate Giolo's tale and described his tattoos as both a regional map and a kind of literati document in the spirit of Temple's *sharawadgi*: "Those nations who use that Art, never permitting the inferior sort to see any regular forms in their Bodies."[86] Pointer, a collector of mostly natural curiosities, took Giolo's skin from his body before burial, and preserved it as a kind of unreadable text in his broader collection of butterflies, eggs, and shells in the Musaeum Pointerianum.[87] In theory, as Hyde suggested, this was a text that could someday be read with careful study; in reality Giolo's remains had become a perverse kind of curiosity, another deeply pessimistic emblem of seventeenth-century achievements in translation.

THE NEWTONIAN SYSTEM

The apogee and breakdown of translation in Hyde's various projects in the years surrounding the Glorious Revolution as well as the skepticism of Temple and Hooke suggested the need for a radically new "system" of exchange, one that recognized complex lines of relations rather than simply dualities, acknowledged the need for a simultaneously open-ended and systematic approach to knowledge, and finally comprehended a certain limit inherent in the potential for translation to move between the phenomena of languages. In April 1686, Vincent, who apparently by this point had given up on his Confucian project, presented the manuscript of the first volume of Newton's *Philosophiae Naturalis Principia Mathematica* to the Royal Society. He described it as a "demonstration of the Copernican hypothesis as proposed by Kepler" that "makes out all of the phaenomena of the celestial motions by the only supposition of a gravitation towards the center of the sun decreasing as the squares of the distances therefrom reciprocally."[88] Copernicanism had been forestalled in England in part because the English translator of Galileo Thomas Salusbury died of plague and most of the copies of his two volume *Mathematical Collections and Translations* (1661, 1665), containing among other English translations Galileo's *Dialogo* and *Discoursi*, burned in the Great Fire. In the aftermath of the fire, in 1667 George Sawbridge only reissued volume one, which translated the *Dialogo sopra I due massimi sistemi del mondo* (1632) or *Systema cosmicum* (1635), a work that focused more specifically on earthly motions rather than describing the Copernican system as a whole.[89] Newton's work would bring the mathematization of Copernicanism back to life, directly challenging the cosmographic synthesis that the Jesuits were putting forward in Paris, Beijing, and Ayutthaya. For Vincent, Newton doubly addressed the failure of the Galilean synthesis of Copernicanism and the Jesuit synthesis of Confucianism to translate in London as "systems." Three weeks later the Royal Society resolved to print the *Principia*.

Almost immediately there was a dispute about whose "system of the world" was being printed, a phrase taken from Galileo but used by Hooke since 1666 to describe his own theories of attraction. Halley, who offered to pay for the printing, wrote to Newton suggesting that he credit Hooke with ideas about "impulse or attraction" in the preface.[90] Newton resented this and almost withheld book three, "De mundi systemate," where he tried to show through phenomena (observations in agreement with computations) that universal gravitation governed the movement of comets. Newton suppressed his stronger view, articulated in a 1685 manuscript also entitled "The System of the World," that a global conspiracy of priests had covered up the fact that ancient Chaldean philosophers (Hyde's Zoroastrians) had known the universe to be heliocentric and largely consisting of a vacuum.[91] Newton chaffed at the provincial credit networks of the Royal Society, circles of witnesses of experiments. Especially in the case of Hooke, he thought of many of their activities as borrowing and claiming Continental ideas that had not been translated in print. He also despised the priestcraft symbolized by James II's "Catholic modernity," which had enabled absolutist formulations of divine right. He instead contended that a true universalism and correct knowledge had been obscured through such activities. A number of societies around the world had begun experiments with absolutist strategies; these strategies were killing the long-term efforts (since the sixteenth century) in London to develop complex decentralized strategies of translation. Although scholars since the nineteenth century have emphasized the hidden or private radical Protestant dimensions of Newton's thinking, which verged on Socinianism of the kind developed by the Polish Brethren, in fact, more attention should be given to the Cabot-like strategies of Newton, who described relationships in which the "center" is defined relative to a conventional framework.

The conceptualization of physics in the *Principia* begins with a localized mathematics defined in relation to absolute space and time rather than directly entering the Galilean debate about geo- or heliocentrism. Galileo's point had been about relativity—velocity is in reference to some object or reference point, a center as it were or common frame of reference (for Galileo the hold of a ship) that is recognized by two or more observers.[92] This had been an important principle for the experimentalism of Boyle and the

Royal Society. Newton began quite differently. Proposition one, that "the areas which bodies made to move in orbits described by radii drawn to an unmoving center of force lie in unmoving planes and are proportional to the times," is usually thought of as the cornerstone of the *Principia*, although it has been criticized my many historians of science in recent years as mathematically and logically flawed.[93] The formulation tried to show mathematical and physical reasons for Kepler's description of the elliptical orbit of Mars and had been suggested by Hooke. Hooke had written to Newton in November of 1679 about "compounding" the celestial motions of planets from both direct (linear) motion along a tangent to the orbit and "attractive motion towards the centrall body," what he elsewhere called a new "system of the world."[94] Computing this was possible in local terms, that is with one center of gravity and one vector force, but it became extremely challenging to consider it globally, that is with multiple centers of gravity (the "three-body" or "n-body problem"). Using triangles, both Hooke and Newton by 1685 were working on a solution to connect the area of the plane of the elliptical orbit, using the radius separating the center of gravity from the orbiting object, to the forces involved in orbital motion. But Hooke, famously credited for "boldness" and "intuition" by Koyré for this insight (acts of genius with no historical precedent), had neither the mathematical models nor the empirical data to support such a systemic approach, while Newton in his notes still struggled with this as a local problem in relation to a single "immovable center." For Newton, the only solution to these epistemological dilemmas ultimately was to develop a new mode of translating between centers through the comparative use of mathematical models employing the convention of absolute space and time to describe phenomena of motion (differential equations) and using the collection of data around the globe to derive solutions from the patterns exhibited in open-ended bodies of data.[95] This kind of intellectual solution needed to be coupled with a more democratic approach to religion and politics required for such data gathering and mutual recogmiton of common standards of space and time as previous work by Cabot, Cavendish, Selden, Ogilby, and Hyde suggested in different ways.

A key distinguishing aspect of Newton's "System of the World" was its use of data on a global scale rather than relying upon Galilean (and Royal Society) relativity. Hooke's interests were largely formal—he had little desire to engage in either large-scale acquisition of data or translating data into conclusions through calculations. In his own work, Newton was preceded by John Flamsteed, who as head of Charles II's Royal Observatory at Greenwich had since 1675 been making detailed star maps and collecting data on planets and comets in order to demonstrate Kepler's laws and the Coperni-can system.[96] Charles II had been worried that current maps were not ac-curate enough for ship navigation on a global scale. Data collection needed to be accompanied, as Selden suggested, by better techniques of measure-ment in an effort to yield greater dominion over the ocean. The project was also at least in theory comparative, and Siam must have at least to Hyde now seemed a missed opportunity. Edward Bernard combed through the Bodleian to give Flamsteed citations from Greek, Hebrew, Persian, and Ara-bic authors about observational data, and the Chinese star maps that Hyde had collected offered similar potential in the summer of 1687.[97] Newton relied on Bernard and Flamsteed's data as well as French data assembled by Giovanni Cassini at Louis XIV's Paris Observatory (est. 1671) for the calcu-lations in his "System of the World."[98] But the centralized data collection of the Royal Observatories and the Bodleian were only one element of New-ton's ability to make claims based on masses of translated data.

Newton also leveraged commercial networks, the most far flung of which was the data gathered by Francis Davenport, a mariner from Boston who ended up at the new East India Company factory at Tonkin in 1672, which had been established to help leverage the Zheng network coming out of Taiwan. He had been ordered by the Company to survey tides at Red River between May and July 1678 because of the oddity of only one flood and ebb tide per day occurring there. While such data might not compete with that of the Jesuits at Beijing, it could perhaps prove of interest for mari-time shippers, given Zheng Jing's interest in calendars, as well as for other polities like Japan or Siam relying on different but interlinked temporal and navigational data sets. Davenport's observations were confirmed in 1683 by Robert Knox and again by William Dampier in 1688, but Davenport was also seen as unreliable, especially in 1688 and 1689, when George White tried to blame him for the disastrous Siamese war. As Simon Schaffer has recently argued, Newton needed reliable global informants and information networks if he was going to show that the moon's gravitation had an effect on tides, while the Company needed such people to convince Asian polities and merchants of their commercial and technical prowess.[99]

Davenport's data was substantially correct, but an error arose from New-ton's theoretical interpretation of it. Newton tried to argue that the tide phenomenon occurred because of the interference of contrary tides coming from the South China Sea and the Indian Ocean. Newton could not measure the movements of the Indian Ocean and South China Sea to rule out his own incorrect solution nor did he understand how the more complex global motions of tidal forces play out locally in specific geographic features. Newton here faced what Francis Bacon, in a discussion about comparing tides in Peru and China, called "An Instance of the Fingerpost," erring by failing to follow Bacon's injunction

> When in the investigation of any nature the understanding is so balanced as to be uncertain to which of two or more natures the cause of the nature in question should be assigned on account of the frequent and ordinary concurrence of many natures, let us inquire whether any such motion be found in nature, or whether it be not rather a thing invented and supposed for the abbreviation and convenience of calculation.[100]

Newton recognized that phenomena like gravity in certain cases required impossible measurements of objects either too large or too distant to allow precision, and more importantly multiple centers of gravitation and vectors of force led to increasingly complex differential problems. Bernard Cohen has argued that Newton used the language of "hypothesis" extensively in the *Principia*, and it was only in the 1713 edition that there is a significant shift to the language of "phaenomena" or appearances.[101] Phenomena saved Newton from Leibnizian formulas of relationality and the need to take into account action at extreme distance as well as absent data and small errors in measurement, what the first edition of the *Principia* already called "insensibles minutias." The language of phenomena in 1687 was still largely suspicious, as when Halley in his dedicatory poem explained that because of Newton one no longer has to be terrified by the strange paths (*via flexa*) of comets, these "bearded astral phenomena" (*barbate Phaenomena Astri*). Newton himself used the concept of "Phenomena" in the 1686 preface in skeptical relation to the leaps of moderns who "rejecting substantial forms and occult qualities—have undertaken to reduce the phenomena of nature to mathematical laws."[102] But the language of phenomena also allowed Newton in this context to argue that "straight lines" and circles were not givens but "problems." (*problemata*)

Newton came to embrace phenomena as his predecessors working with Asian data sets and collections had, famously in his "hypothesis non fingo" (I frame no hypothesis) of the "General Scholium" to the second edition of the *Principia* (1713),

> I have not as yet been able to discover the reason for these properties of gravity from phenomena, and I do not feign hypotheses. For whatever is not deduced from the phenomena must be called a hypothesis; and hypotheses, whether metaphysical or physical, or based on occult quali-

ties, or mechanical, have no place in experimental philosophy. In this philosophy particular propositions are inferred from the phenomena, and afterwards rendered general by induction.[103]

It was not the experiment that needed to be witnessed but phenomena; these were then converted or translated into data framed in terms of absolute time and space. Newton's was not a formula for a center of calculation along the lines of one of the Royal Observatories that centralized all data production in one place.[104] He instead tried to decentralize as much as possible, through the process of translating phenomena, the networks of translation, media, and contracts necessary for framing knowledge. Unlike Bacon's emphasis on the singularity, Newton's method depended on data, lots of it translated around the world by a variety of people and methods. Like Selden and unlike natural law philosophers, Newton was quite ready to accept that better data might actually shift the phenomena described, thus potentially shifting the laws.[105] Thus he broke the dependent relationship between hypothesis (or Latin *suppositiones*) and phenomenon, which had been established by Ptolemy and Euclid, and in doing so opened science to universal translation rather than cultural or local practices of witnessing as practiced by Boyle and the Royal Society.[106] No phenomena were ever more than appearances, temporary clusters of observations and translations into data to be replaced when better ones came along. For Newton, there was no hypothetical center, no clearly evident natural truth, only phenomena generated by apparent conjunctions of forces that had the potential to generate data.

The last person to look at the Selden Map carefully was Edmond Halley in 1705, and he dismissed it precisely in terms of obsolete data about the pheonomenon of magnetic declination. Halley had collected data aboard the HMS *Paramore* (1698–1700) about magnetic variation and published a celebrated map "General Chart of the Variation of the Compass" (1701) of his findings in the North and South Atlantic. In November 1705, Halley would be shown the Selden Map after his appointment as Savilian Professor at Oxford (1703). He had recently published his predictions of the return of the comet that now bears his name, based on historical research of the dates of its appearance. There was historical data about declination on the Selden Map, which Hyde's notes on the Chinese compass made translatable, and presumably Bodley's librarian John Hudson thought could be of use for new editions of Halley's chart. A quick look, however, told Halley that although the map took magnetic variation into account, the variation was fixed, and "Mr. Hally having taken a view of it, concluded it to be full of faults, from some wch he knew to be so from his own observations."[107] This was not a question of dismissing knowledge gained through translation—Halley was learning Arabic in order to finish Edward Bernard's translation of Apollonius's *Conics* from manuscripts in the Bodleian and at Leiden—but rather an assessment of the value of a certain set of data by the editor of the *Principia* along the lines outlined by Newton's methods.

By taking into account radical differences in language, the mediated character of knowledge and the relational properties of systems, Temple's, Hooke's, and Hyde's theories of translation had helped lay the phenomenological ground for this shift in understanding, even if the various instantiations of Newton's *Principia* were not "caused" by these new theories of translation. The broader ground for Newton's "System of the World" came out of much longer-term and more global translation processes shaping London and the universities over the previous century. As the new Newtonian definition of phenomena suggested, these gravitic pulls were precisely the influences that were difficult to see and perhaps impossible to prove in terms of causality. In a more statist and nationalist age, David Hume would even suggest that such broad and inductive processes should be dismissed from the realm of causes, which require contiguity and conjunction.

By the age of Hume, greater efficiencies in the gathering and translation of data also helped reinforce a sense of provincial and nationalist overconfidence. Certainly some in London in 1687 felt that the sheer size of their city, the growth of their economy and trading networks, had indeed made them central. William Petty, Cromwell's great data gatherer in Ireland, argued in his fourth essay on political arithmetic "that London, for aught appears, is the greatest and most considerable city of the world, but manifestly the greatest emporium." He had already noted the year previously in his *Two Essays on Political Arithmetic*, "As for Pekin in China, we have no account fit to reason upon; nor is there anything in the description of the two late voyages of the Chinese emperor from that city into East and West Tartary, in the years 1682 and 1683, which can make us recant what we have said concerning London." Delhi, Agra, Cairo, and Istanbul were similarly dismissed. Petty did not even consider Paris a "great emporium," a status which he ascribed only to London (pop. 696,000), Amsterdam (187,090), Venice (134,000), and Rouen (66,000).[108] Although Petty's data about Asian cities and London itself was wrong, and a radical reduction of the populations reported by Cavendish, the comparative and global phenomenon of London's emporial connections had become apparent and a tool for political self-understanding.

One of Newton's most famous statements seems to suggest a profound awareness of this odd mix of confidence and provincialism that started to emerge from London's global success: "I do not know what I may appear to the world, but to myself I seem to have been only like a boy playing on the seashore, and diverting myself in now and then finding a smoother pebble or a prettier shell than ordinary whilst the great ocean of truth lay all undiscovered before me." The great strengths of the emerging British state after the Glorious Revolution—Parliamentary sovereignty, a national bank (1694), the lapsing of the Licensing Act (1695) and censorship, political union of the crowns (1707), a unified East India trade under a new company (1709), a slew of new joint-stock companies, and a fiscal-military state funded by an efficient network of excise taxes—came out of a greater sense of the nation itself as a phenomenon, an internally comprehensible, rationalized, and unified system that could be measured through data. In many ways, the new successes of London and its new institutions after 1689 hid that "ocean of truth," the complex and decentralized practices of translation that Londoners had engaged at a global level over the course of the previous two centuries to achieve success. In response to increasing global integration and complex and unmanageable relations of translation, media, and contracts in London, the Glorious Revolution reinvented translation as a national phenomenon that occurred through the sovereign authority of Parliament rather than the king, the church, or some imagined conception of divine right. Newton's occult gravitic forces, like the invisible hand of God or in Smith's version the abstract market, replaced the kinds of explicit if contrived images of global connections imagined by Ogilby for Charles II or Dapper for Johan de Witt. The political "revolution," however, like Newton's scientific one, was not simply a national or even European revolution but a global one, emerging from global forces and in turn having a global impact by constructing a particularly powerful and translatable version of scientific "modernity" based on relative certainty in the face of complex and unknowable phenomena. The Newtonian and Glorious revolutions ultimately sealed off the question of causality by relations at a distance as unknowable in favor of more provincial claims to truth, but the dual revolutions were the culmination of efforts over a century and a half to build London into a global city through the creation of practices and institutions for translation, interpretation, and measurement. This process both commenced and came to a head because of changes in the Indian Ocean and East Asia.

Asia and the Making of Modern London

A number of scholars have described the events of 1687 to 1689 in London and the British Isles as a whole as a truly revolutionary moment, the "birth of the modern world."[1] The translation of Copernicus and Galileo into the mathematical models and gravitational system of Newton known as the "Newtonian Revolution," the emergence of the political model of Parliamentary supremacy and sovereignty in the "Glorious Revolution," and the coalescing of the fiscal-military state that ultimately became in 1707 "Great Britain" all suggest a momentous turn of events. Unlike the French and American Revolutions, which have long pedigrees that are often traced back to Locke and Newton or even the Machiavellian moments of the seventeenth century, the Glorious Revolution always seems more immediate and sudden. Efforts to link the revolutions of the 1680s to the early soundings of "radical enlightenment" in Holland, the Protestant translations of Galileo, the radical stirrings of the English Interregnum, a spurt of hard work and "industriousness," urbanization, or the triumph of a particular reading of Baconian induction all certainly belong in this story. Other recent accounts describe the "great divergence" between Britain and in particular East Asia as almost a *lusus naturae*, a freak of nature that created a Leviathan through fortunate access to resources like coal, forests, cod, and land depopulated by disease.[2] The convergence and emergence of the global in London through engagement with Asian urbanism, science, and maritime trade has been the subject of this book. It was the effort in London to translate these diverse forces into a notion of "revolution" that made what followed, the "British Enlightenment" in Roy Porter's phrase, so flexible, cosmopolitan, engaging, and productive.

In the standard "Whig" version of the Glorious Revolution of 1688–89, which Child was already challenging at the end of the Exclusion Crisis in 1681, the Revolution represented a domestic and local triumph of Protestant values and Parliamentary sovereignty against Catholic-leaning absolutism—English interests as against French ones. For some revisionist scholars like J. G. A. Pocock, the concept of a commercial order played little role in the 1680s, as conceptions of the city remained rooted in the neo-Machiavellian concept of civic humanism involving the domestic virtue of the citizen soldier. Recently Steve Pincus has made a "neo-Whig" argument that the Glorious Revolution was political and revolutionary in the sense that there were multiple approaches—Whigs in the 1680s advocating a labor-centered theory of value and Tories, including the director of the East India Company, Sir Josiah Child, arguing for a land-centered one—and that both sides had to articulate their political programs more fully in "modern" and European terms in order to resist French-style Catholic and absolutist modernity. He has joined with James Robinson to argue that this shift towards party politics had the effect of pushing forward a program of economic modernization tied to the emerging institutions of the state and Parliament after the Revolution.[3] This, however, downplays the global character of what was happening in London in the sixteenth and seventeenth centuries and makes the modern into a relatively localized phenomenon.

Those who first tried to articulate a sharp break for the "modern" in seventeenth-century London frequently thought of themselves as making more universal and abstract claims, as when Newton in his prefatory remarks to the 1687 *Principia* defined modern concepts of science against ancient ones. One claim of this book has been that the "early modern" emerging in London from the 1540s to the 1680s was in fact in some ways a late early modernity, coming after the prominence of not only the Portuguese, Spanish, and Dutch but also the Ming, the Ottomans, the Safavids, and the Mughals. Sanjay Subrahmanyam proposed Amir Timur Gurkani's collecting activities at Samarqand (1370–1405) in the aftermath of the fall of the Mongol Empire as a launching point for Asian "early modernity," while at the same time warning that such a move may stretch the concept beyond its productive limits.[4] London's inordinate global role from the eighteenth and especially the nineteenth century, the period of C. A. Bayly's "birth of the modern world," as well as its successful hegemony over the Qing, the Mughals, and even the Safavids and Ottomans may simply be a function of this lateness.[5] This does not mean that historians should simply describe "multiple modernities" and cultural divergences. That nonlinear and decentralized strategy too often seems designed to avoid the problem of translation and to endlessly proliferate exceptionalisms somewhat like Salman Rushdie's tragic midnight's children or Bacon's singularities.[6] At this point, it can become tempting to take a cue from Bruno Latour and toss out the category of the modern altogether as some inevitable hybrid mixture of nature and society, accident and essence, emerging from the very limited view of present successes.

But the concept of the modern still holds wide and cross-cultural popular valence when defined in terms of political popular sovereignty, the rule of law and the freedom to make contracts, the free exchange of ideas and information, and a verifiable scientific method leading to durable claims about truth—that bundle of ideas often described as the Enlightenment. To understand the profound translatability of these practices, which to some have even appeared as "natural laws," the historical understanding of the seventeenth century and the emergence of the modern needs reframing to reveal global processes coming together to produce political, economic, and social change rather than of local ones being exported outward to exert dominance. Any definition of the "early modern" has to be polycentric—political capitals from Paris and St. Petersburg to Beijing and Edo all saw intensified efforts at translation in an effort to comprehend and pull together emerging and shifting global exchange patterns. It has been and remains precisely the subnational—the local, the regional, and especially the urban—that has been the historical site of globalization, the place where multiple global circuits intersect.[7]

While the nation-state may try to disguise the process of translation through linguistic uniformity in order to "imagine" a community and a coherent localized history, the global cities of the early modern period created cultures of interpretation and translation based on the complex techniques and instruments required by exchanges across languages, through networks of people and among political, economic, and legal systems.[8] The outlines of the paradox that emerges when trying to describe processes of mutual recognitions as a single coherent and rational enterprise could already be clearly seen in the differing positions of Selden and Grotius over the constitution of maritime law. The conceptual difficulties appeared again during the age of high imperialism in the contrast between J. R. Seeley's emphasis on the pull of social forces outside of England and John Hobson's capitalist push for markets and raw materials. In the post–World War II period, Amartya Sen's inductive ideas of the struggle to build capabilities and substantive freedoms coming out of his understanding of the trauma of the Bengal Famine of 1943 contrast sharply with the deductive strategy of exceptional intervention by the sovereign proposed by Carl Schmitt, for whom globalization was an imperial and indeed Latin function of Alexander VI's *Inter caetera divinae* and the Treaty of Tordesillas (1494).[9] To avoid the kinds

of conceptual clichés made by Grotius, Hobson, or Schmitt, clichés that
have proven profoundly damaging in the twentieth century, requires a comparative methodology of translation that often runs counter to convenient
narratives of global division into East and West, Europe and China as well as
distinct nation-states. London became a successful city not only because of
its role in Britain, Europe, or the Atlantic world but also because it engaged
in and was indeed shaped in novel ways by economies, polities, and knowledge produced in the Indian Ocean and East Asia.

The potential this approach has for demystifying historical processes can
be seen in how Sir Josiah Child, the director of the London East India Company in the late seventeenth century, understood the "Protestant ethic."
Even more than John Locke, who understood global forces from the perspective of the Carolinas and the Atlantic, or William Petty, who did so through
his surveys of Ireland, Josiah Child, after decades of reading the "India ink"
of the East India Company's correspondence, had a sense of the impact that
global forces and in particular Asia had on London. It is hard to imagine
a thesis quite as radically anti-Weberian as Child put forward in his 1681
Treatise, which tried to claim paradoxically that the "East India trade is the
most National of all Foreign Trades." He attempted to show that London
had not succeeded by exporting its own Protestant values in relation to trade
and commerce but instead that Protestant values had appeared and become
coherent because of London's linkage to Eastern Asia.

Child's reversal of causality is what remains so striking about his argument. The first Whigs like the Earl of Shaftesbury, so important for the later
Glorious Revolution, had come to believe providentially that "Trade thrives
in Protestant Countries, Therefore the Protestant Religion is the cause of
our so great increase in Trade and Navigation, and not the Trade of the East-
Indies." As with concept of the sun revolving around the earth, this was a
commonsensical approach often backed by a theologically coherent world
picture. Child, however, contended:

> First, That the great increase in Trade, is not a constant and infallible
> consequence of the Protestant Religion; because it proves not so in all
> Protestant Countreys: But what ever Nation increaseth in the East-India
> Trade, never fails proportionably to increase in other Foreign Trade and
> Navigation. Secondly, admit that our Reformation to the Protestant Reli-
> gion, were one principal cause at first of our Advance in trade and Navi-
> gation; yet now it is manifest, that the increase of our Trade and Naviga-
> tion, is a great means, under God, to secure and preserve our Protestant

Religion: Foreign Trade produceth Riches, Riches power, Power preserves our Trade and Religion, they mutually work one upon and for the preservation of each other.[10]

The Protestant religion or "spirit," as Weber would later call it, did not make trade and navigation, but instead the rise of trade and navigation, the confessional victory of Protestantism, and later the emergence of Parliamentary democracy had all been enabled because of the development of relations with Asia and in particular the trading systems of the Indian Ocean and South China Sea. Child's argument leaves no ground for distinguishing between "our" values and "their" values, a distinction that would be increasingly reasserted over the eighteenth and nineteenth centuries.

This book has shown that at crucial turning points in British history— the definition of Protestantism in the late 1540s and early 1550s, the decisive and national break with Spain in the 1580s and 1590s, the resistance to following Dutch commercial, legal, and imperial strategies from the 1620s to the 1650s, the emergence of English "absolutism" and strategies of imperial monarchy from the 1660s to 1680s, and finally the years leading up to the Glorious Revolution itself—substantial and global effort went into processes of translation that helped resolve apparently local and English disputes. Londoners translated, and they did not translate alone. Exchanges took place, translation practices developed, books were printed and transcribed, and the physical traces of these processes and sometimes the people themselves made their way back to London. And for every case where some artifact survives, hinting like the Selden Map at the complexity of the process and the wide range of people involved, hundreds of other texts and objects have been lost to memory. But there are still many other such stories in the archive. What better time to try to make visible these collective, cooperative, conflicted, and frequently violent labors to build global networks of exchange? What better time to find historical meaning in a book, a map, or even a game that was sent across the world and kept for centuries with the hope that one day it might be reactivated and translated anew? They tell a story that is after all revolutionary.

ACKNOWLEDGMENTS

In some ways this is a book about the importance of libraries and collecting in mapping the world. It came together after discovering the Selden Map while working my way through the Bodleian's Chinese collection under the guidance of David Helliwell. It took shape at the British Library with the help of Annabel Teh Gallop, Frances Wood, and Ursula Sims-Williams. It relied on the generosity of librarians and staff at the Clark, Getty, UCLA, and Huntington Libraries in Los Angeles, including a summer fellowship at the Huntington, where I have spent most of my vacation time. It gained depth through the libraries at the National Archives of Britain, the Library of Congress, the British Museum, the University of Glasgow, the University of Leiden, Trinity College Dublin, the Royal Society, the Bibliothèque nationale de France, the Archivo General de Indias in Seville, Cambridge University Library, Oxford's St. John's College, the Beinecke and Sterling at Yale, Stanford University, UC Berkeley, the ARSI in Rome, and Archbishop Marsh's Library. Librarians preserve not simply the past but the potential for translation.

I wrote the book in Savannah, Georgia, halfway between the Spanish settlement of Saint Augustine (1565) and the relocated Charleston (1680), and I have particularly benefited from a group of young scholars whom I would like to think of a kind of southern school, namely Tonio Andrade, Holly Brewer, Mi Gyung Kim, Jacob Selwood, Philip Stern, and Nick Wilding. I heartly recommend their works past and forthcoming. Their conferences hosted at Georgia State, Emory, Duke, and North Carolina State proved invaluable to shaping my work. I would also like to thank my colleagues in the History Department at Georgia Southern. In the absence of expensive databases and extensive collections, the librarians at Georgia Southern's Zach Henderson Library did heroic work getting books, microfilms, and in one case a poster-sized printout of sources in Chinese, Japanese, Sundanese, and a host of European languages in which they were not fluent. Georgia Southern supported my yearly trips to the BSECS conference in Oxford as well as research trips to Washington, D.C., Beijing, London, Leiden, Glasgow, Paris, and Dublin.

I hope the influence of a number of people from my years in California and before is apparent in this work, including Daniel Baugh, John Brewer, Haun Saussy, David Sabean, Peter Reill, Felicity Nussbaum, Benjamin Elman, Kirti Chaudhuri, Carolyn Lougee, Peter Stansky, Paula Findlen, and Tim Lenoir. The concepts informing the book came out of work at conferences and in edited volumes produced by Felicity Nussbaum (*The Global Eighteenth Century*, 2001), John Brewer and Frank Trentmann (*Consuming Cultures, Global Perspectives*, 2006), Peter Mancall and Daniela Bleichmar (*Collecting Across Cultures*, 2011), and Tonio Andrade and Xing Hang (*Sea Rovers, Silk, and Samurai*, 2014). I also owe debts of learning to Peter Borschberg, Lucille Chia, Charles Crouch, Stephen Davies, Catherine Delano-Smith, Vera Dorofeeva-Lichtmann, Go Bon Juan, Michael John Gorman, Michelle Haberland, Roger Hart, Maryanne Horowitz, Anne McLaren, Robert Minte, John Moffett, Hyunhee Park, Sumathi Ramaswami, Patricia Seed, Peter Shapinsky, Laura Shelton, Richard Smith, Sarah Tyacke, Richard Tuck, Geoff Wade, Charles Wheeler, and certainly to many friends, students, fellow conference participants, and teachers over the years, too many to mention.

Finally, my work in London would not have been possible without the generosity and companionship of Michèle Cohen, Alan Schechner and Ariel Katz, Josh Atkins and Amy Gilbert, and Brad Greve, all of whom at various points provided me a home away from home. John Brewer introduced me to London, Haun Saussy to translating the Chinese language, and Kirti Chaudhuri to the East India Company archives and the Bodleian, thus in many ways planting the seeds of this entire project. Miles Ogborn and Reed Malcom gave me early advice about publication, and Christie Henry, Abby Collier, Michael Koplow, Robert Nashak, Nick Wilding, John Wills, Jonathan Rabb, and the anonymous reviewers of the manuscript helped see it through. Robert and Shirley Batchelor, Barbara and Edward Gilbert, and Sari Gilbert and Nate and Sage Batchelor put up with and supported this all too ambitious enterprise and deserve my gratitude the most.

The Bodleian Library's collection of Chinese books acquired in the seventeenth-century is in the process of being recataloged by David Helliwell. My initial guide to these collections was the union manuscript catalog of Edward Bernard. A few remain in their original collections but most have been reclassified in the "Sinica" series. The Sinica numbers up to 63 (including reclassifications of books from the collections of Laud, Sincia 41–46, and Selden, Sinica 51–54) were donated mostly by 1661. Many (but not all) of the Sinica numbers above 65 were acquired in the Golius auction in 1696 for Archbishop Marsh, and can be seen in the auction catalog, *Catalogus insignium in omni facultate linguisque Arabica, Persica, Turcica, Chinensi &c. Librorum MSS quos doctissimus clarissimusque* (Lugduni Batavorum: Joannem du Vivé, 1696). I also made use of the Bodleian's Javanese materials mostly acquired in the 1620s (Java b.1–3, 6–7 (R)).

The India Office Records, including the bulk of the surviving papers of the East India Company, are now all in the British Library. For this study, I have consulted the "Charters" (A); "Court Books, 1599–1699" (B/1–41); "Original Correspondence, 1602–1693" (E/3/1–49); "Despatches to the East," or "Letterbooks, 1626–1691" (E/3/84–92); "Factory Records" (Bombay, G/3/1–3; China and Japan, G/12/1–4, 9–10, 13, 15–17; Madras G/19/1–6, 26–31, 36; Java G/21/1–7; G/26; G/40); and the journals of the "Marine Records" (L/MAR/A; L/MAR/B).

A number of other collections within collections were very useful, including the Cotton manuscripts (British Library); the State Papers and Colonial Office Papers (The National Archives); the Pococke manuscripts (Bodleian Library); the Bodley manuscripts (Bodleian Library); the Ashmole manuscripts (Bodleian Library); the Marsh manuscripts (Bodleian Library); the Marshal manuscripts (Bodleian Library); the Laud Collections (Bodleian Library and St. John's College, Oxford); the Bayer Collection (Hunter Manuscripts, Glasgow); the Scaliger Collection (Leiden University); the Golius Collection (Leiden University); and the Royal Society Collection (Royal Society).

In addition to leaving personal notations in many of the Bodleian's books, Thomas Hyde left a group of papers (Sloane 853, Asian) as well as books in his personal study at his death that made their way into the British Library through Sloane's original collections (see the inventory at Sloane 3323, f. 270–72). The books are listed in figure 1. Some of Hyde's manuscripts were purchased by the Bodleian in 1692, and are listed in the Bernard catalog (Bodleian MS Hyde).

John Selden's manuscripts are located in several places although most of his large library went to the Bodleian in 1659. I have consulted for this study the manuscripts in the Bodleian (Selden Supra 105, 108, 109, 111, 112, 117, 152; Arch Seld. A.1, A.2, A. 38, A.72(3), B.3); Lincoln's Inn Library (Hale MSS 11, 12, 84, 86; Maynard MSS 47, 53); the Inner Temple Library (Petyt 529); the Lambeth Palace Library (Fairhurst Papers MS 3513, 3530, 3472, 3474, 4267); and the Clark Library, Los Angeles (Selden 1).

INTRODUCTION

1. The story is told in Anthony Wood, *Life*, ed. Thomas Hearne (Oxford: Clarendon, 1772), 358–62; on Shen Fuzong, who also took the Christian name "Michael"; see Batchelor, "Shen Fuzong," October 2006, Oxford Online Dictionary of National Biography. The work done by Hyde and Shen is described in chapter 5. The records of Hyde and Shen's work are to be found in inscriptions in books in the Bodleian and British Libraries, the manuscript records of their work British Library (hereafter BL) Sloane 853, the printed versions of these in Gregory Sharpe, "Appendix de Lingua Sinensi, Aliisque Linguis Orientalibus Una Cum Quamplurimis Tabulis Aeneis, Quibus Earum Characteres Exhibentur," in *Syntagma Dissertationum Quas Olim Auctor Doctissimus Thomas Hyde* 2 (Oxford: Clarendon Press, 1767), 516–26, and the copperplate engravings that follow, and in the manuscript catalog of Edward Bernard et al., *Catalogi Librorum Manuscriptorum Anglia et Hiberniae* (Oxford: Sheldon Theater, 1697).

2. There is no good contemporary record of this, and the somewhat hyperbolic folk memory is recorded by Anthony Wood, *The History and Antiquities of the Colleges and Halls in the University of Oxford*, 2, pt. 2 (Oxford: John Gutch, 1796), 919; and Gerard Langbaine, "Preface" to John Cheeke, *The True Subject to the Rebel* [1549] (Oxford: Leonard Lichfield, 1641). The rumors of sales to shoemakers ("to scoure their Candlesticks and some to rub their boots") seem to be a misreading of Langbaine quoting John Bale about destruction of manuscripts in 1549 more generally. The neglect may have occurred under the absentee Chancellor Sir John Mason (1552–56), but it is certain that by the time of Cardinal Pole (1556–58), virtually everything was gone.

3. See Strickland Gibson, ed., *Statuta antique universitatis oxoniensis* (Oxford: Oxford University Press, 1931), 342–43; James McConica, "The Catholic Experience in Tudor Oxford," in *The Reckoned Expense*, Thomas McCoog, ed. (Woodbridge: Boy-dell Press, 1996), 42–44; Claire Cross, "Oxford and the Tudor State," in *The History of the University of Oxford III: The Collegiate University*, ed. James McConica, 3 (Oxford: Clarendon Press, 1986), 133–39; A. Vere Woodman, "The Buckinghamshire and Oxfordshire Rising of 1549," *Oxoniensia* 22 (1957), 79–82.

4. Johann Ulmer to Heinrich Bullinger, August 7, 1549, *Original Letters Relative to the English Reformation*, Hastings Robinson, ed. (Cambridge: Cambridge University Press, 1847), 391, translating the Greek original. The break was of course not so dramatic, see Alexandra Walsham, *The Reformation of the Landscape: Religion, Identity and Memory in Early Modern Britain and Ireland* (Oxford: Oxford University Press, 2011); and Eamon Duffy, *The Stripping of the Altars: Traditional Religion in England, c. 1400–1580* (New Haven: Yale University Press, 1992).

5. Bodley to Singleton, London, November 5, 1611, reproduced in Wood, *History and Antiquities*, 2:2:933.

6. The first book acquired was a popular text of selections from Confucius and Mencius, Bodleian Sinica 2. The Sandys purchase included the medical texts by Li Chan (李梃 ca. 1573–1619) *Bian zhu yixue rumen* [編註醫學入門 "Introduction to Medicine, Edited and Annotated"] (1579), Bodleian Sinica 3b, 1–2; Wang Bing (王冰, *Chong guang bu zhu huangdi neijing su wen* [重廣補注黃帝內經素問 "The Yellow Emperor's (Huangdi) Inner Canon"] (late Ming edition of 762 CE text), Bodleian Sinica 8, 1–7 (two missing volumes were acquired later as Sinica 9 and Sinica 10 through the donations of Francis Vere and Owen Wood); and the similarly recollected Gong, Xin, Gong Tingxian (1522–1619), et al., *Xinkan gujin yijian* [新刊古今醫鑑 "Newly Amended Mirror of Ancient and Modern Medicine"] (after 1589), Sinica 12/3 (with Sinica 12/1 and 12/2 acquired through donations by Charles Danvers and Francis Cleere). For a comprehensive list see David Helliwell, "Chinese Books in Europe in the Seventeenth Century," http://www.bodley.ox.ac .uk/users/djh/17thcent/17theu.htm; A. F. L. Beeston, "The Earliest Donations of Chinese Books to the Bodleian," *Bodleian Library Record* 4:6 (December 1953), 306. On Katherine Sandys, see Theodore Rabb, *Jacobean Gentleman* (Princeton: Princeton University Press, 1998), 49.

7. Bodley to Thomas James, June 24, 1607, in G. W. Wheeler, *Letters of Sir Thomas Bodley to Thomas James* (Oxford: Clarendon Press, 1926), 168. Two were donated by Matthew Chubbe, seven from Sir Francis Vere, four from Dean Owen Wood, four from Lord Lumley, two from Sir Francis Cleere, and eleven from Bodley himself.

8. "Indos, Babylonios, Aegyptios atque Gracos, Arabai et Latinos eam pertransisse iam cernimus, iam Athenas deservit, iam a Roma recessit, iam Parisius praeterivit, iam ad Britaanniam insularum insignissimam quin potius microcosmos [in Gk.], accessir feliciter, ut se Graecis et barbaris debitricem ostendat." Thomas James, ed., *Philobiblon* (Oxford: Josephus Barnesius, 1599), 38. On *translatio imperii* see Jacques Le Goff, *La civilisation de l'Occident médiéval* (Paris: Arthaud, 1964), esp. part 2, ch. 6. On James's donation see Batchelor, "Crying a Muck: Collecting, Domesticity and Anomie in Seventeenth-Century Banten and England," in *Collecting Across Cultures*, ed. Daniela Bleichmar and Peter Mancall (Philadelphia: University of Pennsylvania Press, 2011), 122–23. The texts were the Old Javanese *Rosa Carita*, Bodleian MS Java b.1(R) and the Old Sundanese *Bujangga Manik*, MS Java b.3(R). Anthony Wood records the transfer of de Bury's books to Duke Humfrey's and to George Owen, who purchased the site of Durham College. Wood, *History and Antiquities*, 2:2:911.

9. The "megacity" figure from the mayor's 2004 "London Plan," http://www.london .gov.uk/thelondonplan/thelondonplan.jsp. The 2009 revision focuses more on the metro-

politan figure of just under eight million people. "Shaping London," http://www.london
.gov.uk/shaping-london/london-plan/docs/london-plan.pdf.

10. Ferdinand Braudel, *Civilization and Capitalism: Volume III: The Perspective of
the World*, trans. Siân Reynolds (Berkeley: University of California Press, 1992), 143.

11. The most important works by Frisius in this regard are "Libellus de locorum
describendorum ratione," in Petrus Apianus (Peter Apian), *Cosmographia liber* (Antwerp:
A. Birckman, 1533), and *Arithmeticae Practicae Methodus Facilis* (Antwerp: Gregorio
Bontio, 1540), the former in at least twenty-eight and the latter in at least seventy-three
subsequent editions. See N. D. Haasbroek, *Gemma Frisius, Tycho Brahe, and Snellius and
Their Triangulations* (Delft: Meinema, 1968); and F. van Ortroy, *Bibliographie de l'Oeuvre
de Pierre Apian* (Amsterdam: Meridian, 1963). On the Mercator debates see Mark Monmonier, *Rhumb Lines and Map Wars* (Chicago: University of Chicago Press, 2004).

12. See Jeremy Boulton, "London, 1540–1700," in *The Cambridge Urban History of
Britain: 1540–1840*, Peter Clark, David Palliser, and Martin J. Daunton, eds. (Cambridge:
Cambridge University Press, 2000), 316. The population model of London's importance
was put forward by E. A. Wrigley, "A Simple Model of London's Importance in Changing
English Society and Economy, 1650–1750," *Past and Present* 37 (July 1967), 44. Wrigley
suggested that in 1550 there were only 40,000 people, but subsequent scholarship including the suburbs has proposed figures between 75,000 and 120,000. Roger Finlay and Bea-
trice Shearer, "Population Growth and Suburban Expansion," in *London 1500–1700*, A. L.
Beier and Roger Finlay, eds. (London: Longman, 1986), 37–59; J. Landers, "The Population
of London, 1550–1700: A Review of Published Evidence," *London Journal* 15:12 (1990),
111–28. Britain was the first country in the world to have a majority urban population
according to the categories of 1851 census. D. M. Palliser, ed., *The Cambridge Urban History of Britain: Volume I, 600–1540* (Cambridge: Cambridge University Press, 2000), 3.

13. Patricia Fumerton, *Unsettled: The Culture of Mobility and the Working Poor
in Early Modern England* (Chicago: University of Chicago Press, 2006); Miles Ogborn,
Spaces of Modernity: London's Geographies 1680–1780 (Guilford, 1998); Raymond Wil-
liams, *The Country and the City* (Oxford: Oxford University Press, 1973), 155.

14. Caroline Barron, *London in the Later Middle Ages* (Oxford: Oxford University
Press, 2004), 90–91, 102.

15. See Richard Helgerson, *Forms of Nationhood* (Chicago: University of Chicago
Press, 1992); Colin Kidd, *British Identities Before Nationalism* (Cambridge: Cambridge
University Press, 1999); David Armitage, *The Ideological Origins of the British Empire*
(Cambridge: Cambridge University Press, 2000); Karen Newman, *Cultural Capitals: Early
Modern London and Paris* (Princeton: Princeton University Press, 2007).

16. David Ormrod, *The Rise of Commercial Empires: England and the Netherlands*
(Cambridge: Cambridge University Press, 2003), xiii-iv, 337, 341. Key texts in this inter-
pretation are Theodore Rabb, *Enterprise and Empire* (Cambridge: Harvard University
Press, 1967); Kenneth Andrews, *Trade, Plunder and Settlement* (Cambridge: Cambridge
University Press, 1984); Rabb, *Jacobean Gentleman* (Princeton: Princeton University
Press, 1998). On mercantilism see E. F. Heckscher, *Mercantilism*, rev. ed, ed. E. F. Soder-
lund (London: Allen & Unwin, 1955), 128–36. Heckscher later argues that England was to
a great extent exceptional, cf. E. F. Heckscher, "Mercantilism" in *Revisions in Mercantil-*

ism, D. C. Coleman, ed. (London: Meuthen, 1969), 22–23, discussed in Ormrod, 17–27. For an example of the emphasis on the Atlantic in the eighteenth century see David Hancock, *Citizens of the World* (Cambridge: Cambridge University Press, 1995).

17. Brenner and Isett, "England's Divergence from China's Yangzi Delta: Property Relations, Microeconomics, and Patterns of Development," *Journal of Asian Studies* 60:2 (May 2002), 610.

18. Good examples of this vast literature can be seen in Miles Ogborn, *Global Lives: Britain and the World, 1550–1800* (Cambridge: Cambridge University Press, 2008); Alden Vaughan, *Transatlantic Encounters* (New York: Cambridge University Press, 2006); J. H. Elliot, *Empires of the Atlantic World* (New Haven: Yale University Press, 2006); Jack Greene and Philip Morgan, *Atlantic History: A Critical Appraisal* (New York: Oxford University Press, 2008); Bernard Bailyn, *Atlantic History: Concept and Contours* (Cambridge: Harvard University Press, 2005).

19. Robert Markley has identified a general and anti-ecological "fantasy of infinite productivity and profit" at work here. Markley, *The Far East and the English Imagination, 1600–1730* (Cambridge: Cambridge University Press, 2006), 5. But he rejects the idea of any mutual constitution of "England" through Asian encounter let alone "enrichment" of the "Far East" in this process.

20. On the balance of trade with Europe by the late seventeenth century, see the illustrative tables 5.1 and 5.2 in Ogborn, *Global Lives*, 113–14, both adapted from seminal articles by Ralph Davis—"English Foreign Trade, 1660–1700, *EHR* 7:2 (1954), 78–98; "English Foreign Trade, 1700–1774," *EHR* 15:2 (1962), 285–303—which were refined in subsequent works by the same author. See more broadly the new "imperial" history approach in Nicholas Canny, *The Oxford History of the British Empire: Volume I: The Origins of Empire* (Oxford: Oxford University Press, 1998). The classic account of the shift in the 1550s in terms of a search for new markets was given by F. J. Fisher, "Commercial Trends and Policy in Sixteenth Century England," *EHR* 10 (1940), 105–7; and G. D. Ramsay, *The City of London in International Politics at the Accession of Elizabeth Tudor* (Manchester: Manchester University Press, 1975). Instead of the search for markets, revisionists emphasized the formation of an emporial economy see Davis, "English Foreign Trade"; David Fischer, "Development and Organization of English Trade to Asia, 1553–1605," (University of London, PhD Thesis, 1970); Kenneth Andrews, *Trade, Plunder and Settlement* (Cambridge: Cambridge University Press, 1984); Robert Brenner, *Merchants and Revolution* (Princeton: Princeton University Press, 1993); Bruce Lenman, *England's Colonial Wars 1550–1688* (Harlow: Longman, 2001).

21. Child to the Bombay Council, 3 August, 1687, BL, India Office Records (hereafter IOR), E/3/91 f. 209. See Chaudhuri, *The Trading World of Asia*, 316–17.

22. For some broad assessments of this see David Crystal, *English as a Global Language* (Cambridge: Cambridge University Press, 2003); David Bellos, *Is that a Fish in Your Ear? Translation and the Meaning of Everything* (New York: Faber and Faber, 2011), 17–22.

23. Martin Jay, *Songs of Experience* (Berkeley: University of California Press, 2005), 216–17. Compare the opposite tendency in Husserl's comments on Galileo's appropriation of the 'ready-made' of geometry in "Der Ursprung der Geometrie als intentional-historisches Problem," in *Revue international de philosophie* 1:2 (1939).

24. R. G. Collingwood, *The Idea of History* (New York: Oxford University Press, 1956), 158; Jay, *Experience*, 216–19, 235; Bernard Stiegler, *Technics and Time* 3 (Stanford: Stanford University Press, 2011), 36–37. For new scholarship about translation see generally Haun Saussy, ed., *Comparative Literature in an Age of Globalization* (Baltimore: Johns Hopkins University Press, 2006); Saussy, "In the Workshop of Equivalences: Translation, Institutions, and Media in the Jesuit Re-formation of China," in *Great Walls of Discourse* (Cambridge: Harvard University Press, 2001), 15–16; Emily Apter, *The Translation Zone* (Princeton: Princeton University Press, 2005); Lydia Liu, ed., *Tokens of Exchange: The Problem of Translation in Global Circulations* (Durham: Duke University Press, 1999). On the importance of making translation "visible" and the resistance of England as a national space to translation see Lawrence Venuti, *The Translator's Invisibility: A History of Translation*, 2nd ed. (London: Routledge, 2008), 13.

25. For exoticism see Ros Ballaster, *Fabulous Orients: Fictions of the East in England, 1662–1785* (Oxford: Oxford University Press, 2005); Markley, *The Far East and the English Imagination*. On Chinese in particular see David Porter, *Ideographia: The Chinese Cipher in Early Modern Europe* (Stanford: Stanford University Press, 2001), which focuses largely on England; and Nicholas Dew, *Orientalism in Louis XIV's France* (Oxford: Oxford University Press, 2009). On Orientalism as simple appropriation see Robert Irwin, *For Lust of Knowing: The Orientalists and Their Enemies* (London: Allen Lane, 2006), 54–140. Between these is the concept of an "emergent readership" replacing exoticism in Victor Segalen, *Stèles*, Timothy Billings and Christopher Bush, eds. (Middletown, CT: Wesleyan University Press, 2007), 25. The best of this literature describes travel writing as a kind of transcultural experience that was transformative for European culture. See in particular Joan-Pau Rubiés, *Travel and Ethnology in the Renaissance: South India through European Eyes* (Cambridge: Cambridge University Press, 2000); and Mary Pratt, *Imperial Eyes: Travel Writing and Transculturaltion* (New York: Routledge, 1992).

26. The *Bujangga Manik* is Bodleian MS Jav.b.3. It was donated in 1627 along with *Rosa Crita (Carita)* (MSS Jav.b.1), a cosmological manuscript mostly in Javanese but with some Sundanese. See J. Noorduyn in *Three Old Sundanese Poems*, ed. J. Noorduyn and A. Teeuw (Leiden: KITLV, 2006), 241–76; "Bujangga Manik's Journeys Through Java: Topographical Data from and Old Sundanese Source," *Bijdragen tot de taal-, land- en volkenkunde* 138 (1982), 413–42; "The Three Palm-Leaf MSS from Java in the Bodleian Library and Their Donors," *Journal of the Royal Asiatic Society of Great Britain and Ireland* (1985), 58–64; and Batchelor, "Crying a Muck: Collecting, Domesticity, and Anomie in Seventeenth-Century Banten and England," in *Collecting Across Cultures*, 122–23. On the gender tropes see Aihwa Ong and Michael Peletz, eds., *Bewitching Women, Pious Men: Gender and Body Politics in Southeast Asia* (Berkeley: University of California Press, 1995).

27. On the cosmopolis as a unique space of translation produced under the rubric of scribal traditions especially in South India and Southeast Asia see Sheldon Pollock, *The Language of the Gods in the World of Men: Sanskrit, Culture, and Power in Premodern India* (Berkeley: University of California Press, 2006); and Ronit Ricci, *Islam Translated: Literature, Conversion, and the Arabic Cosmopolis of South and Southeast Asia* (Chicago: University of Chicago Press, 2011), 22, 31–65; as well as Kojin Karatani's notion of an 'intercrossing space' in *Architecture as Metaphor: Language, Number, Money*, trans.

Sabu Kosho, ed. Michael Speaks (Cambridge: MIT Press, 1995), xxvi; and Denys Lombard, *Le Carrefour javanais: Essai d'histoire globale* (Paris: L'École des Hautes Études en Sciences Sociales, 1990).

28. What might be loosely called a Calcutta school of historians (despite residence elsewhere)—Bose, Chaudhuri, and Ashin Das Gupta—have all emphasized shifting sets and topologies, comparisons and interactions as a methodology more appropriate than globalism to the Indian Ocean. See Kirti Chaudhuri, *Trade and Civilisation in the Indian Ocean* (Cambridge: Cambridge University Press, 1985); Ashin Das Gupta and N. M. Pearson, eds., *India and the Indian Ocean 1500–1800* (Calcutta: Oxford University Press, 1987); Kenneth McPherson, *The Indian Ocean* (New Delhi: Oxford University Press, 1993); and Sugata Bose, *A Hundred Horizons* (Cambridge: Harvard University Press, 2006), 5. For East Asia, see also the developing critique of the simplified notion of the Chinese tribute system by John Wills, introduction to *China and Maritime Europe, 1500–1800* (Cambridge: Cambridge University Press, 2011), 3–11; Leonard Blussé, *Visible Cities: Canton, Nagasaki, and Batavia* (Cambridge: Harvard University Press, 2008); Victor Lieberman, *Strange Parallels: Southeast Asia Global Context, c. 800–1830* (Cambridge: Cambridge University Press, 2003, 2009).

29. Janet Abu-Lughod, *Before European Hegemony: The World System A.D. 1250–1350* (New York: Oxford University Press, 1989), 116–17, 220–21; Li Guo, *Commerce, Culture, and Community in a Red Sea Port* (Leiden: Brill, 2004); D. Panzak, "Le contrat d'affrètement maritime en Méditerranée: droit maritime et pratique commercial entre Islam et Chrétienté," *Journal of Economic and Social History of the Orient* 45:3 (2002), 342–62; and M. Çizakça, *A Comparative Evolution of Business Partnerships: Islamic World and Europe* (Leiden: Brill, 1996).

30. See James Scott, *The Art of Not Being Governed: An Anarchist History of Upland Southeast Asia* (New Haven: Yale University Press, 2009), xiv. On diasporas along these lines see Engseng Ho, *The Graves of Tarim: Genealogy and Mobility across the Indian Ocean* (Berkeley: University of California Press, 2006); and Francesca Trivellato, *The Familiarity of Strangers: The Sephardic Diaspora, Livorno, and Cross-Cultural Trade in the Early Modern Period* (New Haven: Yale University Press, 2009); Sanjay Subrahmanyam, ed. *Merchant Networks in the Early Modern World* (Aldershot: Variorum, 1996). On brokers and cross-cultural trade see Kapil Raj, *Relocating Modern Science: Circulation and the Construction of Knowledge in South Asia and Europe* (Basingstoke: Palgrave Macmillan, 2007); Alida Metcalf, *Go-Betweens and the Colonization of Brazil* (Austin: University of Texas Press, 2005); Claude Markovits, *The Global World of Indian Merchants, 1750–1947: Traders of Sind from Bukhara to Panama* (Cambridge: Cambridge University Press, 2000). On the figure of the "outsider" see Subrahmanyam, *Three Ways to Be Alien* (Boston: Brandeis University Press, 2011). See also Richard White's emphasis on "middle grounds" in *The Middle Ground: Indians, Empires, and Republics in the Great Lakes Region, 1650–1815* (Cambridge: Cambridge University Press, 1991). For a parallel network and broker theory of translation see Anthony Pym, "Alternatives to Borders in Translation Theory," (1993) in *Translation Translation*, Susan Petrilli, ed. (New York: Rodopi, 2003), 451–63.

31. An argument put forward in Chaudhuri, *The Trading World of Asia and the En-*

glish East India Company (Cambridge: Cambridge University Press, 1978), 135–45, which critiques here the landmark work of J. C. van Leur from the 1930s. Van Leur argued that Asian trading ports were overburdened with a multiplicity of currency, weights, measures, and customs and at the same time information poor because of the extreme localization of markets; see J. C. van Leur, *Indonesian Trade and Society* (The Hague: W. van Hoeve, 1955), 132–35, 201–7, 214. Van Leur's arguments proved influential for Niels Steensgaard, *Carracks, Caravans, and Companies* (Copenhagen: Studentlitteratur, 1973); and Clifford Geertz, "The Bazaar Economy: Information and Search in Peasant Marketing," *American Economic Review* (1978), 29–30.

32. On house societies see Claude Lévi-Strauss, *The Way of the Masks*, trans. Sylvia Modelski (London: Cape, 1983), 184; Lévi-Strauss, *Anthropology and Myth* (Oxford: Blackwell, 1987), 153–55; Charles Macdonald, ed., *De la hutte au palais: sociétés "à maisons" en Asie du sud-est insulare* (Paris: CNRS, 1987). David Sabean uses the phrase "ideologies of the house," emphasizing how the focus on the house idealizes certain models of masculinity. Sabean, *Property, Production, and Family in Neckarhausen, 1700–1870* (Cambridge: Cambridge University Press, 1990), 88–116, 249, 429. For houses more generally as sites of collection see the work of Craig Clunas on the Ming, *Superfluous Things: Material Culture and Social Status in Early Modern China* (Urbana: University of Illinois Press, 1991); and *Fruitful Sites: Garden Culture in Ming Dynasty China* (Chicago: University of Chicago Press, 2004); and on the Dutch, Simon Schama, *The Embarrassment of Riches* (Berkeley: University of California Press, 1988); and Anne Goldgar, *Tulipmania* (Chicago: University of Chicago Press, 2007).

33. *Bujangga Manik*, Bodleian Library, Oriental, Palm Leaf MS Jav.b.3, lns. 287–90, trans. J. Noorduyn, *Three Old Sundanese Poems*, 247.

34. The inspiration here is the comparative essay work of Qian Zhongshu. See Ronald Egan, trans., *Limited Views: Essays on Ideas and Letters* (Cambridge: Harvard University Press, 1998) as well as Qian's "China in the English Literature of the Seventeenth Century," *Quarterly Bulletin of Chinese Bibliography* 1 (1940); and *Xie zai renshengbian shang* [写在人生边上 "Written in the Margins of Life"] (Shanghai: Kaiming shudian, 1946). The idea of "contrapuntal reading" is in Edward Said, *Culture and Imperialism* (New York: Knopf, 1993), 51.

35. Sociology of the book was influential in this regard. See for example Adrian Johns, *The Nature of the Book* (Chicago: University of Chicago Press, 1998); Miles Ogborn, *Indian Ink: Script and Print in the Making of the English East India Company* (Chicago: University of Chicago Press, 2007); and Simon Schaffer, "The Asiatic Enlightenments of British Astronomy," in Simon Schaffer et al., *The Brokered World: Go-Betweens and Global Intelligence, 1770–1820* (Sagamore Beach: Science History Publications, 2009), 49–104.

36. Bodleian Ashmole 1787 (the calendar is now Sinica 88). The bound miscellany included fragments of Cheng Dawei's *Suanfa Tongzong* [算法统宗 "General Source of Computational Methods"] (1592), the Hirado factor Richard Cocks's list of the revenue of the Japanese nobility (ca. 1614), a page from a Chinese medical text, and several silhouettes and paper samples from Europe. According to Yang Yongzhi, *Ming Qing shi qi Tainan chu ban shi* (Taipei: Taiwan xue sheng shu ju, 2007), 16, a 1667 and 1683 copy

also survive in the Kanda Kiichiro collection; see David Helliwell, "Southern Ming Calendars," *Serica*, January 2, 2012, http://oldchinesebooks.wordpress.com/2012/01/02 /southern-ming-calendars/, accessed January 27, 2012.

37. "The Books and Papers of the Late Dr Hyde," BL, Sloane MSS, 3323, f. 270–72

38. The best introduction to the concept of *tu* is Francesca Bray, introduction to *Graphics and Text in the Production of Technical Knowledge in China*, ed. Francesca Bray, Vera Dorofeeva-Lichtmann, and Georges Métailié (Leiden: Brill, 2007), 1–5 and passim; Vera Dorofeeva-Lichtmann, "The Political Concept behind an Interplay of Spatial 'Positions'" *Extrême-Orient, Extrême-Occident* 18 (1996), 9–33; and Michael Lackner, "Die Verplanung des Denkens am Beispiel der *tu*," in H. Schmidt-Glintzer, ed., *Lebenswelt and Weltanschauung in frühneuzeitlichen China* (Stuttgart: Franz Steiner Verlag, 1990), 134–56.

39. Hyde had a Fujianese almanac (Print 15298 a.32; formerly Royal 16 B X), which he took notes in and about (Sloane 853, f. 1), as well as the extensively annotated Yongli Calendar (Print 15298 a.30, formerly Royal 16 B.XI). Sloane 853, f. 23 has translations of the Selden Map, while f. 37 appears to copy Bodleian Sinica 92.

40. Benjamin Elman, *On Their Own Terms: Science in China, 1550–1900* (Cambridge: Harvard University Press, 2005); R. Bin Wong, *China Transformed: Historical Change and the Limits of European Experience* (Ithaca: Cornell University Press, 1997); Geoffrey Lloyd and Nathan Sivin, *The Way and the Word: Science and Medicine in Early China and Greece* (New Haven: Yale University Press, 2002). Because of its significance, I composed a memorandum about the map, which Helliwell sent to Elman, Timothy Brook, and Haun Saussy in February 2008; this resulted in securing funding to restore the map.

41. The work of and discussions with Saussy and Chaudhuri in different ways helped me rethink this. See in particular Haun Saussy, "*Impressions de Chine*, Or How to Translate from a Nonexistent Original," in Victor Segalen, *Stèles*, ed. Timothy Billings and Christopher Bush (Middletown: Wesleyan University Press, 2007); Kirti Chaudhuri, *Trade and Civilisation in the Indian Ocean* (Cambridge: Cambridge University Press, 1985). On entanglement see Nicholas Thomas, *Entangled Objects: Exchange, Material Culture and Colonialism in the Pacific* (Cambridge: Harvard University Press, 1991).

42. G. R. Elton, *The Tudor Revolution in Government* (Cambridge: Cambridge University Press, 1962); Douglas North and Robert Thomas, *The Rise of the Western World* (Cambridge: Cambridge University Press, 1973), 153–56.

CHAPTER ONE

1. The funding structure of the Bristol voyages remains unclear, but recent archival discoveries suggest Italian-style partnerships with nominal royal support. Some financial backing came from the Augustine Friars in London (and their "Lombard Hall"), which historically and especially through the Milanese envoy and mathematician Friar Giovanni Carbonaro had connections with the Lombard merchant-bankers in London. See Evan Jones, "Alwyn Ruddock: John Cabot and the Discovery of America," *Historical Research* 81 (2008), 231–36; Richard Hakluyt, *Principall Navigations, Voiages and Discoveries of the English Nation* (London: George Bishop and Ralph Newberie, 1589), 517;

Jones, "Henry VII and the Bristol Expeditions to North America: The Condon Documents," *Historical Research* 83:221 (August 2010). On Bristol see David Sacks, *The Widening Gate: Bristol and the Atlantic Economy* (Berkeley: University of California Press, 1993).

2. The second John Cabot voyage in 1498, which possibly made it to Coquibaçoa on the Venezuelan coast, definitively confirmed that Cabot had not found Cathay. See James Williamson, *The Cabot Voyages and Bristol Discovery* (Cambridge: Hakluyt Society, 1962), 107–12, 233; Jones, "Ruddock," 244–45.

3. Richard Pynson's *The Boke of John Maundvyle* (London: 1496) survives in a single copy, BL G.6713. William Caxton (d. 1491) obtained the St. Albans copy of Mandeville manuscript in 1490 (BL MS Egerton 1982) presumably with the intention of producing a printed edition in imitation of the 1481 Augsburg one by Anton Sorg. Wynkyn de Worde published Sorg as *A lytell Treatise or Booke, named John Mandevyl, Knyht, borne in Englande* (Westminster: 1499, 1503). Sebastian Cabot himself only cites Polo on the map.

4. Donald Ostrowski, *Muscovy and the Mongols: Cross-Cultural Influences* (Cambridge: Cambridge University Press, 2002), 37–40; and William Hung, "The Transmission of the Book Known as The Secret History of the Mongols," *Harvard Journal of Asiatic Studies* 14 (1951), 433–92. On both the map and a pamphlet made to accompany it Cabot refers to "Tartaroru" as the realm of the "magnu Can." Sebastian Cabot, *Declaratio chartae novae navigatoiae domini almirantis* (Antwerp?: ca. 1544–49), C, Huntington 5680.

5. The classic account of Columbus's sources is Valerie Flint, *The Imaginative Landscape of Christopher Columbus* (Princeton: Princeton University Press, 1992). There is a sense that the distinctiveness of the joint-stock corporation came out of its continuing legal definition along neo-medieval lines. See generally Philip Stern, *The Company State* (Oxford: Oxford University Press, 2011).

6. "What ties the ship to the wharf is a rope, and the rope consists of fibres, but it does not get its strength from any fibre which runs through it from one end to the other, but from the fact that there is a vast number of fibres overlapping." Ludwig Wittgenstein, "The Brown Book," *The Blue and Brown Books* (New York: Harper and Row, 1960), 87.

7. General Archive of Simancas, Libro de Camara, 1546–48, f. 122–23; Cristobal Péres Pastor, "Sebastian Caboto en 1533 y 1548," *Boletin Real Academia Historia — Madrid* 22 (April 1893), 350–51.

8. Richard Hakluyt, *Principall Navigations* (London: 1589), 519–20; John Dasent, *Acts of the Privy Council of England* (London: Stationery Office, 1890–91) 2:137, 320, 374; 3:55. For the response to the emperor see Privy Council to Sir Philip Hoby, Greenwich, April 21, 1550, BL Harleian MSS 523, f. 6–7.

9. For the ambiguities over Cabot's birth in either Bristol or Venice see Richard Eden, *The Decades of the new worlde or west India* (London: William Powell, 1555), 255; and George Best, *A True Discourse of the late voyages of discoverie for the finding of a passage to Cathaya, by the Northeast* (London: Henry Bynnyman, 1578), 1. b.16. On his efforts to claim Venetian birth and citizenship from ca. 1522 and then again in the 1550s see Dispatch from the Council of Ten to Gasparo Contarini, Venice, September 27, 1522, State Archives, Venice, Capi del Consiglio dei X, Lettere Sottoscritte, Filza N. 5, 1522; Dispatch to Contarini, April 28, 1523, Marciana Library, Venice, It. Cl. VII., Cod., MIX. Cart. 294; Peter Vannes to the Privy Council, September 12, 1551, Venice; Dispatch from

the Council of Ten to Giacomo Soranzo, Sept 12, 1551, State Archives Venice, Consiglio dei Dieci Parte Secrete, Filza N. 8, 1551–54. John Cabot, born in Genoa, was naturalized as a Venetian citizen in 1476. "Letters of Naturalization," March 28, 1476, State Archives, Venice, Senato Terra, 1473–77, vii. 109; Henry Harrisse, *Jean et Sébastien Cabot leur origine et leurs voyages* (Paris: E. Leroux, 1882), 387–89. For Ramusio's efforts on Cabot's behalf see Ramusio, "Discorso Sopra Varii Viaggi Per Liquali Sono State Condotte Et Si Potrian Condurre Le Spetierie," *Navigationi et Viagi* 1 (Venice: Giunti, 1550), 401–2. My thanks to Nick Wilding for transcribing references from the Venetian archives.

10. According to Adams, the company was supposed to have 240 investors in the joint-stock, at £25 each for a total of £6000, and money came in from 201 of them. Royal patent, February 6 (26), 1555, the National Archives (hereafter TNA) Patent Rolls, C 66/883/31–32; "The Charter of the Marchants of Russia, graunted upon the discoverie of the saide Countrey, by King Philip and Queene Mary," February 6, 1555, in Hakluyt, *Principall Navigations* (1589), 304–9; and generally T. S. Willan, *The Early History of the Russia Company* (Manchester: Manchester University Press, 1956).

11. Clement Adams's report is in Hakluyt, *Principall Navigations* (1589), 282.

12. John Stow gives a good if late account of these in *The Survay of London*, 3rd ed. (London: George Purslowe, 1618), 161.

13. Cf. Pierre Garcie, *The Rutter of ye Sea*, trans. Robert Copland (London: Richard Bankes, 1528); *Booke of the Sea Carte*, n.d., BL Add 37,024; David Waters, *The Art of Navigation in England in Elizabethan and Early Stuart Times* (London: Hollis and Carter, 1958), 78–81; N. A. M. Rodger, *The Safeguard of the Sea* (New York: W. W. Norton, 1997), 156–57, 164.

14. Paul Griffiths, *Lost Londons* (Cambridge: Cambridge University Press, 2008) 11; A. L. Beier, "Foucault Redux? The Roles of Humanism, Protestantism, and an Urban Elite in Creating the London Bridewell, 1500–1560," *Criminal Justice History* 17 (2002), 33–60.

15. "Society" is Clement Adams's term ("Every man willing to be of the society . . ."). On the "society" see Frederic Maitland, introduction to Otto Gierke, *Political Theories of the Middle Age* (Cambridge: Cambridge University Press, 1900), xxiii; George Unwin, *The Gilds and Companies of London* (London: Meuthen, 1908).

16. Hakluyt, "The new Navigation and discovery of the kingdome of Moscovia, by the Northeast, in the yeere 1553: Enterprised by Sir Hugh Willoughbie knight, and perfourmed by Richard Chanceler, Pilot maior of the voyage,," *Principall Navigations* (1589), 280.

17. Williamson, *The Cabot Voyages*, 49–53, 204–5.

18. Hakluyt, *Principall Navigations* (1589), 264–65.

19. Such "Ordinances" were often issued at the Casa de la Contratación; see for example those issued to Cabot for his 1526 voyage in José Medina, *El Veneciano Sebastián Caboto al Servico de España* 2 (Santiago de Chile: Universidad de Chile, 1908), 29–40.

20. Hakluyt, *Principall Navigations* (1589), 259–63; TNA State Papers (hereafter SP) 12/196, 50–59.

21. Magna Carta (9 Henry III c. 9) gave citizens and freemen of London as well as their corporate organizations the ability to hold property in mortmain (i.e. inalienably with license from the king). The claim to mortmain by "guilds, fraternities, communalities, companies, or brotherhoods," especially charitable religious houses, had been the

target of the Act against Superstitious Uses, 23 Henry VIII, c. 10 (1531), which exempted cities and towns corporate. 37 Henry VIII c. 14 (1545), designed to help pay for wars in France and Scotland, extended such confiscations and called into question the mortmain rights of London corporations. The Act for the Dissolution of Charities, 1 Edward VI, c. 14 S.R. 24, IX (1547), exempted "corporations, guilds, fraternities, companies, and fellowships of mysteries and crafts" not dedicated to "superstitious uses" from such confiscations.

22. See the classic formulation in Stephen Greenblatt, *Renaissance Self-Fashioning* (Chicago: University of Chicago Press, 1980).

23. "che venghono cum questa mercantia da luntani paesi ad casa sua altre caravane, le quale ancora dicono che ad loro sono portate da altre remote regioni. Et fa questo argumento che se li orientali affermanno ali meridionali che queste cose venghono lontano da loro, et così da mano in mano presupposta la rotundita della terra, è necessario che li ultimi le tolliano al septentrione verso l'occidente" and "in Londres magior fondaco de speciarie che sia in Alexandria." Abbé Raimondo de Soncino to the Duke of Milan, December 18, 1497, quoted in *Annuario Scientifico ed Industriale, anno second 1865* (Milan: Biblioteca Utlie, 1866), 701.

24. Williamson, *Cabot Voyages*, 175–89; E. M. Carus-Wilson, *The Overseas Trade of Bristol in the Later Middle Ages* (Bristol: Bristol Record Society, 1937), 157–58, 161–65, 218–89. On Columbus, Henry VII, and Bristol, see David B. Quinn, "Columbus and the North: England, Iceland, and Ireland," *William and Mary Quarterly* 49:2 (April 1992), 278–97.

25. The short 1505 document, literally an "authority" gave the "governor of the merchants adventurers" in Calais the authority by privy seal "to call courts and congregations of the said merchants within the city of London" January 24, 1505, TNA C/66/598/17. A second document by privy seal gave them rights to trade in Holland, Zealand, Brabant, and Flanders under the authority of Castile, license dated June 14, 1507, TNA Patent Rolls, C 66/603/19. Douglas Bisson, *The Merchant Adventurers of England: The Company and the Crown, 1474–1564* (Newark: University of Delaware Press, 1993); A. Sutton, "The Merchant Adventurers of England: Their Origins and the Mercers' Company of London," *Historical Research* 75 (2000), 25–46; E. M. Carus-Wilson, *Medieval Merchant Venturers* (London: Methuen, 1954), 115, 173–74; William Scott, *The Constitution and Finance of English, Scottish, and Irish Joint-Stock Companies to 1720*, 1 (Cambridge: Cambridge University Press, 1912), 11.

26. Caroline Barron, *London in the Later Middle Ages* (Oxford: Oxford University Press, 2004), 107, 109, 117.

27. In 1555, Eden (*Decades*, 277–89) translated an account of Russia by Jovius, *Libellus de legatione Basilii magni Principis Moschoviae ad Clementem VII* (Rome: 1525), and gleaned from Demetrius, the papal envoy to Moscovia in 1525.

28. Eden, *Decades*, 252r.

29. Only a transcript of the charter survives, Guildhall Library Ms 11894, as the company records burned in the Great Fire. The court minutes thus start in 1666/7, Guildhall Library MS 11741/1.

30. Philip and Mary to Ivan, April 28, 1557, TNA 20/60, 1–2. Four years later Elizabeth sent a letter requesting permission for Anthony Jenkinson to trade in Russia.

Elizabeth to Ivan, April 25, 1561, TNA 22/60, 3. Elizabeth continued corresponding with Ivan, settling trade issues through letters and ambassadors.

31. On the charter as a model see Willan, *Russia Company*, 8–9.

32. The Vulgate reads, "Porro ego audivi de te, quod possis obscura interpretari, et ligata dissolvere: si ergo vales scripturam legere, et interpretationem ejus indicare mihi, purpura vestieris, et torquem auream circa collum tuum habebis, et tertius in regno meo princeps eris."

33. Stephen Alford, *The Early Elizabethan Polity: William Cecil and the British Succession Crisis, 1558–1569* (Cambridge: Cambridge University Press, 1998), 15–18; G. Lloyd Jones, *The Discovery of Hebrew in Tudor England: A Third Language* (Manchester: Manchester University Press, 1983), 101.

34. *A Fruteful and pleasaunt worke of the beste state of a publyque weale, and of the newe yle called Utopia: written in Latine by Syr Thomas More knyght, and translated into Englyshe by Ralphe Robynson Citizein and Goldsmythe of London at the procurement, and earnest request of George Tadlowe Citezein and Haberdassher of the same Citie* (London: Abraham Nell, 1551).

35. Susan Brigden, *London and the Reformation* (Oxford: Clarendon Press, 1989), 6–10, 26.

36. Thomas More, *Dialogue of Comfort against Tribulation* in William Rastell, ed., *The Workes of Sir Thomas More, Knyght* (London: John Cawod, et. al., 1557) sec. II. On More and Gardiner see Quentin Skinner, *The Foundations of Modern Political Thought* (Cambridge: Cambridge University Press, 1978), 90–98.

37. Stephen Gardiner, *Letters of Stephen Gardiner*, ed. James Muller (Westport: Greenwood Press, 1970), 369–74.

38. See Michael Riordan and Alec Ryrie, "Stephen Gardiner and the Making of a Protestant Villain," *Sixteenth Century Journal* 34:4 (2003), 1039–63.

39. Cf. 2 Kings 17:18–23. On this theme see John Day, "A Table of the principall matters conteyned in the Byble, in which the readers maye fynde and practice many commune places," *The Byble, that is to say all the holy scripture* (London: 1549), BB3–6; Alford, *Kingship*, 33.

40. On this notion of interdependence rather than mutual exclusivity of norms as key to cross-cultural trade see Avner Grief, *Institutions and the Path to the Modern Economy: Lessons from Medieval Trade* (Cambridge: Cambridge University Press, 2006); and Francesca Trivellato, *The Familiarity of Strangers: The Sephardic Diaspora, Livorno, and Cross-Cultural Trade in the Early Modern Period* (New Haven: Yale University Press, 2009). The conceptualization of the firm here comes from Oliver Williamson, *Economic Institutions of Capitalism* (New York: Free Press, 1985), 32.

41. Maitland, introduction to Otto von Gierke, *Political Theories*, xviii; Gierke, *Genossenschaftsrecht*, III, 279; Janet Coleman, *Ancient and Medieval Memories: Studies in the Reconstruction of the Past* (Cambridge: Cambridge University Press, 1992), 501–37. On Hobbes's subsequent attempt to extend the *persona ficta* concept to the state see Mark Neocleous, *Imagining the State* (Philadelphia: Open University Press, 2003), 72–97.

42. C. W. R. D. Moseley, "Introduction," *The Travels of Sir John Mandeville* (London: Penguin Books, 1983), 10; and Moseley, "The Metamorphoses of Sir John Mandeville," *Yearbook of English Studies* 4 (1974). Stephen Greenblatt argues "the fictive nature of the

performing self" in Mandeville would not have been recognized at the time, but he seems unaware of the notion of the *persona ficta*. Greenblatt, *Marvelous Possessions* (Chicago: University of Chicago Press, 1991), 165.

43. For an example of this distinction see "Discourse of Corporations" (ca. 1587–89): "it is the site and place where every town or city built which is the chief cause of the flourishing of the same or else some special trade, and not the incorporation thereof." *Tudor Economic Documents* 3, R. H. Tawney and Eileen Power, eds. (London: Longmans, 1924), 273–77.

44. M. C. Seyour, ed., *The Defective Version of Mandeville's Travels* (Oxford: Oxford University Press, 2002), 89, 91, using Queen's College Oxford MS 383. The "defective" version translated into English an earlier Anglo-Norman text and was the basis for the 1496 printed edition.

45. Bodleian Library MS e Musaeo 160, f. 113v. The dualist notion of two universals goes back to Augustine, *De Civitate Dei*, XV, 1, referring to Cain and Abel as well as 1 Corinthians 15:46.

46. Cf. Karl Burmeister, *Sebastian Münster: Versuch eines biographischen Gesamtbildes* (Basel: Helbig & Lichtenhahn, 1963), 120–21; Matthew McLean, *The Cosmographia of Sebastian Münster* (Aldershot: Ashgate, 2007), 170–74.

47. See especially *Cosmographia* (1550), 56–73. Jakob Fugger defended Münster in an open letter dated May 8, 1542; see Burmeister, *Münster*, 170–80.

48. July 26, 1550, *Calendar of Patent Rolls, 1549–1551* (London: Stationery Office, 1929), 314.

49. Sebastian Münster, *Cosmographia* (Basel: Petri, 1552), 1083.

50. "Preface to the Reader" in the 1555 *Decades*, d.iii (f).

51. The Portuguese had first gone to Japan in 1543 from islands off Ningbo in Zhejiang Province that they called Liampo. Cabot, making his map in 1544, would obviously not have known this.

52. Bibliothèque nationale de France (hereafter BNF), cartes et plans, rès ge. AA 582. The pamphlet published in Latin and Spanish is Cabot, *Declaratio chartae novae navigatoiae*, which in the "Epilogus" names Cabot as the "auctor huius chartae" and dates his system to 1544 (Cii, in Spanish Fii). Nathan Chytraeus thought they were important enough to reprint the Latin legends from the London map in his *Variorum in Europa itinerum deliciæ* (Oxford: 1594), which differ somewhat from the "Tabula Prima" and "Tabula Secunda" that surround the BNF map. On July 9, 1548, Cabot was granted six months leave "to go to Germany," when Charles V was residing at his palace in Brussels. For the relevant documents see José Medina, *El Veneciano Sebastián Caboto* 1, 394–98.

53. Humphrey Gilbert, *A discourse of a Discoverie for a new Passage to Cataia* (London: Henry Middleton, 1576), l. sing. Diii (written before 1566); Richard Willes, *The History of Travayle in the West and East Indies* (London: Richard Jugge, 1577), 232; Samuel Purchas, *Purchas his Pilgrimes* 3 (London: William Stansby, 1625), 807. Willes writes of a manuscript map by Cabot at Chenies, Buckinghamshire, the house of Francis Russell, Second Earl of Bedford, on which Cabot's "owne discourse of navigation you may reade in his carde drawen with his owne hande" (235). The manuscript map given to Edward VI is last recorded ("Sebastian Gabots maps") in an inventory of twenty-eight manuscript and forty printed maps at Whitehall, all of which burned in the fires of 1691 and 1698.

Royal MSS Appendix 86, f. 94–96. See R. A. Skelton, "The Royal Map Collection of England," *Imago Mundi* 13 (1956), 181. The "Mappa Mundi cortado por el equinocio," sent by Cabot to Charles V from London in November 1553 is now lost as well. Cabot to Charles V, London, November 15, 1554 [*sic*: 1553], *Coleccion de Documentos Inéditos para la Historia de España* 3 (Madrid: 1843), 512–14; and Harrisse, *Cabot*, 283, citing a memorial from Juan Bautista Gesio (Madrid: Real Academia de la Historia, 1843), September 20, 1575.

54. Hakluyt's, "Discourse of Westerne Planting" (1584) notes that "the day of the moneth is also added in his owne ['Gabota'] mappe which is yn the Queenes privie gallorie at Westminster, the copye whereof was sette oute by Master Clemente Adams and is in many marchants houses in London." David Quinn and Allison Quinn, eds., *A Particuler discourse . . . known as Discourse of Western Planting* (London: Hakluyt Society, 1993), 95; and Hakluyt, *Principall Navigations* 3 (London: 1599–1600), 6.

55. E. G. R. Taylor, *Tudor Geography* (London: Meuthen, 1930).

56. See also Richard Pace, *De Fructu quae ex doctrina percipitur liber* (Basel: Froben, 1517), who contrasted the ease of getting to Utopia with the difficulty of actual geography and maps [ed. Frank Manley and Richard Sylvester (New York: Ungar, 1967), 108–9.

57. David Quinn, *Sebastian Cabot and Bristol Exploration* (Bristol: Historical Association, 1968), 25–26.

58. On Dieppe mapmakers see Gayle K. Brunelle, "Images of Empire: Francis I and His Cartographers," in *Princes and Princely Culture*, E. Gosman, ed. (Leiden: Brill, 2003), 81–102.

59. On a copied and translated section of a Javanese map (lost in a December 1511 shipwreck) sent by Afonso de Albuquerque to King Manuel from Maluku on April 1, 1512 see Armando Cortesão and A. Teixeira da Mota, *Portugaliae monumenta cartographia* 1 (Lisbon: Nacional Casa da Moeda, 1960), 79–80; J. H. F. Sollewijn Gelpke, "Afonso de Albuquerque's Pre-Portuguese 'Javanese' Map, Partially Reconstructed from Francisco Rodrigues' Book," *Vijdragen tot de Taal-, Land- en Volkenkunde* 151:1 (1995), 76–99.

60. "a navegaçam dos chines e gores, com suas lynhas ey caminhos deretos por omde as naos hiam, e ho sertam quaes reynos comsynavam huns cos outros." Afonso de Albuquerque to King Manuel, April 1, 1512, *Cartas de Affonso de Albuquerque, seguidas de documentos que as elucidam* 1 (Lisbon: Academia Real das Sciencias, 1884), 64–65.

61. "ver verdadeiramente os chins donde vem e os gores, e as vossas nãos ho caminho que am de fazer pêra as ilhas do cravo, e as minas do ouro omde sam, e a ilha de jaoa e de bamdam, de noz nozcada e maças, e a leira deirrey de syam, e asy ho cabo da terra da navegaçam dos chins, e asy para omde volve, e como daly a diamte nam navegam." Albuquerque, *Cartas*, 64–65.

62. Clusters of dots like this begin to appear on Portuguese portolans from the 1530s.

63. Wang Dayuan (汪大淵), *Daoyi zhilue* [島夷誌略, "Summary Record of the Foreign Islanders,"] (Quanzhou: 1349), Su Jiqing, ed. (Beijing: Zhonghua shuju, 1981).

64. On Cabot's transition to Spanish service see Antonio Barrera-Osorio, *Experiencing Nature: The Spanish American Empire and the Early Scientific Revolution* (Austin: University of Texas Press, 2006), 40. Conflicting accounts appear in Venetian Ambassador Gasparo Contarini, Dispatch for the Court of Spain to the Venetian Senate, Valladolid, December 31, 1522, Marciana Library, Venice, it. Cl. VII, Cod. MIX, Cart. 281–83, who

reports that Cabot was also negotiating with Cardinal Wolsey for the command of a fleet, and Richard Eden, *A treatyse of the newe India* (London: Edward Sutton, 1553), 1 sig. aa. iiii, who claimed that Cabot commanded an abortive voyage with Thomas Perte in 1517.

65. Barrera-Osorio, *Experiencing Nature*, 35–37; José Pulido Rubio, *El piloto mayor de la Casa de Contratación* (Seville: Escuela de Estudios Hispano-Americanos de Sevilla, 1950). On Portugal see John Law, "On the Methods of Long Distance Control, Vessels, Navigation and the Portuguese Route to India," in Law, *Power, Action and Belief* (London: Routledge, 1986), 234–63; G. Beaujouan, "Science livresque et art nautique au XV siècle," in *Les Aspects internationaux de la découverte océanique*, M. Mollat du Jourdin and P. Adam, eds. (Paris: École Practique des Hautes Études, 1966), 13–14.

66. Alison Sandman, "Spanish Nautical Cartography in the Renaissance," in David Woodward, ed., *Cartography in the European Renaissance* (Chicago: University of Chicago Press, 2007), 1095–1142; Sandman, "Mirroring the World: Sea charts, Navigation and Territorial Claims in Sixteenth-Century Spain," in. *Merchants and Marvels*, Pamela Smith and Paula Findlen, eds. (New York: Routledge, 2001), 83–108; Ricardo Padrón, "Sea of Denial: The Early Modern Spanish Invention of the Pacific Rim," *Hispanic Review* 77:1 (February 2008), 1–27.

67. Martín Fernández de Enciso, *Suma de Geographia que trata de todas las partidas y provincias del mundo* (Sevilla: 1519, 1530, 1549).

68. Ricardo Padrón, *The Spacious Word: Cartography, Literature, and Empire in Early Modern Spain* (Chicago: University of Chicago Press, 2004), 84–91. Two subsequent translations of Enciso in the 1530s and 1540s for Henry VIII did not make any significant impact on the English court. See Jean Maillard, "Le premier livre de la Cosmographie en rethorique Francoyse contenant la description des ports et Isles de la mer," British Library MSS Royal 20 B XII, f. 4r; and Roger Barlow, BL Royal 18.B.28, f. 1. Robert Wyer's 1535 *Mappa Mundi . . . the Compasse, and Cyrcuet of the Worlde* copied the commonplace book of the Antwerp merchant Richard Arnold that had first been published in 1503. See also "Cosmographia" (1530, composed ca. 1510), Royal 13 E VII; and another copy in Cotton Nero E. IV.

69. On the Ribeiros see Cortesão and da Mota, *Portugaliae monumenta cartographia* 1, 87–106, and plates 37–40; Surekha Davies, "The Navigational Iconography of Diogo Ribeiro's 1529 Planisphere," *Imago Mundi* 55:1 (2003), 103–12.

70. Richard Hakluyt, ed., *Divers voyages touching the discoverie of America* (London: Thomas Woodcocke, 1582), B3. See also the MSS copies, the oldest dating from 1539 at Hatfield House MSS, no. 29; "The booke made by the worshipfull Master Robert Thorne in Anno 1527," BL, Lansdowne MSS Codex 100/7, f 65–80; and John Dee's damaged copy from 1577, BL, Cotton MSS, Vitellius E VII, used by Hakluyt.

71. See Metcalf, *Go-Betweens and the Colonization of Brazil, 1500–1600*, 76–77.

72. João de Lisboa, "Tratado da Agulha de Marear" (1514), *Livro de Marinharia*, Jacinto de Brito Rebello, ed. (Lisbon: Libanio da Silva, 1903), 18–24; A. R. T. Jonkers, "Parallel Meridians: Diffusion and Change in Early-Modern Ocean Reckoning" in *Noord-Zuid in Oostindisch perspectief*, J. Parmentier, ed. (The Hague: Walburg, 2005), 17–42, esp. table 1.

73. David Woodward, "Roger Bacon's Terrestrial Coordinate System," *Annals of the Association of American Geographers* 80:1 (March 1990), 109–22; O. A. W. Dikce, "The Culmination of Greek Cartography in Ptolemy," *The History of Cartography* 1:177–200.

74. Medina, *Arte de navegar* (Valladolid: Francisco Fernandez de Cordua, 1545). See David Turnbull, "Cartography and Science in Early Modern Europe: Mapping the Construction of Knowledge Spaces," *Imago Mundi* 48 (1996), 5–24; Ursula Lamb, "Science by Litigation: A Cosmographic Feud," *Terrae Incognitae* 1 (1969), 40–57; Lamb, "The Spanish Cosmographic Juntas of the Sixteenth Century," *Terrae Incognitae* 6 (1974), 51–64.

75. In 1542, a pilot from Dieppe, Jean Rotz, presented Henry VIII with a book of charts and a navigational treatise. Jean Rotz, "Boke of Idrography," BL Royal MSS 20.E.IX. On John Rut's 1527 voyage out of Plymouth see Irene Wright, *Spanish Documents Concerning English Voyages to the Caribbean, 1527–1568* (Utrecht: 1929), 29.

76. The *Barbara* captured a Spanish ship in Caribbean along with its charts, and a merchant Grene (who died) "had very excellent goodly carde" of "all the empirours Indions, so alonge the Newe founde lande, with dyvers and other straunge places," bought by William Hare of Berkyng for nine shillings. R. G. Marsden, ed. "The Voyage of the Barbara to Brazil," *The Naval Miscellany* 2 (1912), 61–62.

77. See Eric Ash, *Power, Knowledge, and Expertise in Elizabethan England* (Baltimore: Johns Hopkins University Press, 2004), 91; Alison Sandman and Ash, "Trading Expertise: Sebastian Cabot between Spain and England," *Renaissance Quarterly* 57:3 (Autumn 2004), 813–46.

78. Hakluyt, "The voyage of M. Roger Bodenham with the great Barke Aucher to Candia and Chio, in the yeere 1550," *Principall Navigations* II:1 (1599–1600), 99–101; Jehan Scheyfve to Mary, Queen Dowager of Hungary, June 24, 1550, and the same from 1551 in Royall Tyler, ed., *Calendar of State Papers, Spanish* 10 (London: Public Record Office, 1914), 115, 214.

79. On maps making local and heterogeneous knowledge mobile and commensurable see David Turnbull, *Masons, Tricksters, and Cartographers* (London: Taylor and Francis, 2000); Turnbull, "Rendering Turbulence Orderly," *Social Studies of Science* 25 (1995), 9–33.

80. Adams, "The newe Navigation and discovery of the kingdome of Moscovia," ca. 1554 in Hakluyt, *Principall Navigations* (1589), 280. See also the account by Thomas Edge, "A briefe Discoverie of the Northerne Discoveries of Seas, Coasts and Countries," in Samuel Purchas, *Hakluytus Posthumus, or Purchas his Pilgrimes* 3:3 (London: H. Fetherston, 1625), 462.

81. Kenneth Arrow, "Toward a Theory of Price adjustment," in Moses Abramovitz et al., eds., *The Allocation of Resources* (Stanford: Stanford University Press, 1959), 47, cited in Williamson, *Economic Institutions*, 9. Dennis Flynn and Arturo Giráldez describe the silver cycle as more generally a condition of economic disequilibrium; see "Cycles of Silver: Global Economic Unity through the Mid-Eighteenth Century," *Journal of World History* 13.2 (2002), 394–95.

82. J. D. Gould, *The Great Debasement* (Oxford: Oxford University Press, 1970), 118–26.

83. Correspondence of William Dansell, April 20, April 25, May 17, and June 9, 1549, in Robert Lemon, ed., *Calendar of State Papers Foreign, Edward, Mary and Elizabeth* (London: Stationery Office, 1856)

84. Brenner, *Merchants and Revolution*, 6–9.

85. Some have attributed the fact that it was not published to 1581 to John Hales or

Thomas Gresham. Cf. Raymond de Roover, *Gresham on Foreign Exchange* (Cambridge: Harvard University Press, 1949), 297–98. On Smith's authorship see Mary Dewar, "The Memorandum 'For the Understanding of the Exchange': Its Authorship and Dating," *EHR* 2 ser. 18 (April 1965), 476–87; Daniel Fusfeld, "On the Authorship and Dating of 'For the Understanding of the Exchange,'" *EHR* 20 (April 1967) 145–50; and Gould, *Great Debasement*, 161–64.

86. Smith, *A Discourse of the Commonweal*, ed. Elizabeth Lamond (Cambridge: Cambridge University Press, 1893), 16–17, 44, 104–5. On this as a source for ideologies of "British Emprie" over Scotland and Ireland see Armitage, *Ideological Origins*, 47–51.

87. W. K. Jordan, ed., *The Chronicle and Political Papers of King Edward VI* (Ithaca: Cornell University Press, 1966) 129, 167.

88. Thomas Gresham, April 12, April 16, 1553, *Calendar of State Papers Foreign, Edward*, 273; R. H. Tawney and E. Power, *Tudor Economic Documents* 2 (1924), 146–49, 153.

89. Jack A. Goldstone, "Urbanization and Inflation: Lessons from the English Price Revolution of the Sixteenth and Seventeenth Centuries," *American Journal of Sociology* 89 (1984), 1122–60; Goldstone, "Monetary versus Velocity Interpretations of the 'Price Revolution': A Comment," *Journal of Economic History* 51 (March 1991), 176–81; Goldstone, *Revolution and Rebellion in the Early Modern World* (Berkeley: University of California Press, 1991). The population theory was first espoused by Y. S. Brenner, "The Inflation of Prices in Early Sixteenth Century England," *EHR* 14 (1961), 225–39.

90. Classically Earl Hamilton, *American Treasure and the Price Revolution in Spain 1501–1650* (Cambridge: Harvard University Press, 1934). The thesis was adopted by Ferdinand Braudel and Annales school, Braudel and F. Spooner, "Prices in Europe from 1450 to 1750," in *Cambridge Economic History of Europe* 4, E. E. Rich, ed. (Cambridge: Cambridge University Press, 1967), 374–486.

91. Peter Spufford, *Money and its Use in Medieval Europe* (Cambridge: Cambridge University Press, 1988), 319–38; John Munro, "The Monetary Origins of the 'Price Revolution': South German Silver Mining, Merchant-Banking, and Venetian Commerce, 1470–1540," in Dennis Flynn, Arturo Giráldez, and Richard von Glahn, eds., *Global Connections and Monetary History, 1470–1800* (Aldershot and Brookfield, VT: Ashgate Publishing, 2003), 1–34 and table 2-3.

92. Richard Ehrenberg, *Capital and Finance in the Age of the Renaissance: A Study of the Fuggers and their Connections* (Fairfield, NJ: Augustus Kelly, 1985), 74; Peter Spufford, *From Antwerp to London: The Decline of Financial Centers in Europe* (Wassenaar: Netherlands Institute for Advanced Study, 2005), 16; Herman van der Wee, *The Growth of the Antwerp Market* 2 (The Hague: Martinus Nijhoff, 1963), 109–200.

93. Carlo Cipolla, *Conquistadores, piratas, mercaderes: La saga de la plata española* (Buenos Aires: Fondo de Cultura Económica, 1999), 57.

94. Richard von Glahn, *Fountain of Fortune* (Berkeley: University of California Press, 1996), 83–104, 114; Takeshi Hamashita, "The Tribute Trade System and Modern Asia," *Memoirs of the Toyo Bunko* 46 (1988), 17.

95. On the rise of the *wokou* from the same period see Peter Shapinsky, "With the Sea as Their Domain: Pirates and Maritime Lordship in Medieval Japan," in *Seascapes, Littoral Cultures and Trans-Oceanic Exchanges*, Jerry Bentley, et. al. eds. (Honolulu: Uni-

versity of Hawaii Press, 2007), 221–38; Takeo Tanaka, *Wako* (Tokyo: Kyoikusha Rekishi Shinsho, 1994); C. R. Boxer, *Fidalgos in the Far East, 1550–1770* (Oxford: Oxford University Press, 1968). On Japanese silver mining see Nagahara Keiji and Kozo Yamamura, "Shaping the Process of Unification," *Journal of Japanese Studies* 14 (1988), 77–109; and von Glahn, *Fountain of Fortune*, 88–97, 128–33.

96. See Flynn and Giráldez, "Cycles of Silver," 395; the more subtle distinctions in this article seem to go back to Flynn's earlier balance of payments approach, "A New Perspective on the Spanish Price Revolution: The Monetary Approach to the Balance of Payments," *Explorations in Economic History* 15 (1978), 388–406. For the data on bi-metallic ratios see von Glahn, *Fountains of Fortune*, 57, 127; William Atwell, "International Bullion Flows and the Chinese Economy circa 1530–1650," *Past and Present* 95 (1982), 82.

97. Sanjay Subrahmanyam, *The Portuguese Empire in Asia* (London: Longmans, 1993); C. R. Boxer, *The Portuguese Seaborne Empire* (New York: Knopf, 1969).

98. Bailey Diffie and George Winius, *Foundations of the Portuguese Empire, 1415–1580* (Minneapolis: University of Minnesota Press, 1977), 411–15.

CHAPTER TWO

1. C. R. Boxer, "Three Historians of Portuguese Asia (Barros, Couto and Bocarro)," *Instituto Portugués de Hongkong* 1 (1948), 19–20; Donald Lach and Edwin J. Van Kley, *Asia in the Making of Europe: Volume III, A Century of Advance* (Chicago: University of Chicago Press, 1993), 738. See Fernão Lopes de Castanheda, *História do descobrimento e conquista da India pelos Portugueses* (Coimbra: João de Barreyra and J. Alvarez, 1551–1561), the first book was translated into English by N.L. as *The first booke of the historie of the discouerie and conquest of the East Indias* (London: Thomas East, 1582); and João de Barros, *Décadas da Asia* 1–3 (Lisbon: Germão Galherde, 1552–53; João de Barreira, 1563), the first two of which were published in an Italian-language edition in Venice in 1562.

2. On the dynamicism of the Ming see Timothy Brook, *The Confusions of Pleasure* (Berkeley: University of California Press, 1998); and *The Troubled Empire* (Cambridge: Harvard University Press, 2010).

3. For the classical roots of Chinese efforts "to convey political ideas in terms of conceptualizing space," see Vera Dorofeeva-Lichtmann, "Political Concept behind an Interplay of Spatial 'Positions,'" *Extrême-Orient, Extrême-Occident* 18 (1996), 9–33.

4. Richard Hakluyt, *Principall Navigations* (1589), 815.

5. On the joint-stock as common identity see Stephen Greenblatt, *Shakespearean Negotiations* (Berkeley: University of California Press, 1988), 12. On print and nation see Richard Helgerson, *Forms of Nationhood: The Elizabethan Writing of England* (Chicago: University of Chicago Press, 1992), who frames his account (176) with Rabb, *Enterprise and Empire*, and Andrews, *Trade, Plunder, and Settlement*, to define a developing national Elizabethan dynamic between moderate gentry and merchants.

6. See especially Deborah Harkness, *The Jewel House* (New Haven: Yale University Press, 2007), who emphasizes the increasing diffusion of scientific and technical effort across the entire city.

7. Emily Bartels, *Speaking of the Moor* (Philadelphia: University of Pennsylvania

Press, 2008), 100–117; and more generally Jacob Selwood, *Diversity and Difference in Early Modern London* (Burlington: Ashgate, 2010), 1–17.

8. On the limited concept of a British empire in this period see Armitage, *Ideological Origins*, 105–8, and in particular John Dee's "Brytanici Imperii Limites" (1576), BL Add MSS 59681. Dee was also committed to linking this to joint-stock enterprises with Cathay through cartography, something he shared with Martin Frobisher and Michael Lok. Cf. for example BL Cotton Augustus I.1.1; D. B. Quinn, "Simao Fernandes, a Portuguese Pilot in the English Service ca. 1573–1588," *Congreso Internacional de historia de los descobrimentos, Actas III* (1961), 449–64. See also the Dee polar projection in the Free Public Library of Philadelphia, ca. 1582, made for Humphrey Gilbert, which clearly labels Cathaia as the destination. E. G. R. Taylor, "John Dee and the Map of North-East Asia," *Imago Mundi* 12 (1955), 103–6.

9. *Calendar of State Papers, Spanish* 4 (London: Stationery Office, 1899), 491–92.

10. The report printed in Hakluyt was entitled "The prosperous voyage of the worshipful Thomas Candish of Trimley in the County of Suffolk, Esquire, into the South Sea, and from thence round about the circumference of the whole earth, begun in the year of our Lord 1586, and finished 1588."

11. See generally Peter Mancall, *Hakluyt's Promise* (New Haven: Yale University Press, 2007); Anthony Payne, *Richard Hakluyt* (London: Quaritch, 2008). Hakluyt published the second full edition of Pedro Mártir de Anglería's eight "decades" in Paris in 1587, *De orbe novo Petri Martyris Anglerii decades octo, labore et industria Richardi Hakluyti* (Paris: G. Avray, 1587); the first complete edition was (Alcalá de Henares: Michael Eguia, 1530).

12. See David Armitage, "Literature and Empire," in William Roger Louis et al., eds., *The Oxford History of the British Empire: The Origins of Empire* (Oxford: Oxford University Press, 1998), 115. Richard Barbour elaborated on the expression of this problem as an Elizabethan theatrical trope in terms of the opposition between "domestic representations" and "foreign negotiations." See *Before Orientalism: London's Theater of the East* (Cambridge: Cambridge University Press, 2003), 6.

13. "Certaine notes or references taken out of the large Mappe of China, brought home by Master Thomas Candish 1588," Richard Hakluyt, *Principall Navigations* (1589), 813–15. The phrase "large" suggests that this may have been a large printed sheet map or a manuscript map, possibly derived from a variety of sources. A number of maps derived from Luo's in the late sixteenth century, including Wang Fengyu and Wang Zuozhou's *Guang yu kao* [廣輿考] (1594). Often manuscript copies were labeled "Da Ming" ("Great" or "Large" Ming) as *Da Ming guang yu kao* (1610) BL Or 13160. Population figures and administrative divisions were often printed at the bottom of sheet maps.

14. Hakluyt, *Principall Navigations* (1589), 814.

15. Luo Hongxian, *Guang yutu quanshu* [廣輿圖全書, a facsimile reprint of the 1579 edition] (Beijing: Guo ji wen hua chu ban gong si, 1997), 4r–5r. Luo's data were also reprinted in Zhang Tianfu, *Huang yu kao* (皇輿考), author's preface 1557, 2nd edition with postface 1588, which was a supplement to Supplement to Li Xian et al, *Da Ming yi tong zhi* [大明一统志 "Comprehensive Gazetteer of the Ming Empire"] (1461). Zhang also gives the population of Bei Zhili as 3,413,254. The household and people formula *hukou* (戶口) derives from the *Ming shilu* [明實錄 "Veritable Records of the Ming Dynasty," com-

piled up to the seventeenth century], the census compiled from 1381 and annually from 1391 (although only updated on a decade basis). During the Jiajing reign (1522–66) this was only done decade by decade, and data on individual provinces were published only in the initial editions of 1381 and 1391. Data for the individual provinces was included in Li Dongyang, *Da Ming huidian* [大明會典 "Collected Statues of the Ming Empire"] (1502) for 1491 but (to remain with the Beijing example) it lists 394,500 households and thus cannot be the source of the map's data. Mendoza gives for "Paguia" (Beijing) 47 cities, 150 towns, 2,740,000 paying tribute, 2,155,000 footmen, and 400,000 horsemen, figures for which I have found no corresponding source but might correspond to the mid-fifteenth-century data. See Otto van der Sprenkel, "Population Statistics of Ming China," *Bulletin of the School of Oriental and African Studies* 15:2 (1953), 289–326.

16. The second edition of the OED gives the first use of the word "datum" in English in a non-mathematical sense as 1646, "from all this heap of data . . . it would not follow that it was necessary," Henry Hammond to Francis Cheynell, November 4, 1646, in Hammond, *The Workes* 1 (London: Elizabeth Flesher, 1674), 248. The English translation of Euclid's *Elements* by Henry Billingsley (later Lord Mayor of London) appeared in London in 1570 with a preface by John Dee, but Dee did not translate the concept of data even though Billingsley's Latin edition of Euclid's works contained that text. *Euclidis Megarensis mathematici clarissimi Elementorum geometricorum libri XV . . . His adiecta sunt Phaenomena, Catoptrica & Optica, deinde Protheoria Marini, & Data* (Basil: Joannem Heruagium and Bernhard Brand, 1558). See R. C. Archibald, "The First Translation of Euclid's Elements into English and Its Source," *American Mathematical Monthly* 57:7 (August–September 1950), 447. On the concept of data see Luciano Fiordi, "Philosophical Conceptions of Information," in *Formal Theories of Information: From Shannon to Semantic Information Theory*, ed. Giovanni Sommaruga (Berlin: Springer, 2009), 17–18.

17. The use of precise latitudes here—16° for Hai Pho (海浦 actual 15°53′) and 14° degrees for Qui Nhon (actually 13°46′)—rules out the possibility that this was based on maps derived from Ortelius and Ricci. The Portuguese traded and established a mission at Hoi An ("Faifo") from 1535 and worked with the large Japanese and Chinese trading populations after Nguyen Hoang came down from the north to rule in 1558. See Chingho Chen, *Historical Notes on Hôi-An (Faifo)* (Carbondale: Center for Vietnamese Studies, 1974), 12–16; Anthony Reed, *Southeast Asia in the Age of Commerce* 2 (New Haven: Yale University Press, 1993), 19.

18. Wolfe reprinted with slight modification the 1586 Venetian edition in Italian by Francesco Avanzi sometime in 1587. In April 1587, the printer Thomas East also published a translation of the account of Antonio de Espejo that had been attached to the 1586 Madrid edition of Mendoza's account of China, which detailed the 1583 expedition's interactions with the Zuni, Navajo, and Tano in the "15 provinces" of New Mexico. Mendoza, *New Mexico. Otherwise, The voiage of Anthony of Espeio who in the yeare 1583. with his company, discouered a lande of 15. Prouinces . . . Translated out of the Spanish copie printed first at Madrid, 1586, and afterward at Paris, in the same yeare*, trans. Thomas East (London: Thomas Cadman, 1587).

19. Robert Parke, "To the Right Worshipfull and Famous Gentleman M. Thomas Candish," in Juan González de Mendoza, *The historie of the great and mighty kingdom of China and the situation thereof* (London: J. Wolfe for Edward White, 1589), ¶2f–3r.

20. Hakluyt, *Principall Navigations* 3 (1600), 817. The text mistakenly named the pilots as "Thomas de Ersola" (confusion with the captain Tomas de Alzola) and "Nicholas Roderigo."

21. William Barlow, *The Navigators Supply* (London: G. Bishop, R. Newberry, and R. Barker, 1597), A4.

22. Barlow, "Epistle Dedicatory to . . . Robert, Earl of Essex," *The Navigators Supply*, B2, and J4.

23. John Dee requested on May 15, 1580, that Arthur Pett and Charles Jackman "come by some charts or maps of the country, made and printed in Cathay or China; and by some of these books likewise, for Language &c." in "Mr Dee's book for the Cathay Voyage anno 1580," BL Lansdowne 122 (Burghley Collection), f. 30r. See also BL Cotton Otho. VIII, f. 78–79; and Hakluyt, "Instructions for the North-East Passage by Richard Hakluyt, Lawyer, 1580," *Principall Navigations* (1589), 492–93. The version by Hakluyt's cousin (the lawyer) only refers to "Cambalu or Quinsay, to bring thence the mappe of that country," suggesting that unlike Dee, he still understood the voyage purely in terms of Cathay. These kinds of efforts were relatively widespread in Europe in the 1570s. See for example the account of making the Chinese map for the Guadaroba Nova in the Medici Palazzo Vecchio (1575–76) in Francesca Fiorani, *The Marvel of Maps* (New Haven: Yale University Press, 2005), 115–26.

24. Richard Willes, *The History of Travayle in the West and East Indies* (London: Richarde Jugge, 1577), Aiir. My thanks to Peter Mancall for calling my attention to this text, and see his *Travel Narratives from the Age of Discovery* (Oxford: Oxford University Press, 2006), 26–27, 156. Compare John Dee's effort to define terms using a series of French, Italian, and Latin texts in "Of the famous and rich discoveries of the Great Asia's southerly and easterly coasts" (February–May 1577), Cotton Vitellius C VII, f. 220–21.

25. Sheltco à Geveren, *Of the ende of this worlde, and second coming of Christ: a comfortable and most necessarie discourse, for these miserable and daungerous days,* Thomas Rodgers, trans. (London: Thomas Dawson, 1577). It was a popular seller; additional English editions appeared in 1578, 1580, 1582, 1583, and finally 1589.

26. Willes, *History,* 236. The reference to Gemma Frisius is presumably to his "Charta Cosmographica" in *Cosmographia Petri Apiani, per Gemmam Frisium apud Louanianses Medicum & Mathematicum insignem* (Cologne: 1574). Frisius (d. 1555) worked at Louvain and taught Mercator, but his maps for Apian show a much shorter northwest passage.

27. Willes dedicated the book to Elizabeth Morison, the second daughter of Bridget Hussey by her first marriage to Sir Richard Morison, a Protestant firebrand who died in exile under Mary after having been sent by Edward VI's Privy Council as ambassador to Charles V. During 1575, she had learned both cosmography and Italian with Willes. Pereira's account had been published in Michele Tramezzino, ed., *Diversi avisi dell'Indie di Portogallo* (Venice: 1565). The conventions for translating the geographic names "China" and "Giapan" seem to have been established by Ramusio in his third edition of *Delle Navigationi et Viaggi* 1 (Venice: Giunti, 1563); cf. "Sommario di Tutti Li Regni, Città et popoli oriental, con li traffichi et mercantile, che sui si trovano, cominciando dal mar Rosso fino alli popoli della China" (324); and "Informatione dell'isola nova-

mente scoperta nella parte di settentrione chiamata giapan" (377). Both were Portuguese accounts. "La Relatione dell' isola Giapan" is also advertised on the title page.

28. See Eden, *Decades*, 231, 317. Willes reprinted both passages.

29. Willes, *History*, 237. The translation mostly follows an abridgement of Pereira's narrative written after his escape from prison in Fujian in 1553, which was appended to a collection of Jesuit letters, *Nuovi Avisi Delle Indie di Portogallo . . . et tradotti dalla lingua Spagnola nella italiana* (Venice: Michele Tramezzino, 1565).

30. On the subsequent life of the association between the Chinese and cruelty see Jonathan Spence, *The Chan's Great Continent* (New York: Norton, 1998), 20–31; Timothy Brook, Jérome Bourgon, and Gregory Blue, *Death by a Thousand Cuts* (Cambridge: Harvard University Press, 2008); and Eric Hayot, *The Hypothetical Mandarin* (Oxford: Oxford University Press, 2009), 15n17.

31. The continuing dynamicism of both Wu as well as Fujianese Min in relation to translating languages around East Asia can be seen in the Japanese-Wu translations in Hou Jigao "Jihpen feng tu ji" [日本風土記 "Record of Japanese Customs"] in *Quan zhe bing zhi kao* [全浙兵制考 "The Entire Military System of Zhejiang Examined"] (16th century) 5 (Jinan: Qi lu shu she chu ban she, 1997).

32. Willes called Maffei his "old acquaynted friend" (253). Willes may have used Frois's original Italian manuscript because of the way he spells "Giapan." The initial letters from Japan were translated into Latin *Epistolae Japanicae* (Louvain: Rutgerum Velpium, 1569, 2nd ed. 1571) and *Epistolae Indicae et Japanicae* (Louvain: J. de Witte, 1570) while Willes studied at Louvain. The text also follows the copy in Maffei, *Rerum a Societate Jesu in oriente gestarum* (Dillingen: Sebaldum Mayer, 1571; Paris: 1572; Naples: 1573; Cologne: 1574). For the letter by da Costa criticizing Maffei's translation of his work see Josef Wicki, ed. *Alessandro Valignano: Historia del principio y progresso de la Compañía de Jesús en las Indias Orientales (1542–64)* (Rome: Institutum historicum S.J., 1944), 486–89.

33. Willes, *History*, 253.

34. Zhu Wan, "Shihuo zhi" [食貨志 "Treatise on Food and Money"] in *Piyu zaji* [甓餘雜集], *Siku quanshu cunmu congshu* 78:4 (Jinan: Qilu shushe chuban, 1997).

35. The "Biography of Zhu Wan" in the *Mingshi* accuses Zhu of "taking the law too literally," a classic critique from the Wang Yangming school. Zhang Tingyu et. al., *Mingshi* (Beijing: Zhonghua shuju, 1974), 5403–5, 5424. See also C. R. Boxer, *South China in the Sixteenth Century* (London: Hakluyt Society, 1953), xxviii-xxxi; Tien'tse Chang, *Sino-Portuguese Trade from 1514 to 1644* (Leiden: Brill, 1933), 82–84.

36. See Fernão Mendes Pinto to Baltasar Dias, S.J., November 20, 1555, in Rebecca Catz, *Cartas de Fernão Mendes Pinto e Outros Documentos* (Lisbon: Editorial Presença, 1983), 60–66; Paul Pelliot, "Un ouvrage sur les premiers temps de Macao," *T'oung Pao* 31 (1934–35): 58–94.

37. In this regard, Peter Shapinsky, "Piracy and Cartographic Exchange in Sixteenth-Century East Asia," a paper delivered at the Renaissance Society of America, San Diego, 2013, cites Kuroda Hideo, "Gyōkishiki 'Nihonzu' to wa nanika," in Kuroda Hideo et al., eds., *Chizu to ezu no seiji bunkashi* (Tokyo: Daigaku Shuppankai, 2001), 3–77.

38. On the Gyoki maps see Mary Berry, *Japan in Print* (Berkeley: University of Cali-

fornia Press, 2006), 69. The Homem map is British Library, Additional MSS 5415A; see Armando Cortesão and A. Teixeira da Mota, *Portugaliae monumenta cartographia* 2 (Lisbon: 1960), 5–8, 13; Peter Barber, *The Queen Mary Atlas* (London: Folio Society, 2005).

39. Zheng Ruozeng, *Chouhai tubian* [籌海圖編 "Ocean Plan: Compilation of Maps"], ed. Hu Zongxian (1624 ed., originally 1561). A more thorough multivolume ethnography of Japan, Zheng Shungong's *Riben Yijian* (Mirror of Japan, 1565) composed by the Ming emissary to Minamoto Yoshishige, circulated less widely and also contained Gyoki and route maps.

40. "Instructions for the two Masters Charles Jackman and Arthur Pett, given and adventured to them, at the Court day, gotten at the Moscovy Court, the 17th of May, Anno 1680," BL Lansdowne 122 No 5, f 30, later reprinted by Hakluyt.

41. See Vicente Rafael, *Contracting Colonialism: Translation and Christian Conversion in Tagalog Society* (Durham: Duke University Press, 1993).

42. See generally Bronwen Wilson, *The World in Venice* (Toronto: University of Toronto Press, 2005), 215–21; Derek Massarella, "Envoys and Illusions: The Japanese Embassy to Europe, 1582–90, *De Missione Legatorvm Iaponensium*, and the Portuguese Viceregal Embassy to Toyotomi Hideyoshi, 1591," *Journal of the Royal Asiatic Society* 15 (2005), 329–50; Judith Brown, "Courtiers and Christians: The First Japanese Emissaries to Europe," *Renaissance Quarterly* 47:4 (Winter 1994), 872–906.

43. J. A. Abraanches et al., eds., "Les instructions du Père Valignano pour l'ambassade jaaponaise en Europe," *Monumenta Nipponica* 6 (1943), 395–401.

44. Guido Gualtieri, *Relationi della venuta degli ambasciatori giaponesi a Roma fino all partita di Lisbona* (Rome: Francesco Zannetti, 1586), 182–83. Crowds in Florence supposedly cried tears of joy to see the global reach of the "la vera Chiesa." Lodovico Muratori, *Annali d'Italia dal principio dell'era volgare sino all'anno 1750* 10:2 (Rome: Barbiellini, 1754), 324–25.

45. "Brief relation of the oath taken to Prince Philip in Madrid at the church of St. Jerome," November 12, 1584, TNA SP 94/2/22. The idea of "whiteness" is also articulated by Valignano in his "Summary of the Things of Japan." C. R. Boxer, *The Christian Century in Japan* (Berkeley: University of California Press, 1967), 74; Brown, "Courtiers and Christians," 884–85.

46. "News from Rome," Florence, March 6/16th, 1584/5, TNA SP 101/72/12; "News from Rome" March 19/29, 1584/5, TNA SP 101/72/15; "News from Rome and Venice," June 1/11, 1585, TNA SP 101/72/16; "News from Rome and Venice," June 8, 1585, TNA SP 101/72/17; "Newsletter Sent from Venice," June 29, 1585, TNA SP 101/72/18; "News from Divers Parts," TNA SP 101/95/24; "News from Italy," Venice, August 10, 1585, TNA SP 101/72/23. The embassy left Lisbon for Japan on April 8, 1586. Such news also circulated on a European scale, both through letters and at least fifty-five printed works, including several translations of the Latin account of their visit to Rome. *Acta Consistorii publica exhibiti a . . . Gregorio Papa XII, regnum japoniorum legatis Romae* (Rome: Franciscum Zannettum, 1585), later translated into English by Purchas in 1625; Lach, *Asia* 1:2, 688–706.

47. David Quinn, ed., *The Last Voyage of Thomas Cavendish, 1591* (Chicago: University of Chicago Press, 1975); and "The admirable adventures and strange fortunes of

Master Antonie Knivet, which went with Master Thomas Candish in his second voyage to the South Sea, 1591," in *Hakluytus Posthumous or Purchas His Pilgrims* 16, Samuel Purchas, ed. (New York: AMS Press, 1965), 178–79.

48. Von Glahn, *Fountains of Fortune*, 118.

49. The "Gujin xingsheng zhi tu" survives in the Archivo General de Indias, Seville, Mapas y Planos Filipinas, 5. See J. B. Harley and David Woodward, *History of Cartography* 2:2 (Chicago: University of Chicago Press, 1994), 59; Li Xiaocong, *A Descriptive Catalogue of pre-1900 Chinese Maps Seen in Europe* (Beijing: International Culture Press, 1996), 144–46.

50. Emma Blair and James Robertson, *The Philippine Islands* 3 (Cleveland: A. H. Clark, 1903), 276, 284.

51. Hakluyt, *Principall Navigations*, ed. Arber, 3 (1903), 7–23. In the second edition Hakluyt changed "Navidad" to "Acapulco" because the Spanish had shifted their port.

52. Report of the expedition of Miguel Lopez de Legazpi, 1565, in *Coleccion de documentos ineditos . . . de Ultramar*, 2nd ser., vol. 2, "De las Islas Filipinas" (Madrid: Real Academia de la Historia, 1886), document 27, 291–92.

53. On the Fujian printing industry see Lucille Chia, *Printing for Profit: The Commercial Publishers of Jianyang* (Harvard: Harvard University Press, 2002).

54. Escalante's *Discurso de la navegacion que los portugueses hazen à los reinos y provincias del Oriente, y de la noticia q se tiene de las grandezas del Reino de la China* (Seville: Alonso Escrivano, 1577) was quickly translated by the former Spanish merchant John Frampton as *A discourse of the nauigation which the Portugales doe make to the realmes and prouinces of the east partes of the worlde and of the knowledge that growes by them of the great thinges, which are in the dominions of China* (London: Thomas Dawson, 1579), and it attempted to include one Chinese character, *cheng* (city, 城). Frampton also translated a version of Marco Polo, *The Most Noble and Famous Travels of Marcus Paulus* (London: Ralph Newbery, 1579), after the Spanish of Rodrigo Fernández de Santaella. Gaspar da Cruz's *Tractado em que se cõtam muito por estẽso au cousas da China* (Evora: André de Burgos, 1569) was not translated, probably because it was written in Portuguese and not as easily obtained.

55. For the list books brought back by "Frier Herrada" see Juan Gonzáles de Mendoza, *The Historie of the great and mightie Kingdom of China* 1, G. T. Staunton, ed. (London: Hakluyt Society, 1853–54), 134–37; and Lach, *Asia* 1:2, 778–80.

56. The Jesuit Archives in Rome have a 1587 edition of the *Da Ming huidian*, Jesuit Archives, Rome, Japonica-Sinica IV, 13–24. See Albert Chan, *Chinese Books and Documents in the Jesuit Archives in Rome* (Armonk: M. E. Sharpe, 2002), 547.

57. Mendoza, *Historie* 1:23–24, 80–82, 90–91.

58. David Goodman, *Power and Penury: Government, Technology and Science in Philip II's Spain* (Cambridge: Cambridge University Press, 1988), 62–63; Maria Portuondo, *Secret Science: Spanish Cosmography, and the New World* (Chicago: University of Chicago Press, 2009), 261; A. Cortesão and A. Teixeira da Mota, *Portugaliae monumenta cartographica* 2 (Lisbon: Imprensa Nacional Casa da Moeda, 1960), 123–25.

59. On the reverse of Barbuda's map were also a series of Chinese characters from Escalante, to whom Barbuda acknowledged his debts. Barbuda also cites Duarte Barbosa, Andrea Corsali, Barros, Pigafetta, and Jesuit letters, possibly from Ramusio. Items on the

map also suggest that he had seen charts of China by the Portuguese Bartolomeu Velho (1561) and Fernão Vaz Dourado (1568, 1571, 1580). Lach, *Asia*, 1:2, 818–19; Cortesão and da Mota, II, 96; III, 4–7.

60. A map of China later appeared in Cornelius de Jode, *Speculum Orbis Terrae* (Antwerp: Gerardi de Iudaeis, 1593).

61. Portuondo, *Secret Science*, 262; Jan Hendrik Hessels, *Abrahami Ortelii . . . epistolae* (Cambridge: Ecclesia Londino-Batava, 1887–97), letters no. 62, 99, 210.

62. A new single-sheet map of the "Maris Pacifici (quod vulgo Mar del Sur)" appeared in 1589 in Antwerp [Huntington Library, Map, 149499] showing it as a Spanish lake and announcing the presences of "multa Christianorum" in Japan. Likewise in the handbook-sized *Epitome Theatri Orteliani* (Antwerp: Christopher Plantinus, 1589), Cathay is gone and replaced by the 1584 maps of Tartary and China (M4–5).

63. Details of Drake's voyage were largely kept secret until 1592, but rumors abounded. See Harry Kelsey, *Sir Francis Drake* (New Haven: Yale University Press, 1998), 177–79.

64. For the letters between Camden and Ortelius throughout the 1580s cf. BL Cotton MSS Julius C. V, f. 6, 13, 24, 28, 33, 42. Mercator also visited Camden, and this volume includes a 1579 letter from him (f. 5).

65. William Camden to Abraham Ortelius, Westminister, December 1580, BL Cotton MSS Julius C. V, f. 13. The letter begins "amico suo singulari." Dee's precise relationship with Camden in the 1570s and 1580s is somewhat unclear, although Dee supposedly introduced Ortelius to Camden, and Camden certainly made use of the Mortlake library.

66. See Henry Kamen, *Empire: How Spain Became a World Power* (New York: Harper Collins, 2004), 224–26; Carlos Vega, "Un proyecto utópico: La conquista de China por España," *Boletin de la Asociación Española de Orientalistas* (1982), 14–18.

67. Matteo Ricci to Juan Bautista Román, Zhaoqing, September 1584, in Pietro Venturi, ed., *Opere Storiche del P. Matteo Ricci* 2 (Macerata: Filippo Giorgetti, 1913), 36–49. The letter is translated in R. H. Major, introduction to Mendoza, *Historie*, 1:lxxvii–lxxxix. On this letter and the accompanying map see Boleslaw Szczesniak, "Matteo Ricci's Maps of China," *Imago Mundi* 11 (1954), 127–36.

68. Mendoza, *Historie*, 1: 92.

69. *Avvisi del Giapone de gli anni M.D. LXXXII. LXXXIII. et LXXXIV. Con alcuni altri della Cina dell' LXXXIII. et LXXXIV. Cauati dalle lettere della Compagnia di Giesù. Riceuute il mese di dicembre M.D. LXXXV* (Rome: Francesco Zanetti, 1586), 182.

70. Hakluyt, *Principall Navigations* 3 (1600), 821–22. Unlike Cavendish, Drake, who visited southern Java in 1580 for two weeks, does not appear to have thought of it as part of a Chinese sphere of trade. See Kelsey, *Drake*, 201–3.

71. "A Letter of M. Thomas Candish to the right honourable the Lord Chamberlaine," September 9, 1588, in Hakluyt, *Principall Navigations* (1589), 808.

72. Ping-ti Ho, *Studies in the Population of China* (Cambridge: Harvard University Press, 1959), 3–23. On the property market and gentry in the Ming see Timothy Brook, *The Confusions of Pleasure*, 58–60; Valerie Hansen, *Negotiating Daily Life in Traditional China* (New Haven: Yale University Press, 1995).

73. Timothy Brook, "The Spatial Organization of Subcounty Administration" in Brook, ed., *The Chinese State in Ming Society* (New York: Routledge Curzon, 2005), 19–

42; and Kung-Chuan Hsiao, *Rural China* (Seattle: University of Washington Press, 1960), 31–32, 201. The concept of a "localist turn" is too strong, despite the insights behind it. Peter Bol, "The 'Localist Turn' and 'Local Identity' in Later Imperial China," *Late Imperial China* 24.2 (2003), 3–4.

74. Luo Hongxian, *Nianan wenji* (Taibei: Shangwu yingshuguan, 1974), 6.6a–7b; James Tong, *Disorder Under Heaven* (Stanford: Stanford University Press, 1991); David Robinson, "Banditry and the Subversion of State Authority in China: The Capital Region During the Middle Ming Period (1450–1525)," *Journal of Social History* 33:3 (2000), 527–63; Frederick Mote, "The Ch'eng hua and Hung-chih Reigns, 1465–1505," in *Cambridge History of China* 7:1 (Cambridge: Cambridge University Press, 1988), 376–77.

75. See Anne Gerritsen, *Ji'an Literati and the Local in Song-Yuan-Ming China* (Leiden: Brill, 2007), 217–23; Kandice Hauf, "The Community Covenant in Sixteenth Century Ji'an Prefecture, Jiangxi," *Late Imperial China* 17:2 (1996), 1–22; Robert Hymes, "Lu Chiu-yuan, Academies, and the Problem of the Local Community," in *Neo-Confucian Education,* Wm. De Bary and John Chaffee, eds. (Berkeley: University of California Press, 1989), 432–56.

76. Hu Zhi (1517–85), "Nianan xiansheng xingzhuang" (Record of Conduct of Luo Hongxian), *Henglu jingshe canggao* in *Siku quanshu zhenben,* 366–70 (Taipei: 1984), 23/17a; Luo Hongxian, "Ke xiangyue yin" ["An Introduction to the Engraving of the Xiangyue"], in *Nianan Luo Xiansheng* (Taipei: Siku quanshu zhenben collectanea, 1974), 6/6b; Hauf, "Community Covenant," 14, 38–39.

77. On the various forms of *tu* see Francesca Bray, Vera Dorofeeva-Lichtmann, and Georges Métalié, eds., *Graphics and Text in the Production of Technical Knowledge in China* (Leiden: Brill, 2007).

78. Edmund Plowden, *Les commentaires ou les Reportes* (London: Richard Tottel, 1571), 213; cited in Ernst Kantorowicz, *The King's Two Bodies* (Princeton: Princeton University Press, 1957), 7.

79. See Jean Bodin, "De la souveraineté," *Les six livres de la République* 1:8 (Paris: Jacques du Puys, 1576), 85–86; J. H. Franklin, *Jean Bodin and the Rise of Absolutist Theory* (Cambridge: Cambridge University Press, 1973). Bodin at this point knew nothing of the Ming Empire or China. See Lach, *Asia*, 2:2, 312 n350.

80. Giovanni della Casa coined the phrase to distinguish between civil law (ordinary reason) and the superior actions of princes (reason of state). See Peter Burke, "Tacitism, Scepticism, and Reason of State," in *The Cambridge History of Political Thought, 1450–1750,* J. H. Burns, ed. (Cambridge: Cambridge University Press, 1991), 479–80.

81. Robert Peterson, trans., *A treatise concerning the causes of the magnificence and greatness of cities* (London: R. Ockould and H. Tomes, 1606), 3.

82. Richard Etherington composed an abbreviated translation of this text in manuscript, dedicated to Sir Henry Hobart, chief justice of the Court of Common Pleas, which focused on Botero's attack on Machiavelli. BL Sloane 1065.

83. Botero, *The worlde or An historicall description of the most famous kingdomes and commonweals therein . . . Translated into English,* Robert Johnson, trans. (London: Edm. Bollifant, 1601), 168. On Johnson see Andrew Fitzmaurice, "The Commercial Ideology of Colonisation in Jacobean England: Robert Johnson, Giovanni Botero, and the

Pursuit of Greatness," *William and Mary Quarterly* 64:4 (October 2007). On Botero in England see George Mosse, *The Holy Pretense* (Oxford: Blackwell, 1957), 35–38.

84. See Richard Tuck, *Philosophy and Government* (Cambridge: Cambridge University Press, 1993), 116. On the limits of Raleigh's approach see Nicholas Popper, *Walter Raleigh's "History of the World"* (Chicago: University of Chicago Press, 2012).

85. TNA Colonial Office (hereafter CO) 77/1/11–12. In 1592, the number of Turkey merchants more than doubled to fifty-three in the newly chartered Levant Company, many of whom later pushed for the founding of the East India Company in 1599. Cf. Brenner, *Merchants and Revolution*, 64–65.

86. Samuel Purchas, *Purchas his Pilgrimes* 4 (London: William Stansby, 1625), 1181. Two of Cavendish's Spanish charts survive: Biblioteca Nationale Centrale, Firenze, Port. 30, and Algemeen Rijksarchief, The Hague, Leupe Inv.73, the latter possibly acquired by Hakluyt in 1594–95 and sold to the Dutch through Emanuel van Meteren.

87. Robert Hues, *Tractatus de Globis et eorum usu* (London: 1594); John Davis, *The Seaman's Secrets* (London: Thomas Dawson, 1594); and Davis, *The Worlde's Hydrographical Description . . . whereby appears that there is a short and speedy Passage into the South Seas, to China, Molucca, Philippina, and India by Northerly Navigation* (London: Thomas Dawson, 1595). Hues dedicated his volume to Raleigh, giving a chronology about the Cathay-to-China shift from Mandeville through Cabot and down to Cavendish as well as a preface on the importance of instruments and data collection.

To avoid confusion, I should mention that there are three notable people named John Davis in this period. John Davis of Sandridge (ca. 1550–1605, aka Davys) was involved in early voyages searching for a northern passage to China in the 1580s, went with Cavendish in 1591 and published the *Seaman's Secrets* and *Hydrographical Description*, sailed with Cornelius Houtman (1598–1600) and then James Lancaster (1601–3), and died off Borneo in 1605 fighting Japanese sailors. John Davis of Limehouse (d. 1621, aka Davy) first sailed with Lancaster in 1601 and left a rutter of the East Indies, BL Sloane 3959, that was also published by Samuel Purchas, *Purchas his Pilgrimes* 1 (London: William Stansby, 1625), 440–44. Finally, John Davies (1563–1625, aka Davis), who is not discussed in this book, published on mathematics and was involved in the Essex conspiracy.

88. Colm Lennon, *Sixteenth Century Ireland: The Incomplete Conquest* (Dublin: Gill & Macmillan. 1994), 229–34; Nicholas Canny, *Making Ireland British 1580–1650* (Oxford: Oxford University Press, 2001), 146–64.

89. See Vaughan, *Transatlantic Encounters*, 31–35. On the importance of Harriot in defining "geometrical landscapes" see Amir Alexander, *Geometrical Landscapes: The Voyages of Discovery and the Transformation of Mathematical Practice* (Stanford: Stanford University Press, 2002), as well as his manuscript mathematical papers, BL Add 6782–89, demonstrating this approach.

90. E. G. R. Taylor, ed., *The Original Writings and Correspondence of the Two Richard Hakluyts*, 2, 369.

91. *Twelfth Night*, act 3, scene 2, lns. 79–80.

92. There are two versions, with different cartouches in the south Pacific and south Atlantic, although the map remains the same. See for example BL 683.h.5,6 for the south Pacific version and BL 212.d.1–2 for the south Atlantic version.

93. HKUSC, G7400 1590.S54. The copy of this map from the Cotton Library, BL Cotton Augustus, II.ii.45, dates from 1607. The provinces, river systems, coastline, and depiction of the Great Wall and Gobi Desert all derive from Luo, but the long Korean peninsula shows a clear debt to Yu Shi, and the map of Japan does not correspond to that from Luo. In the preface to his *Certaine Errors in Navigation* (London: Valentine Sims, 1599), Wright published excerpts of two letters of apology from Hondius trying to appease Wright's anger over the "Christian Knight" map.

94. Wright's method was first described and attributed in Thomas Blundeville, *M. Blundevile his Exercises, Containing Sixe Treatises* (London: John Windet, 1594).

95. See in particular the Trinity College Dublin MS 387; Wright, *Certaine Errors*, PPP3; Ash, *Power, Knowledge, and Expertise in Elizabethan England*, 166–68; Mark Monmonier, *Rhumb Lines and Map Wars* (Chicago: University of Chicago Press, 2004), 63–74. As Monmonier suggests, Thomas Harriot's approach to this problem from 1585 was actually more elegant in its simplicity.

96. See Peter Barber's emphasis on the lack of interest in maps at court. Barber, "Was Elizabeth I Interested in Maps—and Did It Matter?" *Transactions of the Royal Historical Society* 14 (2004), 185–98. For a well-argued counterpoint see Leslie Cormack, *Charting an Empire: Geography at the English Universities, 1580–1620* (Chicago: University of Chicago Press, 1997), 204. For Elizabeth's failed attempt to write to the "serenissmóque principi," "imperatori," and "Monarchae" of China on behalf of Benjamin Wood and the merchants Thomas Bromfield and Richard Allen in 1596, see TNA CO 77/1/17, which has annotations in Robert Cecil's hand; and Hakluyt, *Principall Navigations* 3 (1600), 853–54.

97. "The names of such persons as have written with their own hands, to venture in the pretended voyage to the East Indies," (September 22, 1599) in *The Dawn of British Trade to the East Indies as Recorded in the Court Minutes of the East India Company 1599–1603*, Henry Stevens, ed. (London: Henry Stevens and Son, 1886), 1–4.

98. "An assembly of the persons heerunder named holden the xxiiith of September 1599," in Stevens, *Dawn*, 4–6.

99. "At an assemblie of the Comitties or ye directors of the viage the xxvth of September 1599," in Stevens, *Dawn*, 7–9. The first news of Dutch success in Java had arrived August 8, 1597 by way of newsletter from Holland, State Papers Foreign, Holland, SP 84; see *Calendar of State Papers Colonial Series, East Indies, China and Japan, 1513–1616* (London: Longman, 1862), 2, 98 no. 253.

100. "An assembly of Committies holden at Mr Aldn Godderd the 4 of october 1599," in Stevens, *Dawn*, 9.

101. The translation of *John Huighen van Linschoten. His Discours of Voyages into ye Easte & West Indies. Devided into Foure Bookes* (London: John Wolfe, 1598) also supported this. William Cecil himself, as Elizabeth's treasurer, had a large collection of maps now at Hatfield House as well as a copy of Ortelius's atlas at Burghley House that also contained manuscript maps by John Dee and Robert Norman. See R. A. Skelton, *A Description of Maps and Architectural Drawings in the Collection Made by William Cecil, First Baron Burghley, Now at Hatfield House* (London: Roxburghe Club, 1971).

102. Gyles van Hardwick to Peter Artson, September 30, 1598, in *Calendar of State Papers Colonial*, 99.

103. Coppin to Robert Cecil, August 9, 1599, HMC Salisbury, 9: 282–83. On the status of the navy and Greville's obliviousness see Rodger, *The Safeguard of the Sea*, 338–39; Ronald Rebholz, *The Life of Fulke Greville* (Oxford: Clarendon Press, 1971), 118–19.

104. Hakluyt's second edition of the *Principall Navigations* had a 1598 title page promising an account of Essex's victory at Cadiz. This was removed in 1599, when Hakluyt dedicated the second and third volumes to Robert Cecil.

105. Earl of Northumberland to Cecil, August 1, 1600, *Calendar of the Manuscripts of . . . Salisbury. The Cecil Manuscripts* 10 (London: Historical Manuscripts Commission, 1904), 260. A manuscript satire in Italian claiming to be a response from "Prince Taico-sama" to Elizabeth's Latin letter to the "King of China" for Benjamin Wood's voyage in 1596 (printed in Hakluyt in 1600) most likely came out of the Essex circle in this period as a critique of Cecil's faction. See "1600 The Emperor of China his letter to the Queene of England," Folger MS V.a. 321, 34v–35r, reprinted in A. R. Braunmuller, *A Seventeenth-Century Letter-Book* (Newark: University of Delaware Press, 1983); Timothy Billings, "The Emperor of China His Letter to Queen Elizabeth," http://www.folger.edu/html/folger_institute/mm/EssayTB.html; Hawkins to Cecil, March 20, 1602–3, HMC Salisbury, 12:697.

106. Cathay did not entirely disappear; see for example Joseph Moxon's map of the northern hemisphere from 1660, BL Map 792.d.8.

107. Bruce, *Annals*, 1: 121–26.

108. TNA CO 77/1/26–27 ca. 1599–1600. Transcribed in Bruce, *Annals*, 1:115–21. The document next to this is a comprehensive list of "Commodities of the East Indies" dated 1600, TNA CO 77/1/30. See also the draft Huntington EL2360 transcribed in Heidi Brayman Hackel and Peter C. Mancall, "Richard Hakluyt the Younger's Notes for the East India Company in 1601," *Huntington Library Quarterly* 67:3 (2004), 423–36. Note that the Huntington draft reads "China extending to Cathaia" (f. 1r).

109. BL Harleian, 306, f. 17–25; reprinted in Bruce, *Annals*, 136–39.

110. Barlow, *The Navigators Supply*, K.

111. *Measure for Measure*, act 2, scene 1, ln. 93.

CHAPTER THREE

1. Wallace Notestein and Frances Relf, eds., *Commons Debates for 1629* (Minneapolis: University of Minnesota Press, 1921), 62.

2. The original (lost) manuscript circulated in the summer of 1619 at the request of Buckingham, the new lord high admiral, but James I wanted it changed, and it was rejected on its second submission. See G. J. Toomer, *John Selden: A Life in Scholarship* 1 (Oxford: Oxford University Press, 2009), 388–432; David Berkowitz, *John Selden's Formative Years* (Washington: Folger Library, 1988), 54–55, 308–9. Selden tells his version of the story (mistakenly remembering it as 1618) in *Vindiciae Joannis Seldeni* (London: Cornelius Bee, 1653), 15–28. Selden first published a revised version during the ship money controversy under Charles I as *Mare Clausum seu De dominio maris* (London: William Stanesbeius, 1635). The English translation appeared in support of the first Navigation Act (1651) as Marchamont Nedham, trans., *Of the dominion, Or, Ownership of the Sea* (London: William Du-Gard, 1652).

3. The Chinese names are from the Selden Map. For a general account see Batchelor, "The Selden Map Rediscovered: A Chinese Map of East Asian Shipping Routes, c. 1619," *Imago Mundi* 65:1 (January 2013), 37–63.

4. For the sense that the EIC and the VOC were increasingly independent but drawing on medieval legal precedent see Stern, *The Company State*; and Eric Wilson, *Savage Republic: De Indis of Hugo Grotius* (Leiden: Martinus Nijhoff, 2008). On the traditional patrilineal aspects embedded in this see Julia Adams, *The Familial State* (Cornell: Cornell University Press, 2005). On the longer-term process of state formation in Southeast Asia in this period as a strange parallel see the first volume of Lieberman, *Strange Parallels* (2003).

5. With him on his return to London from Japan in 1623–24, Richard Cocks brought the history *Azuma Kagami*, one of the first Japanese books to use movable type, which chronicles the years 1180 to 1266; Cocks died along the way. The two surviving volumes of his *Azuma Kagami* were dispersed. Trinity College Dublin, MS 1645, given by Archbishop of Dublin John Parker; Cambridge UL FJ.274.17, donated through the library of Bishop of Ely John Moore in 1715. The Cambridge volume is marked "Oxford 1626" and may have exchanged owners at the same time as the Selden Map. Cocks mentions in 1616 buying fifty-four volumes in Kyoto. See Peter Kornicki, *The Book in Japan* (Leiden: Brill, 1998), 313. He had earlier sent back an almanac for 1615. Bodleian Library Sinica 47, printed in 1614, see Purchas, *Pilgrimes*, I.4.407.

6. Tonio Andrade, *Lost Colony: The Untold Story of China's First Great Victory over the West* (Princeton: Princeton University Press, 2011); and John Wills, "Maritime Europe and the Ming" and Wills and John Cranmer-Byng, "Trade and Diplomacy with Maritime Europe, 1644-c. 1800," in Wills, ed., *China and Maritime Europe, 1500–1800* (Cambridge: Cambridge University Press, 2011), esp. 24–77 and 183–254.

7. This gave it an almost mythic status in American history, as in Charles Andrews, *The Colonial Period of American History: England's Commercial and Colonial Policy* 4 (New Haven: Yale University Press, 1938), 36–37. In 1663, owners of the 1652 edition could purchase a new title page and the old dedication to Charles I for rebinding in their old copies. Pepys did this on April 17, "ashamed to have the other seen dedicated to the Commonwealth."

8. On the novelty of Qing strategies and the comparable approaches of the French and Russians in the late seventeenth century, see Laura Hostetler, *Qing Colonial Enterprise: Ethnography and Cartography in Early Modern China* (Chicago: University of Chicago Press, 2001).

9. On Selden's broader intellectual life, see most recently G. J. Toomer, *John Selden: A Life in Scholarship* (Oxford: Oxford University Press, 2009), who sees *Mare Clausum* as rather weak and politically motivated [I: 388–437]. See also those who focus on sacred history: Jason Rosenblatt, *Renaissance England's Chief Rabbi: John Selden* (Oxford: Oxford University Press, 2006), and Reid Barbour, *John Selden: Measures of the Holy Commonwealth* (Toronto: University of Toronto Press, 2003); and those who focus on natural and common law: Sergio Caruso, *La miglior legge del regno. Consuetudine, diritto naturale e contratto nel pensiero e nell'epoca di John Selden* (Milan: Giuffrè, 2001), and Paul Christianson, *Discourse in History, Law, and Governance in the Public Career of John Selden* (Toronto: University of Toronto Press, 1996).

10. On Hakluyt's translation see David Armitage, ed., *The Free Sea* (Indianapolis: Liberty Fund, 2004).

11. See the work of the St. Andrews professor William Welwood, "Of the Community and Propriety of the Seas," *An Abridgement of All Sea-Lawes* (London: Humfrey Lownes for Thomas Man, 1613), chapter 27, 61–72. The queen, Anne of Denmark, sharing this northern perspective, requested Welwood's more direct response, *De dominio maris, iuribusque ad dominium praecipue spectantibus assertio breuis et methodica* (London: Thomas Creede, 1615). See J. D. Alsop, "William Welwood, Anne of Denmark, and the Sovereignty of the Sea," *Scottish Historical Review* 49 (1980), 171–74. Grotius left his response in manuscript, "Defensio capitis quinti Maris Liberi oppugnati a Guilielmo Welwood" (ca. 1615), in *Mare Clausum*, ed. Samuel Muller (Amsterdam: F. Muller, 1872), 331–61.

12. Inner Temple Library, MS Petyt 529; Armitage, *The Free Sea*, xx–xxi; Taylor, *The Original Writings and Correspondence of the Two Richard Hakluyts*, 2:497–99. Hakluyt was also involved in the translation of Gotthard Arthus, *Dialogues in the English and Malaiane Languages*, trans. Augustine Spalding (London: William Welby, 1614). See F. M. Rogers, "Hakluyt as Translator," *The Hakluyt Handbook* 1, ed. David Quinn (London: Hakluyt Society, 1974), 37–39. On Anglo-Dutch negotiations see G. N. Clark and W. J. M. van Eysinga, "The Colonial Conferences between England and the Netherlands in 1613 and 1615," *Bibliotheca Visseriana* (1940), 15: 1–270; 17: 1–155. An early skeptic was Richard Cocks, who wrote to Thomas Wilson in London, December 10, 1614, that the Dutch habit of seizing Chinese junks endangered trade with both China and Japan. Purchas, *Pilgrimes*, 1.4.408–9.

13. On the Dutch and Spanish struggle over Ternate in 1606 see the extensive account in Bartolomé Leonardo de Argensola, *Conquista de las Islas Malucas* (Madrid: Alonso Martin, 1609); and Antonio de Morga, *Sucesos de las Islas Filipinas* (Mexico: Cornelio Cesar, 1609). Sir Henry Middleton brought back letters in 1606 from the rulers of Ternate, Tidore, and Banten. The Ternate letter mentions Drake and the sultan's predecessor having sent a gold ring to Elizabeth (June 1605), TNA SP 102/4/24, as does Argensola, *Conquista*, 263–64.

14. Sir Thomas Hawkins, trans., *The Cause of the Greatness of Cities . . . With certain observations concerning the sea* (London: Elisabeth Purslowe, 1635), supplanted Robert Peterson's 1606 translation (made perhaps with the help of Raleigh), which did not contain the "Relatione del mare."

15. David Armitage, "Making the Empire British: Scotland in the Atlantic World 1542–1707," *Past and Present* 155 (May 1997), 52–53; Peter Borschberg, *Hugo Grotius, the Portuguese, and Free Trade in the East Indies* (Singapore: NUS Press, 2010), 1–19. Grotius had a limited engagement with the Spanish natural law tradition of the School of Salamanca, notably the Dominican Francisco Vitoria (*De Indis*, 1532; and *De potestate civili*, 1528, from the schoolbook version *Relectiones Theologicae XII*, 1557) and Fernando Vázquez de Menchaca (*Controversiae illustres*, 1572), written in opposition to Portuguese claims as well as the work of more contemporary "Spanish Lawyers" like Diego de Valdés's *De dignitate regum regnorumque Hispaniae* (Granada: Ferdinand Diaz a Montoya, 1602). See *Mare Clausum* (1635), 72–74.

16. Aka *De Rebus Indicanis*. See Borschberg, *Hugo Grotius*, 48–53. Grotius relied

on the news pamphlet *Corte en sekere Beschrijvinghe vant veroveren der rijcke ende gheweledighe kracke comende uyte Ghewesten van China* (Middelburg: Richard Shilders, 1604).

17. Grotius, *De Jure Praedae Commentarius* (Oxford: Clarendon Press, 1950), 298; Borschberg, *Hugo Grotius*, 52; Martine Julia van Ittersum, *Profit and Principle: Hugo Grotius, Natural Rights Theories, and the Rise of Dutch Power in the East Indies, 1595– 1615* (Leiden: Brill, 2006), 50. Tuck notes the more general importance of "sociability" and neighborliness to Grotius's theories as an alternative to theistic natural law. Richard Tuck, *Natural Rights Theories: Their Origin and Development* (Cambridge: Cambridge University Press, 1979), 60, 72.

18. Selden's book was so popular that in April 1636, five months after he wrote his preface in November 1635, Charles I issued a proclamation against importing foreign pirated copies, *A proclamation to forbid the importing, buying, selling, or publishing any forraine edition of a booke lately printed at London by His Maiesties command, intituled Mare Clausum* (London: Robert Baker, 1636). A second edition also appeared in 1636. Tuck notes that the references to Grotius's *De Jure Belli ac Pacis* (Paris: Nicholas Buon, 1625; Amsterdam: William Blaeu, 1631) indicate Selden made revisions between 1618 and 1635. Tuck, *Natural Rights Theories*, 86; see also Tuck, "Grotius and Selden," in *The Cambridge History of Political Thought*, J. H. Burns and Mark Goldie, eds. (Cambridge: Cambridge University Press, 1991), 499–529.

19. Malynes, *Consuetudo, vel lex mercatoria, or The ancient law-merchant Diuided into three parts: according to the essentiall parts of trafficke* (London: Adam Islip, 1622).

20. *Mare Clausum* (1635), 91–97. Ambrose, *Hexameron*, ch. 10, had said there were only land measurers ("Geometram") and not sea measurers ("Thalassometram"). Selden cites John Tzetzes's edition of Lycophron. See Toomer, *Selden*, 1:408. The notion of "hydrography" appears on John Dee's tree graph of mathematical disciplines at the end of his preface to the 1570 edition of *Euclid*.

21. See also the letter by Selden to the Canterbury Orientalist Francis Taylor, June 25, 1646, in Henry J. Todd, *Memoirs of the Life and Writings of Brian Walton* 1 (London, 1821), 41n-43n, in which Selden claims the study of Biblical Oriental languages ("Orientalis disciplinae") is as useful as Galileo's telescope in clearing away obfuscations and revealing starry messengers (*sidereis nuntiis*). On the history of the Arabic font see Toomer, *Eastern Wisedome*, 114–15, 171, 271. The Oxford Arabic fonts were imported from Leiden in 1637 and Selden used them extensively in *De Jure Naturali et Gentium*. For his *Titles of Honour* (London: William Stansby, 1614; 2nd ed., 1631), Selden had Arabic woodcut inserts made.

22. *Mare Clausum* (1635), 72–74.

23. "Sed interea stabilitum est ex Jure Universali Obligativo, quo Pactis standum est et servanda fides." *Mare Clausum* (1635), 16; see Toomer, *Selden*, 1, 397.

24. See Tuck, *Philosophy and Government*, 217; *Natural Rights*, 89, 96–98. The clearest version of this argument is Selden's *De Jure Naturali et Gentium*, 35, referencing Grotius's *De Jure Belli ac Pacis*.

25. Pocock, *The Machiavellian Moment* (Princeton: Princeton University Press, 1975), 333–400; Pocock, *The Ancient Constitution and the Feudal Law* (Cambridge:

Cambridge University Press, 1957), 64, 289; and Alan Cromartie, *Sir Matthew Hale* (Cambridge: Cambridge University Press, 1995), 107–8.

26. Harold Berman, "The Origins of Historical Jurisprudence: Coke, Selden, Hale," *Yale Law Journal* 103:7 (May 1994), 1695. See also the more limited concept of "age of partnership" and "age of companies" in Holden Furber, *Empires of Trade* (Minneapolis: University of Minnesota, 1976); B. B. Kling and N. M. Pearson, eds., *The Age of Partnership* (Honolulu: University of Hawaii Press, 1979); James Tracy, *The Rise of Merchant Empires* (Cambridge: Cambridge University Press, 1990).

27. The 1619 proposal sent to James I for a joint company is TNA CO 77/15 Eliz-Charles II, f. 2–12. On reactions see the correspondence of Batavia's governor general Jan Pietersz Coen, in H. T. Colenbrander, ed., *Jan Pietersz Coen. Bescheiden omtrent zijn bedriff in Indië*: 1 (The Hague: Nijhoff 1919), esp. 543–44.

28. See J. C. Van Leur, *Indonesian Trade and Society* (The Hague: 1955); Anthony Reid, *Southeast Asia in the Age of Commerce, 1450–1680* (New Haven: Yale University Press, 1988, 1993).

29. Like Raleigh, Sandys seems to have been influenced by Botero in this regard rather than Grotius. See Tuck, *Philosophy and Government*, 117; Theodore Rabb, "The Editions of Sir Edwin Sandys' *Relation of the State of Religion*," *Huntington Library Quarterly* 26 (1963), 323–36.

30. Sandys, June 20, 1627, BL IOR B/10 20–29. On Sandys's parallel vagueness in terms of natural law see Rabb, *Jacobean Gentleman*, 30; and Noel Malcolm, "Hobbes, Sandys, and the Virginia Company," *Historical Journal* 24:2 (June 1981), 306. Selden invested in the Virginia Company in 1620 along with Sir Thomas Roe and Sir Robert Cotton, but like Cotton had disassociated himself from the venture by 1624. See Toomer, *Selden*, 1:320.

31. Ogborn, *Indian Ink*, 59–60; Annabel Teh Gallop, "Ottoman Influences in the Seal of Sultan Alauddin Riyat Syah of Aceh (r. 1589–1604)," *Indonesia and the Malay World* 32: 93 (July 2004), 176–90; L. F. Brakel, "State and Statecraft in Seventeenth-Century Aceh," *Pre-colonial State Systems in Southeast Asia*, ed. Anthony Reid (Kuala Lumpur: Royal Asiatic Society, 1975), 56–66; William Foster, ed., *Voyages of Sir James Lancaster* (London: Hakluyt Society, 1940), 124.

32. The English version of Elizabeth's original letter is TNA CO 77/1 1570–1621, 32–33. The second letter is in Samuel Purchas, *Purchas his Pilgrimes* 1:3 (London: 1625), 154–55. Miles Ogborn has emphasized the active elements of interpretation and translation in the delivery of the letter and Robert Markley the sense of cooperation against the "third man" of the Iberians. Ogborn, *Indian Ink*, 62–63; Markley, *Far East*, 39.

33. The trading permit is Bodleian MS Douce OR.e.4,. The most common word for seal in western Malay areas is *cap*, derived from Hindi *chhap* and Persian *chhapa*. Frederick de Houtman's Malay vocabulary (compiled 1599–1601, printed 1603) gives "tjap" for both "seghel" and "zeghel" (143, 176). See Annabel Teh Gallop, *Malay Seal Inscriptions* 1 (Phd. University of London, SOAS, 2002), 56–58. On the complexities of the *farman* as a diplomatic tool see Sudipta Sen, *Empire of Free Trade: The East India Company and the Making of the Colonial Marketplace* (Philadelphia: University of Pennsylvania Press, 1998), 77–80. On Alauddin "the usurper" see Anthony Reid, "Trade and the Problem of

Royal Power in Aceh: Three Stages," in *Pre-colonial State Systems in Southeast Asia*, ed. Anthony Reid and Lance Castles (Kuala Lumpur: MBRAS, 1975), 48–49.

34. On this capture see Sanjay Subrahmanyam, *Improvising Empire* (New Delhi: Oxford University Press, 1990), 42–45, 170; Borschberg, *The Singapore and Melaka Strait* (Singapore: National University of Singapore Press, 2010), 60–61; and Martine Julia van Ittersum, introduction to Hugo Grotius, *Commentary on the Law of Prize and Booty* (Indianapolis: Liberty Fund, 2006), xiii.

35. The Arabic letters with transcriptions and annotated translation are now MS Bodleian Or 575, f. 14–19. Archbishop William Laud donated this letter in 1635 in a small collection containing two Chinese essays (Sinica 42), a Japanese almanac for 1615 (see below), and a series of other Arabic letters, transcriptions, and translations, including three 1615 passes from Zidan al-Nasir, the sultan of Morocco, to trade in sugar. Another copy of the second half of the Arabic is Bodleian MS Douce Or.e.5 f.1f. The Arabic letter also appeared in print in Purchas's 1625 collection, translated into English by William Bedwell, *Pilgrimes* 1:3, 160.

36. Jacob van Heemskerck at Banten to the VOC in Amsterdam, August 27, 1603 (received March 17, 1604), translated in the appendix to Grotius, *Commentary on the Law of Prize and Booty*, 535–36.

37. F. C. Wieder, ed., *De reis van Joris van Spilbergen naar Ceylon, Atjeh en Banten, 1601–4* (The Hague: Martinus Nijhoff, 1933), 82–83.

38. See Johann Theodore de Bry, *Icones seu Gennuinae et Espressae Delineationes Omnium Memorabilium* (Frankfurt: Wolfgang Richter, 1607), images VII and IX.

39. Grotius writing for the directors of the VOC to the sultan of Tidore, winter 1606–7, in *Commentary on the Law of Prize and Booty*, 553.

40. On relation of Zeeland and Moucheron to Caribbean and African interests see van Ittersum, "Mare Liberum in the West Indies? Hugo Grotius and the Case of the Swimming Lion, a Dutch Pirate in the Caribbean at the Turn of the Seventeenth Century," *Itinerario* 37 (2007), 59–94.

41. Chaudhuri, *The English East India Company* (London: Cass, 1965), 21.

42. For the fourth EIC voyage, see "Lycence for transportaion of moneys uncoyned 1607," and Robert, Lord Salisbury to the East India Company, April 11, 1609, *The First Letter Book of the East India Company, 1600–1619*, Sir George Birdwood, ed. (London: Bernard Quaritch, 1893), 224–27, 282. Cecil approved £20000 in 1608, and as of April 1609 £6000 remained for a fifth voyage under David Middleton.

43. TNA 14/44, 62*. Cf. J. Knowles, "Jonson's Entertainment at Britain's Burse: Text and Context," in *Re-Presenting Ben Jonson*, M. Butler, ed. (Basingstoke and London: Macmillan, 1999), 133; Linda Levy Peck, *Consuming Splendor* (Cambridge: Cambridge University Press, 2005).

44. See Leonard Blussé, "No Boats to China: The Dutch East India Company and the Changing Pattern of the China Sea Trade, 1635–1690," *Modern Asian Studies* 30:1 (February 1996), 51–76.

45. See Van Dyke, "The Anglo-Dutch fleet of Defense, 1618–1622," in Leonard Blussé, *About and Around Formosa* (Taipei: SMC Publishing, 2003), 61–81.

46. On Dabhol see Foster, *English Factories, 1618–1621*, 286–89, 296–97, 300; and *English Factories 1622–1623*, 228, 264–71. Niels Steensgaard, *Carracks, Caravans, and*

Companies (Copenhagen: Studentlitteratur, 1972), saw the capture of Hormuz as a the triumph of the new corporations, but Ashin Das Gupta saw its significance as encouraging the rise of Surat and the use of medium-sized merchant ships. Das Gupta, "Indian Merchants and the Western Indian Ocean: The Early Seventeenth Century," *Modern Asian Studies* 19:3 (1985), 487.

47. "Translation of the Joint agreement between the English Company and the Netherlanders for Establishing the Trade of Bantam" (1622), BL, IOR, G/21/3A, v. II, f. 361–62.

48. See the charts made by the English, TNA MPF 189 (formerly part of SP 112/39); and then the derivative chart made by Hessel Gerritsz ca. 1617, TNA MPF 188 (formerly SP 112/38). Sarah Tyacke, "Gabriel Tatton's Maritime Atlas of the East Indies, 1620–1621: Portsmouth Royal Naval Museum, Admiralty Library Manuscript, MSS 352," *Imago Mundi* 60:1 (January 2008), 59.

49. Beginning with pamphlets supporting the English by Patrick Copland and Thomas Knowles, *A Courant of Newes from the East India* (February 8, 1622); *A Second Courant of Newes from the East India in Two Letters* (1622); followed by the pro-VOC pamphlet *The Hollanders Declaration of the Affairs of the East Indies* (Amsterdam [London]: Edward Allde, 1622), translating *Waerachtich verhael, van 't geene inde eylanden van Banda, inden jaere sestien-hondert eenentwintich, ede te vooren is ghepasseer* (1622); which was in turn responded to by Bartholomew Churchman, *An Answer to the Hollanders Declaration, Concerning the Occurents of the East India* (London: Nicholas Okes, 1622). Further pamphlets followed. On the *orang kaya* see J. Kathirithamby-Wells, "Royal Authority and the Orang Kaya in the Western Archipelago, ca. 1500–1800," *Journal of Southeast Asian Studies* 17:2 (September 1986), 256–67; and John Villiers, "Trade and Society in the Banda Islands in the Sixteenth Century," *Modern Asian Studies* 15:4 (1981), 723–50.

50. *Calendar of State Papers East Indies, 1622–24*, 283; and BL IOR B/6/31, May 1624; [Sir Dudley Digges], *A True Relation of the Unjust, Cruell, and Barbarous Proceedings against the English at Amboyna* (London: H. Lownes for Nathanael Newberry, 1624); *Waerachtich verhael van de tijdinghen gecomen wt de Oost-Indien* (Amsterdam: 1624); *A True Declaration of the News that Came out of the East Indies* (London: 1624); *Antwoorde van de Duytsche relatie, aengaende die ghepretendeerde conspiratie vande Enghelschen in Amboyna* (Amsterdam: 1624); John Skinner, *The Answer unto the Dutch Pamphlet, Made in Defence of the Unjust and Barbarous Proceedings against the English at Amboyna* (London: 1624); the ballad "News out of the East India of the Cruell Usage of Our English Merchants at Amboyna" (London: ca. 1625); and a sermon dedicated to the EIC, Robert Wilkinson and Thomas Myriell, *The Stripping of Joseph or the Crueltie of Brethren to a Brother* (London: William Stansby, 1625).

51. Chaudhuri, *English East India Company*, 31, 65.

52. Thomas Mun, *England's Treasure by forraign trade* (London: J. G. for Thomas Clark, 1664), 34.

53. "To the Reader," *A True Relation of the Unjust, Cruell, and Barbarous Proceedings*, A1–4. On the massacre as one of the first truly "public" issues in London in relation to Asia, see Anthony Milton, "Marketing a Massacre: Amboyna, the East India Company, and the Public Sphere in Early Stuart England," in *The Politics of the Public Sphere in Early Modern England*, Peter Lake and Steven Pincus, eds. (Manchester: Manchester

University Press, 2007), 168–90; and Karen Chancey, "The Amboyna Massacre in English Politics, 1624–1632," *Albion* 30:4 (Winter 1998), 583–98.

54. Cf. Toomer, *Selden*, 1:132–33, 154–55; and Selden, *Titles* (1614) 51, 85–111, 374–81; 2nd ed. (1631), 109; Selden to Erpenius, February 17, 1622, Bodleian Library; MS Selden Supra 108, 208. Selden's notebook for the 1631 edition of *Titles of Honour* with notes on collecting also survives as Clark Library, MS Selden 1. For the translations of various letters from princes, many of which came to James I by way of the East India Company, see Lincoln's Inn MSS, Hale 11, 48/221–42, and Hale 12, DD/59/354. The latest date is November 1631, the Safavid Shah Safi (1629–42) to Charles I, for which the Persian original is TNA SP 102/4/10. On Selden's work with Pococke at Oxford on Arabic and Ethiopian (Ge'ez) see Lambeth Palace, Fairhurst Papers MS 3513, f. 32–38. See also Selden's other transcriptions of letters concerning Tidore, Banten, and Spanish transshipments of silk in Lambeth Palace, MS 3472, f. 77, 85–86, 113.

55. "Dedication," *The Workes of Benjamin Jonson* (London: William Stansby, 1616). Selden's introduction to Hebrew, Arabic, and Aramaic probably occurred in 1609 through James Ussher. Cf. Jason Rosenblatt, "Milton, Natural Law, and Toleration," in *Milton and Toleration*, Sharon Achinstein et al., eds. (Oxford: Oxford University Press, 2007), 133.

56. Stow, *The Survay of London*, 29.

57. Purchas's first edition of the *Pilgrimage* (London: William Stansby for Henrie Fetherstone, 1613) contained a dedicatory poem by Selden, and in the second edition Purchas wrote of "my learned friend Mr. Selden of the Inner Temple, whose books and notes have furnished this booke with no few notes in this and other arguments." *Pilgrimage* 2 (1614), 131. On the Inns and Selden see Toomer, *John Selden*, 1: 9–27.

58. Sir John Borough, the keeper of records in the Tower of London, had assembled by 1633 a massive archive of documents supporting British sovereignty over the oceans, posthumously published during debates over the Navigation Acts as *The Soveraignty of the British Seas; Proved by Records, History and the Municipal Laws of this Kingdom* (London: J. Roberts, 1651). On Cotton, Arundel, and others see Jennifer Summit, *Memory's Library* (Chicago: University of Chicago Press, 2008), 136–96; Richard Ovenden, "The Libraries of the Antiquaries (c. 1580–1640) and the Idea of a National Collection," in *The Cambridge History of Libraries* 1 (Cambridge: Cambridge University Press, 2006); Colin Tite, *The Manuscript Library of Sir Robert Cotton* (London: British Library, 1994). In 1629, Charles I ordered the dispersion of Cotton's library, which after 1623 also included Camden's collection. Cf. Toomer, *Selden*, 1: 45–46. Selden's catalog of Arundel's marbles is *Marmora Arundeliana* (London: John Bill, 1629).

59. Buckingham to Tradescant, 1625, TNA SP 16/13/155–56.

60. J. C. T. Oates, *The Manuscripts of Thomas Erpenius* (Melbourne: Bibliographical Society of Australia and New Zealand, 1974); G. J. Vossius, *Oratio in obitum clarissimi ac praestantissimi viri, Thomae Erpenii . . . Item Catalogus librorum Orientalium, qui vel manuscripti, vel editi, in bibliotheca Erpeniana exstant* (Leiden: Johannis Maire, 1625).

61. The Persian manuscript is a Qur'anic commentary, Cambridge MS. Mm.4.15.

62. See Arthur Macgregor, *Tradescant's Rarities* (Oxford: Clarendon Press, 1983), 17–18. On loot as an attempt to show the "possessor's capacity for otherness" see Susan Stewart, *On Longing* (Baltimore: Johns Hopkins, 1984), 148; and Finbarr Flood, *Objects of Translation* (Princeton: Princeton University Press, 2009), 121–26. For Cope's influence

on Hakluyt, see the latter's dedicatory letter to Cecil, October 29, 1601, in the English edition of Antonie Galvano, *The Discoveries of the World* (London: G. Bishop, 1601).

63. Temple, ed., *Travels* 2:1 (1919), 1–3. For an introduction to collecting in this period see Marjorie Swann, *Curiosities and Texts: The Culture of Collecting in Early Modern England* (Philadelphia: University of Pennsylvania Press, 2001).

64. Bainbridge to Ussher, October 3, 1626, in Ussher, *Letters* no. 110, p. 370; Bainbridge to Selden, August 27, 1627, MS Selden supra 108, f. 236r. The copy of the *Minhaj* is now Bodleian MS Laud Or. 278. See Toomer, *Eastern Wisdom*, 72–73.

65. Summit, *Memory's Library*, 197, 208.

66. Francis Bacon, "The Advancement of Learning," book II, in *Francis Bacon: A Critical Edition*, Brian Vickers, ed. (Oxford: Oxford University Press, 1996), 230

67. Bacon, *Novum Organum* 1 (London: 1620), cxxix. The argument about the obscure origins of such technologies had its roots in Polydore Vergil's *De Inventoribus Rerum* (Venice: Christophorus de Pensis, 1499). See Paolo Rossi, *Philosophy, Technology, and the Arts in the Early Modern Era*, trans. Benjamin Nelson (New York: Harper Torchbooks, 1970), 83, 88–89.

68. Summit, *Memory's Library*, 203–4. For the publication date see Vickers, ed., *Francis Bacon*, 786–87. The first draft was possibly made at the time of the colonial conferences (ca. 1614–15), after *Mare Liberum* and before *Mare Clausum*.

69. Ph. S. van Ronkel, "Account of Six Malay Manuscripts of the Cambridge University Library," *Bijdragen tot de Taal-, Land- en Volkenkunde van Nederlandsch-Indie* 6:2 (1896), 1–53.

70. "waerinne dat vergadert zijn diversche woorden in Duyts en de Maleys," Cambridge MS Or. L 1.6.5; the handwriting changes at f. 26. See generally the edition by L. F. Brakel, *The Hikayat Muhammad Hanafiyyah* (The Hague: Martinus Nijhoff, 1975). The other manuscripts were a *Hikayat Yusuf*, Cambridge MS Or. Dd.5.37, and a composite manuscript, Cambridge MS Or. Gg.6.40.

71. *Hikayat Bayan Budiman*, Bodleian MS Pococke 433, acquired by the library in 1693 at Pococke's death. Pococke obtained it ca. 1630–40. The wrapper reads, "This is the Mola tounge Spoke By the Molaianes in the Sou[th] Seases the Coste of Vormeo ["Morneo" crossed out]" and "Malaica quaedam folia imperfect," referring to the fact that this includes only the first three stories. The Bodleian fragment is the oldest extant Malay copy; see R. O. Winstedt, ed., *Hikayat Bayan Budiman* (Kuala Lumpur: Oxford University Press, 1966).

72. *Hikayat Seri Rama*, MS Laud Or. 291. Max Saint, "Laud, Pococke, and Three Malay Manuscripts in Oxford," *Indonesia and the Malay World* 14:41 (November 1986), 45–48.

73. *Caritanira Amir*, Bodleian MS Jav.b.2. It is the earliest surviving copy of a menak. Probably because of Purchas, it was thought at the time to be from the Malabar Coast of southwest India, with an ink notation on the cover "Liber Lingua Malabanca Style ferrio scriptus."

74. Amartya Sen, *The Idea of Justice* (Cambridge: Harvard University Press, 2009), 70.

75. Lambeth Palace, Fairhurst Papers MS 3513, f. 37. By the late 1640s, Selden grew increasingly skeptical of kabalistic and occult theories of language, remarking on the artificiality of number as a "humane Imposicon." Selden, *Table Talk*, 84; Feingold, "John

Selden," 70. For universal language schemes see especially Francis Lodwick, *A Common Writing* (London: Francis Lodwick, 1647); Lodwick, *The Groundwork, or foundation laid (or so intended) for the framing of a new perfect language: and an universal or common writing* (London: s.n., 1652); Cave Beck, *The Universal Character* (London: Thomas Maxey, 1657); as well as the two broadsheets published by George Dalgarno in 1657 in David Cram and Jaap Maat, *George Dalgarno on Universal Language* (Oxford: Oxford University Press, 2001). Lodwick had a substantial collection of Chinese books by the 1650s, loaning Wilkins copies that included a Lord's Prayer he engraved for his *Essay.* Vivian Salmon, *The Works of Francis Lodwick* (London: Longman, 1972), 135; Felicity Henderson and William Poole, *On Language, Theology, and Utopia* (Oxford: Clarendon Press, 2010), 22–23; and John Wilkins, *Essay towards a Real Character and a Philosophical Language* (London: S. Gellibrand, 1668), 450–51.

76. The London *Polyglot* edited by Brian Walton appeared between 1654 and 1657 in nine languages (Charles I had asked Laud for twenty-four) with the help of Edward Pococke, Thomas Greaves, Thomas Hyde, James Ussher, and many others. See Peter Miller, "The 'Antiquarianization' of Biblical Scholarship and the London Polyglot Bible (1653–1657)," *Journal of the History of Ideas* 62:3 (July 2001), 463–82.

77. Selden's student texts, Clark Library Selden MS 10, with its study of cosmology, and MS.1963.007, "A Briefe treatise of a book called Speculum Universe or Universal mirror," May 17, 1605, already show extensive use of Greek and Hebrew as well as tabular logic. Clark MS 1, f. 133, has an early tree diagram by Selden of concepts of legal authority. For the popularity of Ramusian techniques in legal education see Abraham Fraunce, *The Lawiers Logike: Exemplifying the Praecepts of Logic by the practice of the common Lawe* (London: William How, 1588).

78. See Grotius, "The Preliminary Discourse," *The Rights of War and Peace* 1, Richard Tuck, ed. (Indianapolis: Liberty Fund, 2005), 79–81. This was not true of Lodwick as Henderson and Poole argue in *On Language*, x, 49, 201–2. Unlike Selden's historical perspective, Lodwick made this argument in both linguistic and proto-racial terms (the distinct "generation" and generations of black and white children) in a Restoration text that indirectly justified the Royal African Company (MS Sloane 913, f. 363–64).

79. See the inventory made ca. 1654 Bodleian Library MS Selden Supra 111, f. 121v–128r; and Inventory of Books, October 1649, Lambeth Palace, Fairhurst Papers MS 3513, f. 21, with emendations from November and December 1649; and M. Barratt, "The Library of John Selden and Its Later History," *Bodleian Library Record* 3:31 (March 1951), 128–42.

80. The quote is from *Areopagitica* and refers to Selden's *De Jure Naturali et Gentium Iuxta Disciplinam Ebraeorum* (London: Richard Bishop, 1640), dedication 1639. Selden took his epigram "loca, nullius ante trita solo. Iuvat integros accedere Fontes Atque haurire" from Lucretius, *De Rerum Natura*, 926–27. For the story of the requests see David Wilkins, *Opera omnia* 1, xliv; J. Milton French, *The Life Records of John Milton* 2 (New Brunswick: Rutgers University Press, 1949–58), 237.

81. Edward Pococke to Selden, August 23, 1650, Bodleian Library, Selden Supra 108, f. 147; and more generally several letters between Gerard Langbaine and Pococke between 1652 and 1654 in Bodleian Library, MS Selden Supra 109. Pococke and Selden collaborated on the Latin/Arabic text *Eutychii Patriarchae Alexaandrini Aannalium* (Oxford:

Henry Hall, 1654). Greaves used Selden's manuscripts for his *Discourse of the Roman Foot* (London: M. F. for William Lee, 1647) and a book Selden himself wanted published, *Elementa linguae Persicae* (London: Jacob Flesher, 1649). Selden saved Greaves's library from Parliamentary seizure in 1648; see Mordecai Feingold, "John Selden and the Nature of Seventeenth-Century Science," in *In the Presence of the Past*, Richard Bienvenu, ed. (Dordrecht: Kluwer, 1991), 66.

82. Theodor Graswinckel, *Maris liberi vindiciae* (The Hague: Adrian Vlacq, 1652); Selden, *Ioannis Seldeni vindiciae* (London: Cornelius Bee, 1653). Graswinckel had sent Selden a critique in 1635 that Selden did not respond to, which was more directly aimed at Pietro Burgus's *De dominio Serenissimae Genuensis Reipublicae in mari Ligustico* (Rome: Marcianus, 1641), supporting limited Genoese claims to the Ligurian Sea. Two other English pamphlets most likely by Nedham also supported the Parliamentary cause: *Additional Evidences Concerning the Right of Soveraigntie and Dominion of England in the Sea* (London: William Du Gard, 1652); and *Dominium Maris: or the Dominion of the Sea . . . translated out of Italian* (London: William Du Gard, 1652). English translations of Grotius, *Of the law of warre and peace*, appeared in London in 1654 and 1655. See Tuck, *Natural Rights Theories*, 89.

83. The phrase comes from *De synedriois et praefecturis juridicis veterum Ebraeorum liber tertius et ultimus*, 3:14.9 (London: Cornelius Bee, 1655), 304, in reference to a Greek inscription and frieze in the museum reproduced on 305.

84. See Eric Nelson, *The Hebrew Republic* (Cambridge: Harvard University Press, 2010); Adam Sutcliffe, "The Philosemitic Moment? Judaism and Republicanism in Seventeenth-Century European Thought," in Sutcliffe, et. al., *Philosemitism in History* (Cambridge: Cambridge University Press, 2011), 67–92.

85. Thomason's list of these books is Bodleian BB 8(9) Art. Selden; and *Catalogus Librobum Diversis Italiae* (London: John Legate, 1647), 47–56. This broader proposal in Parliament to build a modern public library at Cambridge failed. *Journals of the House of Commons* 5 (London: 1803), 512; Israel Abrahams and C. E. Sayle, "Purchase of Hebrew Books by the English Parliament in 1647," *Transactions of the Jewish Historical Society of England* (1915–17), 63–77; Oates, *Cambridge University Library*, 231–40; Toomer, *Selden*, 2:577.

86. Selden was particular interested in al-Idrisi's *Geographia Nubiensis*; cf. Feingold, "John Selden," 68. Bodley had in his library a surviving copy of Chaucer's "Treatise on the Astrolabe" (1391–93), Bodleian Library MS Bodley 619. On exchanges of such objects see Avner Ben-Zaken, *Cross-Cultural Scientific Exchanges in the Eastern Mediterranean, 1560–1660* (Baltimore: Johns Hopkins University Press, 2010). For the gift of the astrolabe see Toomer, *Selden*, 1:390–92.

87. Selden's astrolabe is Oxford Museum of the History of Science, no. 37527, North Africa, ca. early 17th century. Robert Gunther, *The Astrolabes of the World* 1 (London: Holland Press, 1976), 293–97.

88. Selden's edition was entitled *Xinqin quanxiang dazi tongsu yanyi sanguozhi zhuan* [新鐫全像大字通俗演義三國志傳 "Newly engraved, fully illustrated, great writing, everyday moral narrative *Three Kingdoms History*"] Bodleian Sinica 51/1–6. The 1592 Liu Longtian edition *Xinke anjian quanxiang piping sanguozhi zhuan* [新刻按鑑全像批評三國志傳] also made its way to London, surviving as juan 7–8 (University Library

Cambridge, donated Richard Holdsworth, 1649), juan 9–10 (Württemberg Landesbibliothek), juan 11–12 (partial, Bodleian Sinica 46, donated by Laud), juan 19–20 (BL 15333.e.1). The London booksellers George Thomason and Octavian Pullen had juans 15–20 of what appears to be a different edition *Xinke yanyi quanxiang sanguozhi zhuan* [新刻演義全像 三國志傳] printed at Liguang Pavilion. Bodleian Sinica 55, donated July 6, 1659. The oldest known edition is the 1548 Ye Fengchun obtained through Fujian in the 1570s, Escorial G.IV.24–30. See Helliwell, "Chinese Books in Europe in the Seventeenth Century"; Frances Wood, "Chinese Books in the British Museum," in *The Art of the Book in China*, Ming Wilson and Stacey Pierson, eds. (London: Percival David Foundation, 2006), 223–24; Anne McLaren, "Popularizing *The Romance of the Three Kingdoms*," *Journal of Oriental Studies* 33:2 (1995), 165–83. Hyde may have owned juan 19–20 and the colophon reprinted by Walton since he translated it with Shen in 1687 on the same sheet with notes about the Selden Map, BL Sloane 853, f. 23.

89. "Specimen Characterum Chinensum, ex initio eujusdam libri, eorum typis impressi," in Walton, ed., "Prolegomenon 2," *Biblia Sacra Polyglotta* (London: Thomas Roycraft, 1657), 14. The colophon reads "Wanli renchen zhongxia yue / shu lin Yu shi Shuangfeng tang" [萬曆壬辰仲夏月/書林余氏雙峰堂 "Wanli 29 (1592), second month of summer / Book collection of the Yu family, Shuangfeng Hall (Yu Xiangdou)"]. As noted, *juan* 1 from which the colophon came is no longer extant. Richard Holdsworth, who had *juan* 7 and 8 of the novel, was a signatory along with Walton and Ussher to a moderate Royalist petition in 1647 suggesting Charles I offer temporary toleration for dissenters. On the colophon see Lucille Chia, "Chinese Books in the Philippines," in *Chinese Circulations: Capital, Commodities and Networks in Southeast Asia*, Eric Tagliacozzo and Wen-Chin Chang, eds. (Durham: Duke University Press, 2011), 271–72.

90. Bodleian MS. Arch Selden A.1, ca. 1540, on European paper. Printed in Purchas, *Pilgrimes* 3 (1625), 1065–1117. See H. B. Nicholson, "The History of the Codex Mendoza," in Frances Berdan and Patricia Anawalt, *The Codex Mendoza* 1 (Berkeley: University of California Press, 1992), 1–2. The library catalog is MS Selden Supra 111, f. 121v–128r. B. C. Barker-Benfield suggests that the "liber" was the Codex Selden and the "rotulus" the Selden Roll. Barker-Benfield, "The Bindings of Codex Mendoza," *Bodleian Library Record* 17:2 (October 2000), 100, 104.

91. Purchas, *Pilgrimes* 3 (1625). 1067. The problem of divergent ways of describing historical time had been raised in the sixteenth century by Joseph Scaliger; see Anthony Grafton, *Joseph Scaliger* 2 (Oxford: Oxford University Press, 1993), 394–459.

92. Purchas, *Pilgrimes* 3 (1625), 1065.

93. Bodleian "Selden Roll," MS. Arch. Selden A. 72 (3). The history of the Mixtec was first discussed in Spanish in Antonio de Herrera y Tordesillas, *Historia general de los hechos de los Castellanos en las islas y tierra firme del Mar Oceano* 3 (Madrid: Juan Flamenco, 1601), ch. xii-xiv.

94. Bodleian "Codex Selden," MS. Arch Selden A.2, dated 1560 on cover, with 1556 as the last date according to the Aztec calendar. Elizabeth Hill Boone has suggested that this manuscript is noncartographic despite the footprint motif. Boone, "The House of the Eagle," in *Cave, City, and Eagle's Nest*, David Carrasco and Scott Sessions, eds. (Albuquerque: University of New Mexico Press, 2007), 35.

95. On Codex Bodley, Bodleian MS Mex D 1, see Nicholas Johnson, "Roads as Con-

nectors in Mixtec Pictorial Histories," in *Painted Books and Indigenous Knowledge in Mesoamerica*, Elizabeth Boone, ed. (New Orleans: Middle American Research Institute, 2005), 131. Codex Laud, Bodleian Library MS. Laud Misc. 678, was part of a set of manuscripts called the "Borgia group," made by the Tolteca-Chichimeca alliance between the Mixtecs and Tenochtitlan at Cholula and using a very different style.

96. See James Ussher, "Epistle to the Reader," *Annals of the World* (London: John Crook, 1658), which first appeared in Latin from 1650–54. Selden met Ussher in 1609 and corresponded with him, Bainbridge, and Peiresc on chronology in 1620s and 1630s, when Ussher was in Ireland as archbishop of Armagh (1625–56). Ussher delivered Selden's funeral oration in 1654. The two collaborated with Ralph Cudworth on the work *De anno civili veterum* (Leiden: Pieter van der Aa, 1644), a comparison of Jewish and Karaite calendars. See Feingold, "John Selden," 68; and Cudworth's notes in Selden's edition of Johann Kepler and Tycho Brahe, *Astronomia Nova αἰτιολογητός, seu Physica cœlestis* (Prague and Heidelberg: Gotthard Vögelin for Kepler, 1609), Bodleian A.1.2 Med. Seld. (previously G.1.16 Art. Seld.).

97. "Codicil," June 11, 1653, in David Wilkins, *Works of John Selden* 1 (London: 1726), lv. The compass is now Oxford Museum for the History of Science, 44055. For a full description of the provenance of the Selden Map see Batchelor, "The Rediscovery of the Selden Map: An Early Seventeenth-Century Chinese Map Depicting East Asian Shipping Routes," *Imago Mundi* 65:1 (January 2013), 37–63.

98. A "Sanhedrin" of 70 was the idea of the Fifth Monarchist John Rogers and Major General Thomas Harrison, although for practical reasons of representation Barebone's Parliament, which first met in July, had 140 members on Cromwell's suggestion. Selden's history of the Sanhedrin is *De synedriis et praefecturis juridicis veterum Ebraeorum*, published in three parts, 1650, 1653, and 1655 by Jacob Flesher and Cornelius Bee.

99. On the Rump's commercial policy see J. E. Farnell, "The Navigation Act of 1651, the First Dutch War, and the London Merchant Community," *Economic History Review* (1964), 438–54; Brenner, *Merchants and Revolution*, 577–637; and Blair Worden, *The Rump Parliament, 1648–1653* (Cambridge: Cambridge Univ. Press, 1974), 254–62.

100. See Marchamont Nedham, "To the Supreme Autoritie of the Nation, the Parliament of the Commonwealth of England, November 19, 1652," in John Selden, *Of the Dominion or Ownership of the Sea* (London: William Du Gard, 1652). William Watts made a translation in 1636 that may have been the basis for Nedham's edition. Cf. William Watts to John Selden, July 11, 1636, Bodleian MS Selden Supra 108, 82.

101. Bodleian Laud Or. 145. *Shunfeng xiangsong* is written in Chinese characters on the front flysheet with an interior inscription, "Liber Guil. Laud Archibpi Cant et Cancillar Universit Oxon 1637." Cf. the transcription along with another Bodleian rutter in Xiang Da, ed. *Liang zhong hai dao zhen jing* (Beijing: Zhonghua shu ju, 1961). For conflicting interpretations as to its composition see Roderick Ptak, "Jottings on Chinese Sailing Routes to Southeast Asia, Especially on the Eastern Route in Ming Times," *Portugal e a China. Conferencias nos encontros de historia luso-chinesa* (Lisbon: Fundação Oriente, 2001), 113–17; J. V. G. Mills, trans., *Ma Huan Ying-yai sheng-lan: The Overall Survey of the Ocean's Shores* (Cambridge: Cambridge University Press, 1970); Mills, "Chinese Navigators in the Insulinde about AD 1500," *Archipel* 18 (1979), 69–93; and Tian Rukang, "The First Printed Chinese Rutter—Duhai fangcheng," *T'oung Pao* 68: 1–3 (1982).

102. Purchas, *Pilgrimes*, I: xlvi.

103. See Edmund Scott, *Exact Discourse of the Subtilties, Fashishions [sic], Pollicies, Religion, and Ceremonies of the East Indians, as well Chyneses as Javans, there abyding and dweling* (London: W. White for Walter Burre, 1606).

104. BL Sloane 3668, 3959.

105. Purchas, *Pilgrimes*, I: 4, 385–95, 440–44.

106. Purchas, *Pilgrimes*, III: 2, 361.

107. BL Cotton Augustus, I.1.45. Cotton raided the State Papers with the claim he was going to establish a national library. Sarah Tyacke has suggested Tatton as one possible copyist (personal communication). See the very similar printed map "Sinarum Regni alioruq regnoru et insularu illi adiacentium description," ca. 1597–1607, HKUST, G7400 1590 .S54.

108. The Cotton map claims the Ming has "150 great Cittes, 235 small ones, 1154 townes, villages decayed 211, forts of guard 213." British Library, Cotton Augustus, II.ii.45. The *Mingshi* (Taibei: Guo fang yan jiu yuan, 1963; orig. Beijing: 1739), juan 40, counts 159 prefectures (*qu* 區), 240 subprefectures (*zhou* 州), and 1144 counties (*xian* 縣). Gerritsz's chart is Berlin, Staatsbibliothek Preussischer Kulturbesitz, T.7557. See the reproduction in Tyacke, "Gabriel Tatton," 49. For the Kadoya chart see the reproduction in Nakamura, "The Japanese Portolanos," fig. 4; and Nakamura, *Goshuinsen kokaizu* (Tokyo: Nihon Gakujutsu Shinkokai, 1965), 550–51. Both Ortelius (1587) and Jan Huygen van Linschoten, *Itinerario, Voyages ofte Schippvarert* (Amsterdam: Cornelis Claesz, 1596), between 22–23, included maps of East Asia as well.

109. Translated from the French, "Memoire des commoditéz de l'union et incommoditéz de la diversité des compaaignies traffiquants aux Indies orientales" (1615), Nationaal Archief, The Hague, 1.10.35.02, Collectie Hugo de Groot, Supplement, 40:137–41; see Borschberg, *Hugo Grotius*, 76, 315–16.

110. Richard Cocks, *Diary*, ed. E. M. Thompson, 2 (London: Hakluyt Society, 1883), 42, June 15, 1618. The wealthy and powerful Osaka merchant Sueyoshi Magozayemon Yoshiyasu (1570–1617) had the Japanese red seal licenses for Luzon, Tonkin, and Siam from 1604 to 1617 and owned a portolan chart. Hirosi Nakamura, "The Japanese Portolans of Portuguese Origin of the XVIth and XVIIth Centuries," *Imago Mundi* 18 (1964), 29. Li seems to have taken over those routes by 1618.

111. Chinese traders from Quanzhou and Zhangzhou in Fujian came to Taiwan from at least the 1590s as part of the deerskin trade; see Chen Di (陳第), *Dongfan ji* [東番記, "Eastern Foreigners Record"] (1603). The Selden map shows two resupply base camps on Taiwan, labeled north harbor (北港 *beigang*) near the subsequent Spanish fort San Salvador at Keelung (1626) and *jiali* forest (加里林), just north of what would become the Dutch Fort Zeelandia (1624) at Tayouan. One of these was likely run by the Japanese-born Chinese pirate Yan Siqi; cf. Tonio Andrade, *How Taiwan Became Chinese* (New York: Columbia University Press, 2007), 11; Salvador Diaz, "Relaçao" in José Borao, *Spaniards in Taiwan* 1 (Taipei: SMC, 2001), no. 21. Taiwan begins to appear in records of Li Dan's shipping in February 1618 along with Tonkin; see Iwao Seiichi, "Li Tan, Chief of the Chinese Residents at Hirado," *Memoirs of the Research Department of the Toyo Bunko* 17 (1958), 44–45.

112. Surviving accounts of the *Elizabeth* affair include Léon Pages, *Histoire de la reli-*

gion chrétienne au Japan 1 (Paris: C. Douniol, 1869), 450, transcribing a letter from one of the captured priests; Jacinto Orfanel, *Historia Eclesiastica de los sucessos de la christian-dad de Japon* (Madrid: Alonso Martin, 1633), 141–52; and the diary of the Hirado chief factor Richard Cocks, itself not part of the EIC papers but surviving as BL Additional MS 31,300–1, running from June 1, 1615, to January 14, 1619, and from December 5, 1620, to March 24, 1622. Other documents can be found in Anthony Farrington, ed., *The English Factory in Japan, 1613–1623* (London: British Library, 1991).

113. Cocks, *Diary*, 2: 324.

114. See the April 6, 1625, report of Wang San, an owner of a junk travelling from Quanzhou to Batavia, to the Dutch in *Daghregister gehouden in 't Casteel Batavia* (The Hague: Martinus Nijhoff, 1896), 139–40, which also gives a good account of the junk trade.

115. The figure of eighty is from Bartolomé Martínez, the Dominican provincial who slipped through the Dutch blockade in from Appari in January 1619 to warn Chinese merchants in Guangdong and Fujian. His report on this failed mission is APSR (Avila), Formosa, Tomo 1, f. 371–77, is reprinted in Borao, *Spaniards in Taiwan*, 1:46.

116. Letters by Richard Cocks (December 10, 1614, and February 15, 1617, Hirado), Edmund Sayer (December 5, 1615, December 4, 1616) in Purchas, *Pilgrimes*, 1.4.409–11. Sayer complains that the Chinese pilot they hired in Siam only knew coastal navigation and fell sick, forcing Sayer to muddle through the voyage.

117. When Shen Fuzong saw the Selden Map in 1687, he made a distinction between the lined diagram of the wind or compass rose as a *gepan* (格盤), a strange phrase meaning pattern board, and the actual label on the Selden Map of *luojing* (羅经). BL Sloane 853, f. 23.

118. Laud's *Shunfeng Xiangsong* will say to *zhen* or to needle rather than *ji* or "to plot" when giving navigational instructions. "Watch" (literally "change") indicated both time and distance, and for distance it was very much an approximation, somewhere between fourteen and twenty miles. Mei-Ling Hsu, "Chinese Marine Cartography: Sea Charts of Pre-Modern China," *Imago Mundi* 40 (1988), 112 n2.

119. In December 1650, Gerard Langbaine presented Selden with a Greek manuscript copy of Magdalen College Library's Alypius's "Introduction to Museic," which had complex tables showing the harmonic relationships of notes. Clark Library, MS 3.

120. See Roger Hart, *The Chinese Roots of Linear Algebra* (Baltimore: Johns Hopkins University Press, 2011); and more broadly Jack Goody, *The East in the West* (Cambridge: Cambridge University Press, 1996), 49–81.

121. See Batchelor, "The Selden Map Rediscovered," and in the same issue Stephen Davies, "The Construction of the Selden Map: Some Conjectures," *Imago Mundi* 65:1 (January 2013), 97–105.

122. William Baffin, *The Voyages of William Baffin* (London: Hakluyt Society, 1891), 145, 154. See also Laurens Reael, *Observatien of Ondervindinge an de Magneetsteen, end de Magnetische Kracht der Aerde* (Amsterdam: Spillebout, 1651), who subsequently made similar observations of ranges in the Indian Ocean and East Asia.

123. Henry Gellibrand, *A Discourse Mathematical on the Variation of the Magneti-call Needle* (London: William Jones, 1635).

124. A. R. T. Jonkers, *Earth's Magnetism in the Age of Sail* (Baltimore: Johns Hopkins University Press, 2003), 202. The map plotting this data only shows that collected from 1651 to 1700.

125. See Tyacke, "Gabriel Tatton," 42, 60; Jonkers, *Earth's Magnetism*, 138–41.

126. Fei Xin, *Xing cha sheng lan* [星槎勝覽 "Description of the Starry Raft"] (1436), quoting a popular saying among sailors.

127. Needham, *Science and Civilisation*, 4:1, 286, showing a compass on loan to the Ashmolean. On the concept of *jian* see Batchelor, "A Taste for the Interstitial (間): Translating Space from Beijing to London," in David Sabean and Malina Stefanovska, eds., *Spaces of the Self* (Toronto: University of Toronto Press, 2012), 281–304.

128. See Alexander Wylie, "The Magnetic Compass in China," *Chinese Researches* 3 (Shanghai, 1897), 157, originally *North China Herald*, March 15, 1859; cited by Joseph Needham, *Science and Civilisation in China* (Cambridge: Cambridge University Press, 1971), 4:1, 310. I have not been able to verify these citations by Wylie. See also the figure of 7.5° west recorded in Xu Zhimo (徐之鏌), *Chong juan luojing ding men zhen jian yi tu jie* [重鐫羅經頂門針簡易圖解] (ca. 1580?, preface 1623) reprinted in *Siku quanshu cunmu congshu zibu* 64 (Tainan: Zhuangyan wenhua, 1997). On the problematic nature of Needham's table as a collection of disparate and decontextualized examples see Fu Daiwie, "On Mengxi Bitan's World of Marginalities and 'South-Pointing Needles,'" in Viviane Alleton and Michael Lackner, eds., *De l'un au multiple* (Paris: Editions MSH, 1999), 177–201.

129. The Jesuit response to Gellibrand's discoveries came in Athanasius Kircher, *Magnes* (Rome: L. Orignani, 1641) which argued that magnetism flowed between fixed celestial and terrestrial poles along fibers. In support of this he used Jesuit observations from around the world, including China in the 1654 edition (Rome: V. Mascardi, 1654) of the original 1641 work.

130. *Bianyong Xuehai qunyu* [便用學海群玉, "Convenient to use: Seas of knowledge, mines of jade"], revised by Wu Weizi (Jianyang: Xiong Chongyu, 1607), juan 2. The surviving copy is Leiden University Acad. 226. See Koos Kuiper, *Catalogue of Chinese and Sino-Western Manuscripts in the Central Library of Leiden University* (Leiden: Leiden University Library, 2005), 70–75. See also the similar images in Xiong Chongyu (熊冲宇 aka. Chengye), ed., *Xinke taijian lifa zengbu ying fu tongshu* [新刻太监曆法增補應福通書, "Newly engraved court calendar system supplement and almanac"] (Jianyang: Xiong Chongyu, ca. 1573–1619), Naikaku Bunko, 305.288; Zhang Huang (章潢, 1527–1608), *Tushu bian* [圖書編 "Compendium of Maps and Writings"], ed. Wang Shanglie (1613), which was compiled between 1562 and 1585; and Yu Xiangdou, ed. (余象斗) *Xinke tianxia simin bianlan santai wanyong zhengzong* [新刻天下四民便覽三台萬用正宗 "Santai's newly engraved convenient orthodox instructions for myriad uses among the people of the world"] (Jiangyang: Yu Xiangdou, 1599), Tokyo Daigaku, Toyo Bunka, N3079, reproduced in *Chugoku nichiyou ruisho shusei* 3–5 (Tokyo: Kyuko shoin, 2000). Timothy Brook sees the *Wanyong zhengzong* as the source for the map of the Ming; personal communication, April 22, 2013.

131. On *fenye* versus geomancy or the "form" vs. "compass" schools see especially Richard Smith, *Fortune-tellers and Philosophers: Divination in Traditional Chinese Society* (Boulder: Westview Press, 1991), 67–70, 134–39; and John Henderson, "Chinese Cosmographical Thought," *History of Cartography*, 2:12:210, 216–24. This is the reason that astronomy was so important in Beijing. Purchas in a translation of a letter from a Fujianese official uses a tributary language to state that the emperor of China "is so mightie, that he governeth all that the Moone and Sunne doe shine upon" [*Pilgrimes*, 3.2.309].

132. The key work defining the South China Sea through the eastern and western routes was Chen Dazhen (陳大震), *Dade nanhai zhi* [大德南海志 "Great South China Sea Record"] (ca. 1307). See also Wang Dayuan, *Daoyi zhilue* (1350), and his reference to southern "island peoples." For an indication of what was new ca. 1617 see Zhang Xie, *Dongxiyang kao* [东西洋考, "Inspection of Eastern and Western Oceans"] (Zhangzhou: 1617/1618).

133. Jane Burbank and Frederick Cooper, *Empires in World History* (Princeton: Princeton University Press, 2010), 183. See also Lauren Benton, *A Search for Sovereignty: Law and Geography in European Empires* (Cambridge: Cambridge University Press, 2009); and Sailha Belmessous, *Native Claims: Indigenous Law Against Empire* (Oxford: Oxford University Press, 2011).

CHAPTER FOUR

1. See the lavish printed version meant to circulate outside London. John Ogilby, *The Relation of his Majesties Entertainment Passing through the City of London, to his Coronation* (London: Thomas Roycroft, 1661).

2. See Plato, *The Sophist*, 223d, as cited in Grotius, *Mare Liberum*, ch. 8. On this theme of the royal emporium, which emerged during the trade depression of 1659, see Blair Hoxby, "The Government of Trade: Commerce, Politics, and the Courtly Art of the Restoration," *ELH* 66:3 (1999), 591–627.

3. On the fragility of this settlement see Gary S. De Krey, *London and the Restoration* (Cambridge: Cambridge University Press, 2005), 14–15, 47–63.

4. Barbour, *Before Orientalism*, 68–101; Daniel Vitkus, *Turning Turk: English Theater and the Multicultural Mediterranean, 1570–1630* (London: Palgrave, 2003); Jonathan Burton, *Traffic and Turning: Commerce, Conversion, and Islam in English Drama* (Newark: University of Delaware Press, 2005); Kevin Sharpe, *Image Wars: Promoting Kings and Commonwealths in England, 1603–1660* (New Haven: Yale University Press, 2010).

5. On the depiction of Piazza Navona and the obelisk in print during the 1650s see Rose Marie San Juan, *Rome: A City Out of Print* (Minneapolis: University of Minnesota Press, 2001), 187–217; and more generally Dorothy Metzger Habel, *Urban Development of Rome in the Age of Alexander VII* (Cambridge: Cambridge University Press, 2002). For similar efforts in Paris see Orest Ranum, *Paris in the Age of Absolutism* (University Park: Penn State University Press, 2002). For the importance of engraving to Restoration royalism see the contemporary accounts of John Evelyn, who claimed engraving was a Chinese art, *Sculptura or the history and art of chalcography and engraving in copper* (London: J. C. for G. Beedle, 1662), 32–33, 47; and the account of Charles II's engraver William Faithorne, *Art of Graveing and Etching* (London: William Faithorne, 1662).

6. Brenner, *Merchants and Revolution*, 4.

7. The idea that "tribute" is a concept that could still translate between the Ming and Qing and the Portuguese and Dutch is elaborated in John Wills, *Pepper, Guns, and Parleys* (Cambridge: Harvard University Press, 1974); and Wills, *Embassies and Illusions* (Cambridge: Harvard University Press, 1984).

8. Ogilby, *Entertainments*, 9–10.

9. Chaudhuri, *Trading World of Asia*, 143. For the detailed workings of this in the

Coromandel Coast trade see Sinnappah Arasaratnam, *Merchants, Companies, and Commerce on the Coromandel Coast, 1650–1740* (Delhi: Oxford University Press, 1986).

10. On the need for cooperation in the Restoration see Arnold Sherman, "Pressure from Leadenhall: The East India Company Lobby, 1660–1678," *Business History Review* 50:3 (Autumn 1976), 329–55. On the new emphasis on slavery in the Atlantic colonies and indeed "state regulation over people" and "territory" see Christopher Tomlins, *Freedom Bound: Law, Labor, and Civic Identity in Colonizing English America* (Cambridge: Cambridge University Press, 2010), 426–28; Alison Games, *The Web of Empire* (New York: Oxford University Press, 2008), 291–93.

11. October 9, 1651, in Henry Scobell, *A Collection of Several Acts of Parliament* 2 (London: John Field, 1652), 176.

12. 12 Car. II, Cap. 18, "Charles II, 1660: An Act for the Encouraging and increasing of Shipping and Navigation," *Statutes of the Realm* 5 (London: 1819), 246–50.

13. BL Add. MSS 25,115 f. 81; TNA C.O. 389/1, f. 8; and an incomplete copy TNA C.O. 77/7, no. 90.

14. BL IOR, G/21/4, f. 4–8.

15. "An Act for the Encouragement of Trade" [1663], *Statutes of the Realm* 5:449–52.

16. The model was the Guinea Company ("The Company of Adventurers of London," est. 1618), which because of lack of capital had issued licenses to private traders, as well as a slew of other competing and often short-lived organizations up until 1660. Cf. Holly Brewer, "Slavery, Sovereignty, and 'Inheritable Blood,'" http://www.nyu.edu/pages /atlantic/Lockeslaverynyu.pdf, accessed November 1, 2012.

17. "Articles of Agreement between the Royal Company and the East India Company made the 16th day of October 1662," BL, IOR, E/3/86 f.171; K. G. Davies, *The Royal African Company* (London: Taylor and Francis, 1957), 41–46; Chaudhuri, *Trading World of Asia*, 169–70.

18. See Simon Schaffer, "Golden Means: Assay Instruments and the Geography of Precision in the Guinea Trade," in *Instruments, Travel, and Science: Itineraries of Precision*, Marie-Noëlle Bourguet, et al. (London: Routledge, 2002), 20–50.

19. The trade was complex in that the Dutch made better guns for better prices, prompting complaints by English gunmakers in the 1680s. Often even East Indian goods like cowries were obtained in Holland. See Davies, *Royal African Company*, 173.

20. The Dutch slave trade grew to 64,800 people and the Portuguese fell to 53,700 people from 1651 to 1675 after the Dutch pushed the Portuguese out of the Gold Coast (Ghana) in 1642. See David Eltis, "The Volume and Structure of the Slave Trade: A Reassessment," *William and Mary Quarterly* 58:1 (2001), 43. According to the slave trade database, fifty-five voyages left London under the Company and its successor the Royal African Company between 1663 and 1675. Most of these voyages went to Barbados and Jamaica. Cf. The Trans-Atlantic Slave Trade Database, http://www.slave voyages.org.

21. Bruce Lenman, "The East India Company and the Trade in Non-Metallic Precious Metals from Sir Thomas Roe to Diamond Pitt," in *The World of the East India Company*, H. V. Bowen, et al., eds. (Boydell Press, 2002), 97–109. Robert Boyle in his 1663 "Observations on a Diamond that Shines in the Dark" noted that "diamonds deserve the rather to be enquired into, because the commerce they help to maintain between the western

and eastern parts of the world, is very considerable." Peter Shaw, ed., *The Philosophical Works of the Honourable Robert Boyle* 3 (London: W. and J. Innys, 1725), 144.

22. Gedalia Yogev, *Diamonds and Coral* (Leicester: Leicester University Press, 1978), 82–93.

23. See Oxinden's correspondence in BM Add 40699, 40700, 40701; Larry Neal and Stephen Quinn, "Markets and Institutions in the Rise of London as a Financial Center in the Seventeenth Century," in *Finance, Intermediaries, and Economic Development*, Stanley Engerman et al, eds. (Cambridge: Cambridge University Press, 2003), 19–20.

24. Craig Muldrew, *The Economy of Obligation: The Culture of Credit and Social Relations in Early Modern England* (London: Macmillan, 1998), 115; S. Quinn, "Balances and Goldsmith-Bankers: The Co-ordination and Control of Inter-banker Clearing in Seventeenth-Century London," in *Goldsmiths, Silversmiths, and Bankers: Innovation and the Transfer of Skill, 1550 to 1750*, D. Mitchell, ed. (London: Centre for Metropolitan History, 1995), 53–76.

25. Ward Barrett, "World Bullion Flows," in *The Rise of Merchant Empires*, J. D. Tracy, ed. (Cambridge: Cambridge University Press, 1990), 251. The silver/gold conversion ratio used here is 14.5 to 1.

26. J. Horsefield, *British Monetary Experiments* (Cambridge: Harvard University Press, 1960), xii-xiv.

27. Von Glahn, *Fountains*, 226. On this return to bimetallic equilibrium see Flynn and Giráldez, "Cycles of Silver," 391–427. Goldstone describes this as especially harmful for the Ottomans and Spanish. Goldstone, *Revolution and Rebellion*, chapter 4. On the effects on the Dutch see Leonard Blussé, "No Boats to China: The Dutch East India Company and the Changing Pattern of the China Sea Trade, 1635–1690," *Modern Asian Studies*, 30 (1996), 51–76.

28. See the accounts of expenses for soldiers in Bombay from February 1662 to December 1664, TNA CO 77/8 (1655–62) f. 170.

29. G. Z. Refai, "Sir George Oxinden and Bombay, 1662–1669," *English Historical Review* 364 (1977), 573–81.

30. See "The State of the Ship King Ferdinando's Voyage to East India," TNA CO 77/11, 1668–70, f. 222, a printed broadsheet about fraud and conspiracy in relation to Oxinden's voyage to China. The resulting lawsuit lingered in the House of Lords. "William Love, et al. v. Henry and Sir James Oxinden, et al.," October 30, 1673.

31. Richard Temple, ed., *The Diaries of Streynsham Master* 1 (London: John Murray, 1911), 190–92; and see correspondence in BL Add. MSS 40696–40713, 54332–54334.

32. TNA SP Ext. 8.2 f. 126f is the original Arabic letter, and the English translation is f. 126r. A second series of letters to both Charles II and Christian V of Denmark from 1675 are at f. 45, 46, and 58.

33. Cotton Mather, *A Brief History of the Warr with the Indians in New-England* (Boston and London: John Foster and Richard Chiswell, 1676); Elliot, *Empires of the Atlantic World*, 78, 102, 149–50, 266–67.

34. Peter Heylyn, *Microcosmus or A Little Description of the Great World: A Treatise Historicall, Geographicall, Political, Theological* (Oxford: John Lichfield and James Short, 1621), "Dedication," 11. The presentation copy is Huntington Rare Books 55354.

35. George Vernon, *The Life of the Learned and Reverend Dr. Peter Heylyn* (London:

C. Harper, 1682); and John Barnard, *Theologico-Historicus, or the True Life of the Most Reverend Divine, and Excellent Historian Peter Heylyn* (London: C. Harper, 1683).

36. Foster, *Court Minutes, 1660–3*, 148–57.

37. James Hart, *Justice Upon Petition* (London: Harper Collins, 1991), 246; December 17, 1667, *Journal of the House of Lords* 12 (London: 1767–1830), 172–73.

38. De Kray, *London and the Restoration*, 321–25; TNA Royal African Company Court Minute Book, 1672–78, T/70/100, f. 8–22.

39. See Shapin and Schaffer, *Leviathan and the Air Pump*; Peter Dear, "*Totius in verba*: Rhetoric and Authority in the Early Royal Society," *Isis* 76 (1985), 145–61.

40. Wilkins, *Essay Towards a Real Character*, 10, 13. On Wilkins and Webb see especially Rhodri Lewis, *Language, mind and nature: artificial languages in England from Bacon to Locke* (Cambridge: Cambridge University Press, 2007), 190 and passim; David Porter, *Ideographia*, 26–48. The efforts with regard to Malay were picked up by William Mainston in tandem with the 1682 Bantenese embassy to London, see his "Gramatica Mallayo-Anglica," Bodleian Library MS Ashmole 1808, f. v-x, donated by Elias Ashmole (d. 1692). Another copy is BL Oriental Malay Add. 7043. On the circulation of Mainston's manuscript through Robert Boyle see Mainston to Boyle, December 19, 1682, May 15, 1683, Michael Hunter, ed. *Correspondence of Robert Boyle* 5 (London: Pickering and Chatto, 2001), 369, 411. Brief accounts of Malay as a polite language of merchants appeared earlier in Linschoten, *Itinerario, voyage ofte schipvaert* (Amsterdam: Cornelis Claesz, 1596), 24. See William Poole, *The World Makers* (Oxford: Peter Lang, 2010), 82.

41. John Webb, *An Historical Essay Endeavoring a Probability that the Language of the Empire of China is the Primitive Language* (London: Nathaniel Brook, 1669), 118. A second edition was printed under the title *The Antiquity of China* (London: Obadiah Blagrave, 1678). The dedicatory epistle is dated May 1668. Webb had designed the frontispiece for Walton's *Polyglot Bible* (1657).

42. Beale to Evelyn, July 14, 1668, BL Add. MS 78312, f. 105.

43. On Ogilby see Katherine S. Van Eerde, *John Ogilby* (Folkestone: Dawson, 1976); on Hooke and Ogilby see E. G. R. Taylor, "Robert Hooke and the Cartographical Projects of the Late Seventeenth Century (1666–1696)," *Geographical Journal* 90 (1937), 529–40

44. John Ogilby, *An Embassy from the East-India Company of the United Provinces, to the Grand Tartar Cham Emperour of China* (London: John Macock, 1669), translating Johan Nieuhof, *Het Gezantschap der Neerlandtsche Oost-Indische Compagnie aan den Grooten Tartarischen Cham* (Amsterdam: van Meurs, 1665). On this essay see Leonard Blussé and R. Falkenburg, *Johan Nieuhofs Beelden van Een Chinareis, 1655–1657* (Middelburg: Stichting VOC publicaties, 1987).

45. TNA SP 44/23, Warrants, 1666–67, f. 416–17.

46. Lach and Van Kley, *Asia* 3:1 (1993), 483.

47. Following Ogilby, Elkanah Settle's play, *The Conquest of China by the Tartars* (London: 1676; performed ca. February 1674) depicted the Ming/Qing transition as a legitimate succession, against some more pessimistic Dutch literary accounts. See Edwin Van Kley, "An Alternative Muse: The Manchu Conquest of China in the Literature of Seventeenth-Century Northern Europe," *European Studies Review* 6 (1976), 21–44. News of the conquest had come from Martino Martini's *Novus Atlas Sinensis* (Amsterdam: Blaeu, 1655), translated into Dutch, French, and German, and his history *Sinicae histo-*

riae decas prima (1658), but only the translation of Álvaro Semedo, *The History of that Great and Renowned Monarchy of China* (London: E. Tyler for John Crook, 1655), composed in 1638 before the fall of the Ming, appeared in English before Ogilby's translations.

48. See Benjamin Schmidt, "Accumulating the World," in Lissa Roberts, ed., *Centers and Cycles of Accumulation in and Around the Netherlands* (Munster: LIT Verlag, 2011), 129–54; Friderike Ulrichs, *Johan Nieuhofs Blick auf China* (Wiesbaden: Harrassowitz Verlag, 2003), 43–53; Dawn Odell, "The Soul of Transactions: Illustration and Johan Nieuhof's Travels in China," *De zeventiende eeuw* 17:3 (2001), 225–42; Leonard Blussé and R. L. Falkenburg, *Johan Nieuhofs beelden van een Chinareis* (Middelburg: Stichting VOC Publicaties, 1987).

49. On engravings of China see Ernst van den Boogaert, *Het verheven en verdoven Azië: Woord en beeld in het Itinerario en de Icones van Jan Huygen van Linschoten* (Amsterdam: Het Spinhuis, 2000); Sun Ying, *Wandlungen des europäischen Chinabildes in illustrierten Reiseberichten des 17. und 18. Jahrhunderts* (Frankfurt am Main: Peter Lang, 1996); Richard Strassberg, *China on Paper* (Los Angeles: Getty Research Institute, 2007).

50. Athanasius Kircher, *China monumentis, qua sacris qua profanis, nec non variis naturae et artis spectaculis, aliarumque rerum memorabilium argumentis illustrata* ["*China illustrata*"] (Amsterdam: Jan Janszoon van Waesberge and Eliza Weyerstraet, 1667), with engravings taken from Chinese and Mughal originals, possibly acquired by the Austrian Jesuit Johann Grueber (1623–80). Lack and Van Kley, *Asia* 3:4 (1993), 1737. It was translated for the same publishers into Dutch by J. H. Glazemaker 1668 and French by F. S. Dalquié in 1670. Kircher's engravings of a "bridge" (棧道 *zhandao*) in Shaanxi and the Great Wall appeared first in London in the *Philosophical Transactions* 2:26 (June 3, 1667), 484–88. On his religious syncretism see Dino Pastine, *La nascita dell'idolatria: L'oriente religioso di Athanasius Kircher* (Firenze: La Tuova Italia, 1978). Schall's work was edited for European publication by the Jesuit Johan Foresi, *Historica Narratio de Initio et Progressu Missionis Societatis Jesu apud Chinenses* (Vienna: Matthaei Cosmerovij, 1665).

51. John Adams [Johann Adam Schall von Bell], "A Narrative of the Success of an Embassage sent by John Maatzuyker de Badem General of Batavia Unto the Emperour of China and Tartary, the 20th of July 1655 . . . Written by a Jesuite in those Parts," in Ogilby, *Embassy*, 9 (added to the end and paginated separately).

52. See Haun Saussy, "*China Illustrata*: The Universe in a Cup of Tea," in *The Great Art of Knowing*, Daniel Stolzenberg, ed. (Stanford: Stanford University Libraries, 2001), 107.

53. On such late seventeenth-century imperial image making in Qing China, Russia, and France see Laura Hostetler, *Qing Colonial Enterprise: Ethnography and Cartography in Early Modern China* (Chicago: University of Chicago Press, 2001).

54. George Humble published *The Prospect of the Most Famous Parts of the World* (London: John Dawson, 1627), with some maps augmented or revised by John Speed. The engraving was all done in Holland.

55. Bodleian Western Manuscripts, MSS Wood 658, f. 792.

56. TNA SP 29/173, f. 226, item 109.

57. February 29, 1684, BL IOR E/3/90, f. 254.

58. Arnold Montanus, *De Nieuwe en Onbekende Weereld Beschryving van America en 't Zuidland* (Amsterdam: Jacob van Meurs, 1671) for America; and Olfert Dapper's

Naukeurige Beschrijvinge der Afrikaensche Gewesten (Amsterdam: Jacob van Meurs, 1668); Dapper, *Gedenkwaerdige Gesantschappen der Oost-Indische Maetschappy in't Vereenigde Nederland aen de Kaisaren van Japan* (Amsterdam: Jacob van Meurs, 1669); and Dapper, *Asia, of naukerige beschrijving van het rijk des grooten Mogols en de groot gedeelt van Indien* (Amsterdam: Jakob van Meurs, 1672), for Africa, Japan, and Asia respectively; and, in addition to a new edition of Nieuhof, Dapper's *Gedenkwaerdig bedryf der Nederlandsche Oost-indische Maetschappye op de kuste et in het keizer-rijk van Taising of China*, 2 vols. (Amsterdam: Jacob van Meurs 1670) for China. For the history of these see John Wills, "Author, Publisher, Patron, World: A Case Study of Old Books and Global Consciousness," *Journal of Early Modern History* 13 (2009), 375–433.

59. For Blome's unsuccessful attempt in April 1669 to register to translate and publish the Africa volume, see G. E. Eyre, *A transcript of the registers of the worshipful company of Stationers* 2 (New York: Smith, 1950), 399. Ogilby's *Brittania* (1675) also rivaled Blome's 1673 edition.

60. As has been argued in epistemological and political terms by Wills, "Author, Publisher," 376, 389–90. On the Radical Enlightenment concept in terms of religion see Jonathan Israel, *Radical Enlightenment: Philosophy and the Making of Modernity, 1650–1750* (Oxford: Oxford University Press, 2001); and in terms of commerce see Harold Cook, *Matters of Exchange* (New Haven: Yale University Press, 2007).

61. Ogilby, *Africa*, 424–29.

62. See Temple's very popular *Observations Upon the United Provinces of the Netherlands* (London: A. Maxwell, 1673), written ca. 1668 during Temple's embassy and submitted in 1670 but only published during the war.

63. See David Armitage, "John Locke, Carolina, and the *Two Treatises on Government*," *Political Theory* 32:5 (October 2004), 602–27. The originals of the "Fundamental Constitutions" are July 21, 1669, TNA 30/24/47/3, and as a broadsheet Bodleian Ashmole F4 (42) with the manuscript signature of Shaftesbury, dated March 1, 1670, but probably published between 1671 and 1672.

64. Ogilby, *America*, 206–7.

65. On this synthesis see Wills, "Author, Publisher," 405–13.

66. "The Chineses also keep great Feasts in their Vessels on the River, making merry with varieties of Meat and strong Liquor; in which manner the greatest Mandarins often recreate themselves." Ogilby, *Atlas Chinensis* (1671), 358–59, from Dapper, *Gedenkwaerdig*, 392. On travel and pilgrimages in relation to consumption see Pomeranz, *Great Divergence*, 142; Glen Dudbridge, "A Pilgrimage in Seventeenth Century Fiction: T'ai-shan and the *Hsing-shih yin-yuan chuan*," *T'oung Pao* 77:4–5 (1991), 226–52; and Susan Naquin and Chun-fang Yu, eds., *Pilgrims and Sacred Sites in China* (Berkeley: University of California Press, 1992). On similar images on porcelain see Batchelor, "On the Movement of Porcelains," 104.

67. Liuxi Meng, *Poetry as Power* (Lanham: Lexington Books, 2007), 26.

68. Ogilby, *Atlas Chinensis* (1671), 571. The Dapper versions of these are *Beschry-ving*, 106–11.

69. Ogilby, *Atlas Chinensis* (1671), 543, see also the further lists on 544–46; and Dapper, p. 136–123 (printer's error, actually 136–139).

70. Gu Zuyu, *Dushi fangyu jiyao* (ca. 1630–60), juan 90, gives a figure of 3,300 *li*

rather than 3,340 (Ogilby/Dapper) for Hangzhou (Hangcheu), the capital of Zhejiang province. Juan 80 gives a figure of 5,870 for Changsha, the capital of Hunan, rather than 5,570 for "Junnan" (transposing the province and the capital in this case). Gu's text was not printed until 1811. See also Gu Yanwu's (顧炎武) Tianxia Junguo libing shu (天下郡國利病書 "On the benefits and faults of imperial local administration," ca. 1639–62). For the Shangcheng yilan, see Chia, Printing for Profit, 228–29, 325; and more generally Timothy Brook, Geographical Sources of Ming-Qing History (Ann Arbor: University of Michigan Press, 1988). My thanks to Lucille Chia for providing me with a copy of the text from the Naikaku Bunko 123.0005.

71. Ogilby, Atlas Chinensis, 613.

72. An advertisement concerning the English atlas, with the proposals (London: John Ogilby, February 10, 1671/2), copy in Harvard University Special Collections.

73. Ogilby, Britannia: Vol 1 or an Illustration of the Kingdom of England and Dominion of Wales: By a Geographical and Historical Description of the Principal Roads thereof. Actually Admeasured and Delineated in a Century of Whole-Sheet Copper-Sculps (London: John Ogilby, 1675). It was almost immediately pirated for the new issue of John Speed and Robert White, Theater of the Empire of Great Britain (London: Thomas Basset and Richard Chiswel, 1676).

74. 12 Car. II, cap. 35. See also John Hill's early scheme for a London to York penny post, Penny Post, or a Vindication of the Liberty and Birthright of every Englishman in carrying Merchants' and other Men's Letters against any restraint of Farmers (London: s.n., 1659).

75. James How, Epistolary Spaces (Aldershot: Ashgate, 2003), 52–55; Howard Robinson, The British Post Office (Princeton: Princeton University Press, 1948).

76. Ogilby also issued in 1675 the inexpensively priced Itinerarium Angliae, or a Book of Roads, which appeared in various versions after Ogilby's death from the printer William Morgan, Ogilby's partner and successor as cosmographer to Charles II.

77. In 1680, William Dockwra and Robert Murray patented the penny post, a private postal system for London that delivered mail in a seven-mile radius from the center of the city connected with networks of shops, coffee houses, taverns, and pubs. See The Practical Method of the Penny-Post (London: 1681); Steven Pincus, 1688: The First Modern Revolution (New Haven: Yale University Press, 2009), 72; Mark Knights, Politics and Opinion in Crisis (Cambridge: Cambridge University Press, 2006), 173.

78. The notion of "dispersed knowledges" comes from Michael Fischer, Mute Dreams, Blind Owls, and Dispersed Knowledges (Durham: Duke University Press, 2004), especially part 1 about Zoroastrian ritual and the Shahnama.

79. On the fears associated with this see Nabil Matar, Turks, Moors, and Englishmen (New York: Columbia University Press, 1999), 105–7.

80. On Dryden see Rahul Sapra, The Limits of Orientalism: Seventeenth-Century Representations of India (Newark: University of Delaware Press, 2011). The play reenacts the succession crisis of Shah Jahan (d. 1666) beginning in 1658 as described by François Bernier, Histoire de la dernière revolution des états du Grand Mogul (Paris: Claude Barbin, 1670, 1671); and Évènmens particuliers, ou ce qui s'est passé de plus considerable après la guerre pendant cinq ans . . . dans les États du Grand Mogul (Paris: Claude Barbin, 1670, 1671), translated respectively by the secretary of the Royal Society, Henry

Oldenburg, as *The History of the Late Revolution of the Empire of the Great Mogul* (London: Moses Pitt, 1671, 1676); and *A Continuation of the Memoires of Monsier Bernier concerning the empire of the Great Mogol* (London: Moses Pitt, 1672). On Bernier see Dew, *Orientalism in Louis XIV's France*, 131–67.

81. See the broadsheet Thomas Garraway (Garway), "An exact description of the growth, qualities and virtues of the tea leaf" (London: ca. 1658), British Museum; "That excellent and by all Physitians approved China Drink," *Mercurius Politicus* 435 (September 1658); *A True and Perfect Description of the Strange and Wonderful Elephant Sent from the East-Indies and Brought to London on Tuesday the Third of August, 1675* (London: 1675). Dryden's play premiered on November 17, 1675.

82. In addition to donating a large number of Persian and Arabic manuscripts to the Bodleian, Laud donated in 1639 a collection of eight Arabic and Persian manuscripts to St. Johns, acquired from Kenelm Digby before he went into exile in France. See Emilie Savage-Smith, *A Descriptive Catalogue of Oriental Manuscripts at St. John's College* (Oxford: Oxford University Press, 2005). These included Ulugh Beg's *Ziji-i Jadid-i Sultani* (ca. 1420–38) (this copy dated 1532, St. Johns College, Oxford MS 151). Hyde used the Ulugh Beg manuscript along with Bodleian MS Pococke 226 and MS Saville 46 for his *Tabulae longitudiniis and latitudinis stellarum fixarum ex observatione Ulugh Beighi* (Oxford: Henry Hall, 1665), updating three previous partial editions (1648, 1650, 1652) by John Greaves, who used his own Persian and Arabic copies—Bodleian MS Greaves 5 (Cairo: al-Rifa'i, 1536), and St. Johns MS 91, the latter Arabic copy had been acquired from Laud after 1640 and extensively annotated by Greaves.

83. *Risalat Hayy ibn Yaqzan* (AH 707/1303 CE), Bodleian Library MS Pococke 263. Edward Pococke (ed.) and Edward Pococke, Jr. (trans.), *Philosophus Autodidactus sive Epistola Abi Jaafar ebn Tophail de Hai Ebn Yokdhan. In qua Ostenditur quomodo ex Inferiorum contemplatione ad Superiorum notitiam Ratio humana ascendere possit* (Oxford: H. Hall, 1671). See G. J. Toomer, *Eastern Wisedome and Learning* (Oxford: 1996), 218–23; G. A. Russell, "The Impact of Philosophus Autodidactus," in Russell, ed., *The 'Arabick' Interest of the Natural Philosophers in Seventeenth Century England* (Leiden: Brill, 1994), 224–62; and Avner Ben-Zaken, *Reading Hayy Ibn-Yaqzan: A Cross-Cultural History of Autodidacticism* (Baltimore: Johns Hopkins University Press, 2011).

84. Pococke, *Litvrgiæ Ecclesiæ Anglicanæ, Partes pracipuae . . . in linguam Arabicam traductae* (Oxford: Henry Hall, 1674).

85. Failure is an important theme in Games, *The Web of Empire*, 253, which emphasizes religious fragmentation in England rather than the role of textual production outside of England.

86. Eliot, *Mamusse Wunneetupanatamwe Up-Biblum God* (Cambridge, MA: Samuel Green and Marmaduke Johnson, 1663), was presented to Charles II as well as to Oxford and Cambridge. Boyle also sponsored the distribution of Bibles in Welsh and Irish and was involved in William Seaman's partial *Kütüp-ü paklarin Türkide bir nümudari-yi yahsi Kadis Yuhanna Resulin Türki zebana mütercem olmus üç risalesidir. Specimen Turcicum S.S. Scripturae* (London: James Flesher, 1659), which used the Oxford Arabic fonts.

87. The first letter of Pangeran Ratu of Banten to James I in October 1605 (TNA SP 102/4/8) was in Arabic, but two others to Charles I dated 1628 and 1635 were in Malay (TNA SP 102/4/50; TNA SP 102/4/37). Sultan Abdul Fatah in writing to Charles II ini-

tially used Arabic, as part of his broader program following the lead of Aceh and Johor to connect Banten with the Islamic world—especially the Ottomans and Mughals. January 4, 1665/Jumadilakhir 16, 1075, TNA SP Ext. 8/2 f. 126; n.d. (before 1675) Ext. 8/2 f. 45. On January 31, 1675 CE(Zulkaidah 5, AH 1085, according to the Arabic date on the letter), Sultan Abdul returned to writing in Malay to both Charles II and to Christian V of Denmark. TNA Ext 8/2, f. 46; TNA Ext. 8/2 f. 58. This seems to have been temporary because other than a letter in romanized Malay with an English translation from the "young Sultan" Abul Nasar from 1680 (TNA CO 77/14, f. 22–23), Sultan Abdul sent all of his remaining correspondence in Arabic (one letter from 1680 TNA CO 77/14, f. 38; and three from 1682, CO 77/14, f. 111; CO 77/14 f. 112; CO 77/14, f. 114–15). See Gallop, "Seventeenth-Century Indonesian Letters in the Public Record Office," 412–39.

88. Boyle to Robert Thompson, March 5, 1677, in *The Works of the Honourable Robert Boyle* 4, Thomas Birch, ed. (London: J. and F. Rivington, 1772), 226; Streynsham Masters to Samuel Masters, December 9, 1678, Royal Society Boyle Letters, 4, f. 39, Fol/1. Among Hyde's books were a copy of a letterbook by Hadiki entitled *Makturat Persice* dated Rabi 2, AH 1077 (Surat: 1666), Nasta'liq Persian, BL Asian Reg.16.B.23, with models for writing letters to relations, friends, officials of inferior rank, etc. He also had a glossary to Rumi's *Masnavi* by Abd ul-Latif ben Abd Ullah Kabirryyah, dated AH 1081 (Surat: 1670), Nasta'liq Persian, BL Asian Reg.16.B.19, an important work of synthetic scholarship responding to the Mughal fashion for Sufi poetry. On Hyde's shift towards the study of practical languages after completing the Bodleian catalogue in 1674 see David Vaisey, "Thomas Hyde and Manuscript Collecting at the Bodleian Library," in *The Foundations of Scholarship: Libraries and Collecting, 1650–1750* (Los Angeles: William Andrews Clark Library, 1992), 8.

89. *Jang ampat evangelia derri tuan kita Jesv Christi, daan Berboatan derri jang apostali bersacti: bersalin dallam bassa Malayo = The four Gospels of Our Lord Jesus Christ, and the Acts of the holy Apostles, translated into the Malayan tongue* (Oxford: H. Hall, 1677).

90. Streynsham Master to Samuel Masters, December 9, 1678, *Correspondence of Robert Boyle* 6, appendices. For a good account of the brokered relations in Madras, especially between Streynsham Master and the Telugu-speaking Kasi Viranna see Ogborn, *Global Lives*, 80–93; and Arasaratnam, *Merchants, Companies, and Commerce*.

91. Among these were three copies of the *Saddar* ("Hundred Subjects," ca. 16th century) (Navsari: 1674) BL Asian, Reg.16.B.7, previously Hyde's; (Navsari: 1675) copied by Herbad Hormuzyar, BL Add 6998; Reg.16.B.1, 174b-330; and Reg 16.B.15, a poetical version of the same labeled in Latin "Sudder Nuzzum" by Iranshah ben Malakshah that dated to 1495 in a copy from 1640 CE (Muharram AH 1050), which Hyde translated and described in his *Historia Religionis veterum Persarum* (Oxford: Sheldonian Theater, 1700), 431–88, including the Persian introduction and conclusion. One copy of Zartosht Bharam's *Zaratusht-Nama* (late 13th century), by the scribe Herbad Khorshed ben Isfandiyar ben Rustom (Navsari: 1679) BL Asian, Reg.16.B.8, also arrived in London. Finally, there were two copies of Zartosht Bharam's late 13th-century Persian translation of the *Arda Viraf Nama* (ca. 300–600 CE) in Avestan and Nasta'liq Persian, BL Asian, Reg.16.B.2 and Reg.16.B.1, f. 18–174a; the scribe for both is also Herbad Khorshed ben Isfandiyar ben Rustom (Navsari: 1674). On the flyleaf of the latter is written, "This booke very hard to

be procur'd. For when I had prevailed with the priest to write it for me, he durst not let his owne cast or sect know of it, but wrote it all in the night when all eyes were shut and asleep," suggesting a complex politics of sharing such texts. [B.1] Hyde failed to get a copy of the *Zend-Avesta*. [Hyde to Boyle, November 29, 1677, Boyle, *Works*, 6: 567.] On Avestan generally see the Avestan Digital Archive, http://ada.usal.es/; and Karl Hoffmann, "Avestan Language," *Encyclopedia Iranica*, 3:1 (1987), 47–62. For the religious underpinnings of this on Aungier's part, whose approach to Protestantism seems similar to John Locke's in Carolina, see Stern, *The Company State*, 101–13.

92. See D. L. White, *Parsis as Entrepreneurs in Eighteenth Century Western India: The Rustum Manock Family* (PhD. thesis, University of Virginia, 1979); Jivanji Modi, "Rustam Manock (1635–1721), the Broker of the English East India Company and the Persian Qisseh of Rustam Manock," *Asiatic Papers* 4 (Bombay: 1929), 101–337, which gives an account of the laudatory poem in Persian dated ca. 1711 that begins with Rustom Maneck helping out the Parsis after one of the two sackings of Surat in either 1664 or 1670. On the institution of the *panchayat*, which was new to Parsis in the seventeenth century and modeled after village assemblies (*yat*) in India, see Jesse Palsetia, *The Parsis of India* (Leiden: Brill, 2001), 25–26.

93. BL Asian, Reg.16.B.6. Unfortunately Hyde's Avestan or Persian types, which were once part of the British Museum collection, cannot be located today.

94. BL Asian Reg.16.B.14. See also the description in Hyde, *Historia Religionis Persarum*, 319.

95. On Akbar's shift to Persian and its diffusion as an administrative and literary language see Muzaffar Alam, "The Culture and Politics of Persian in Pre-Colonial Hindustan," in *Literary Cultures in History: Reconstructions from South Asia*, Sheldon Pollock, ed. (Berkeley: University of California Press, 2003), 159–67. The story of migration because of religious persecution was canonized in Behman Kaaikobad Sanjana, *Qisseh-i Sanjan* (Navsari: 1599), a text that survives only in later copies but was well-known in seventeenth-century Surat.

96. The usual date given for the reintroduction for the *jizya* is 1679, but particular groups including Brahmins and Parsis were subject to new head taxes already in 1675. See Surat to Company, November 26, 1669, in Forrest, *English Factories in India, 1668–9*, 190; John Fryer, Bombay, September 22, 1675, in his *New Account of East India and Persia in Eight Letters Being Nine Years Travels Begun 1672 and Finished 1681* (London: Chiswell, 1698), 144.

97. Aungier, "Proposals Touching Bombay Island," (1671) in *Selections from the State Papers preserved in the Bombay Secretariat (Home Series)* 1, George Forrest, ed. (Bombay: Government Central Press, 1887), 51–56. Bantenese trade was going so well in the 1670s (and the Red Sea so poorly) that Gujarati's tried to revive their old trade with Java in 1674, with the encouragement of the Sultan of Banten who wanted to establish ties with the Mughals. Cf. Chaudhuri, *Trading World of Asia*, 197.

98. On knowledge about currency relations in the Indian Ocean in London see Adam Olearius, *Voyages and Travels*, trans. John Davies (London: T. Dring and J. Starkey, 1662), esp. 299–300; and Jean-Baptiste Tavernier, *The six voyages of John Baptista Tavernier*, trans. John Phillips (London: Daniel Cox, 1677).

99. Fryer, *A New Account of East India and Persia*, 64–65.

100. On such networks generally see G. A. Nadri, "The Maritime Merchants of Surat: A Long-term Perspective," *Journal of the Economic and Social History of the Orient* 50:2–3 (2007), 235–58; Sanjay Subrahmanyam, "Iranians Abroad: Intra-Asian Elite Migration and Early Modern State Formation," *Journal of Asian Studies* 51 (1992).

101. Aungier to Bombay, April 26, 1677, in Forrest, *Home Series*, 1:111–13.

102. John Shepherd, *Statecraft and Political Economy on the Taiwan Frontier, 1600–1800* (Stanford: Stanford University Press, 1993); Liu Ts'ui-jung, "Han Migration and the Settlement of Taiwan," in Mark Elvin and Liu Ts'ui-jung, *Sediments of Time: Environment and Society in Chinese History* (Cambridge: Cambridge University Press, 1998), 165–74.

103. See Chen Di, *Dongfanji* (東番記, "Eastern Barbarian Record," 1603); Emma Jinhua Teng, *Taiwan's Imagined Geography: Chinese Colonial Travel Writing and Pictures, 1683–1895* (Cambridge: Harvard East Asian Monographs, 2006), 62–67; Laurence Thompson, "The Earliest Chinese Eyewitness Accounts of the Formosan Aborigines," *Monumenta Serica* 23 (1964), 175.

104. See Tonio Andrade, *Lost Colony* (Princeton: Princeton University Press, 2011); Andrade, *How Taiwan Became Chinese* (New York: Columbia University Press, 2008); Wills, "Maritime China from Wang Chih to Shih Lang," *From Ming to Ch'ing*, 216–17. On the Southern Ming see Lynn Sturve, *The Southern Ming, 1644–1662* (New Haven: Yale University Press, 1984).

105. Henry Dacres at Bantam to George Foxcroft at Madras, April 7, 1670, BL IOR G/21/5E, f. 6. The Taiwan correspondence can also be found in Anthony Farrington, Chang Hsiu-Jung, and Ts'ao Yung-Ho, eds., *The English Factory in Taiwan, 1670–1685* (Taipei: Taiwan National University, 1995).

106. Simon Delboe at Taiwan to EIC, September 16, 1672, BL IOR G/21/4B f. 137.

107. Captain William Limbrey, January 6, 1673, BL IOR G/21/4B f. 100.

108. See Kristof Glamann, *Dutch-Asiatic Trade, 1620–1740* (Copenhagen: Nijhoff, 1955), 54–63.

109. The 1671 copies with their owners are Bodleian Library Sinica 57 (Robert Boyle); Sinica 58 (Henry Aldrich, Dean of Christ Church); Magdalen College, Cambridge, Pepys Library 1914 (Samuel Pepys); Emmanuel College, Cambridge, MS 3.2.17 (fragment); Clare College, Cambridge G1.3.44 (Joseph Mayron); BL 15298.a.30 (Thomas Hyde); BL 15298.a.6(1). A 1676 copy is St John's College, Cambridge S.14 (John Dacres), and for 1677 see Bodleian Sinica 88.

110. See for example Charles Sweeting and Thomas Angeir at Taiwan to Edward Barwell and John Chappell at Amoy, February 24, 1678, BL IOR G/12/16 f. 103.

111. The key shift was the calendar case of 1664, in which Schall and other Jesuits were initially imprisoned for having created a false calendar; see Chu Pingyi, "Scientific Dispute in the Imperial Court: The 1664 Calendar Case," *Chinese Science* 14 (1997), 7–34.

112. The only study of the calendars is Huang Dianquan, *Nanming da tongli* (Tainan: Jing shan shu lin, 1960). See also Jiang Risheng, *Taiwan Waiji* [臺灣外記 "External Records of Taiwan"] (completed 1704, published ca. 1713).

113. Richard Smith, *Fortune-Tellers and Philosophers: Divination in Traditional*

Chinese Society (Westview, CT: Westview Press, 1991), 78–79; Richard Smith, *Chinese Almanacs* (Oxford: Oxford University Press, 1992), 8; Marc Kalinowski, *Divination et société dans la Chine medieval* (Paris: Bibliothèque nationale de France, 2003), 106.

114. Hyde and Shen translated *zhongxing* both literally as "mediae gloriae" and as the imperial title (as opposed to the "name" Yongle), noting that it indicates a range of aspirational concepts. ["se sperando seu optando hoc tempus esse mediae seu sumum gloriae hujuas familiae, non autem finem seu de china."] For the rhetoric of "restoration" as a political goal of reuniting the bureaucracy and army with the throne for those advising Yongli see Jin Bao, "Ling hai fen yu," 5b–7a in *Shiyuan Congshu* (Wuxing: Bing Chen, 1916), cited in Ian McMorran, "Wang Fu-Chih and the Yung-li Court," in Spence and Wills, *From Ming to Ching*, 150; and Qu Shisi, "Bao zhongxing jihui shu," in *Qu Shisi ji* (Shanghai: Shanghai guji chubanshe, 1981), 104–7.

115. Henry Dacres at Banten to Bartholomew Peartree, James Arwaker, Ellis Crisp, and Charles Firth at Taiwan, June 30, 1671, BL IOR G/21/6A f. 19–22. This plan failed when the Dutch told the Japanese that Charles II was married to a Portuguese princess. See Simon Delboe's diary ("Japan Journall"), BL IOR G/21/4 f. 118–30; TNA CO 77/12 f. 232–47, 250–60, and 262–69.

116. Bodleian Library, Ashmole 1787 (1).

117. Ad Dudink and Nicolas Standaert, "Ferdinand Verbiest's *Qionglixue* 窮理學 (1683)," in *The Christian Mission in China in the Verbiest Era*, Noel Golvers, ed. (Leuven: Leuven University Press, 1999), 11–31; Elman, *On Their Own Terms: Science in China, 1550–1900*, 144–46, 397.

118. BL Print 15298 a.30, f. 7v and f. 8r.

119. See in particular Print 15298 a.32 (formerly Royal 16 B X), figure 35 in the present volume. An emblem of the cosmos and more trigrams are on the second and third pages of this seventeenth-century Fujianese almanac. The Bodleian also acquired, possibly in the 1670s, a Qing almanac, *Chen liang jun xuan xinhai nian tongshu bianlian* [陳良駿選辛亥年通書便覽 "Well-explained and selected year 48 almanac and brief guide"] (Guangzhou: 1671–72), dated Kangxi 10, Bodleian, Sinica 96.

120. Contract for English Factory at Taiwan, September 10, 1670, BL IOR G/21/4B, f.54.

121. EIC to Dacres, October 23, 1674, BL IOR E/3/88, f. 136; Dacres to EIC, December 4, 1674, BL IOR G/21/4B, f. 115; Dacres to Captain John Wittey in London, February 8, 1675, BL IOR G/21/4B, f. 143.

122. EIC Court Minutes, BL IOR B/33, f. 365–69.

123. Charles James and Caesar Chamberlain at Surat to the EIC, March 10, 1677, BL IOR E/3/37 no. 4270.

124. John Dacres, Edward Barwell, and Samuel Griffith at Taiwan, August 1675, BL IOR G/12/16 f. 88; John Dacres et al., August 5, 1675, BL IOR G/12/16 f. 87; and John Dacres et al. to Henry Dacres, December 22, 1675, BL IOR E/3/36, no. 4150. The gunners were Edward Pepper, Phillip Bishop, John Baptista, and Anthony Martivan.

125. Edward Barwell at Amoy to Surat, November 2, 1677, BL IOR E/3/38 no. 4293.

126. EIC to Amoy, August 12, 1681, BL IOR E/3/89, f. 372–77; Josiah Child to the King of Amoy and Taiwan [Zheng Jing], August 12, 1681, BL IOR E/3/89, f. 380.

127. EIC to Bantam, London, July 15, 1682. BL IOR E/3/90, f. 6. A marginal heading reads, "The whole S Seas trade has not mainteyned charges these ten years."

CHAPTER FIVE

1. See in particular James II's sympathy with the policies of Louis XIV and the global project of the French Jesuits against Pope Innocent XI, in Pincus, *1688*, 122–31.

2. The Siamese Revolution lasted from May to September of 1688. For this as a revolution rather than a mere dynastic succession, see Pierre Joseph d'Orléans, *Histoire de M. Constance, premier minister du Roy de Siam, et de la dernière revolution de cet Estat* (Paris: Daniel Horthemers, 1690); Jean Vollant des Verquains, *Histoire de la révolution de Siam arrivée en l'année 1688* (Lille: J.-C. Malte, 1691); Marcel Le Blanc, S.J., *Histoire de la revolution du roiaume de Siam arrivée en l'année 1688* (Lyon: Horace Molin, 1692); and most importantly Simon de La Loubère, *De Royaume de Siam* (Paris: Abraham Wolfgang, 1691; English trans. London: F. L. for Thomas Horne et al., 1693). La Loubère was Louis XIV's envoy who added an "Avertissement necessaire" about the expulsion of the French by Phra Phetracha in November 1688. The language of "revolution," which seems to have emerged from the French accounts and publications, also translated into English almost immediately, *A Full and True Relation of the Great and Wonderful Revolution that Happened Lately in the Kingdom of Siam . . . Being the substance of letters writtein in October 1688 and February 1689 from Siam and the Coast of Coramandel. Never before published in any language, and now translated into English* (London: Randal Taylor, 1690). This text appears to have translated a letter from a Dutch source and the diary of a French officer. Elihu Yale wrote of "strange news of the great revolutions at Syam" in a letter to the EIC from Madras, January 1689, BL IOR E/3/47 no. 5658; see also January 30, 1690, E/3/48, no. 5698. See generally E. W. Hutchinson, *1688 Revolution in Siam* (Hong Kong: Hong Kong University Press, 1968). Kirti Chaudhuri and Jonathan Israel explain, "It is not generally appreciated by the historians of early modern Europe that 1688–9 marked not only a 'revolution' in Britain but also one in the Indian Ocean." Chaudhuri and Israel, "The East India Companies and the Revolution of 1688–9," in Israel, *The Anglo-Dutch Moment* (Cambridge: Cambridge University Press, 2003), 407; and for a broader picture John Wills, *1688* (New York: Norton, 2001).

3. Gabriel de Magalhães, *A New History of China* (London: Thomas Newborough, 1688), 254. Claude Bernou translated the Portuguese manuscript of Magalhães that Couplet had brought with him in 1682 as *Nouvelle Relation de la Chine* (Paris: 1688; new editions in 1689 and 1690).

4. The war with the Mughals ended by treaty in February 1690, and at the same time the Company ordered the war with Siam ended after the revolution that ousted both Constantine Phaulkon and the French forces there. See "East India Company to Sir John Child at Bombay," January 31, 1690, BL IOR E/3/92, f. 78–79.

5. On the unprofitable but long-lasting factory in Bengkulu (est. 1685) see Anthony Farrington, "Bengkulu: An Anglo-Chinese Partnership," in *The Worlds of the East India Company*, H. V. Bowen et al., eds. (Woodbridge: Boydell Press, 2002), 111–17.

6. Pincus's emphasis on "Catholic modernity" is helpful here, although his stress

on trade as a "zero-sum" game fails to understand the problem of changes in Asian trade affecting London. Cf. Pincus, *1688*, 372–81.

7. EIC to Elihu Yale, December 12, 1687, in John Bruce, *Annals of the Honorable East India Company*, 591. The figure of 100,000 is from Om Prakash, *European Commercial Enterprise in Pre-colonial India* (Cambridge: Cambridge University Press, 1998), 148.

8. C. H. Philips, "The Secret Committee of the East India Company," *Bulletin of the School of Oriental Studies* 10, no. 2 (1940), 299–315.

9. August 6, 1684, Court Minutes, B/37; Ray and Oliver Strachey, *Keigwin's Rebellion* (Oxford: Clarendon Press, 1916), 116.

10. See the documents reprinted in *An Historical Account of Some Memorable Actions, Particularly in Virginia* (London: J. Roberts, 1716). The phrase "revolution" was used in 1684–85 by Grantham and others during the settlement negotiations; see Strachey and Strachey, *Keigwin's Rebellion*, 142. The governor of Virginia William Berkeley and Grantham framed the 1676 uprising in Virginia as a "rebellion."

11. *Cobbett's Complete Collection of State Trials* 10 (London: Hansard, 1811), 371–554. This account is taken from the manuscript notes made by Samuel Pepys as secretary of the Admiralty, now at Cambridge, Magdalen College.

12. *EIC v. Sandys*, 382, 392, 412, 414. Pincus sees this as an argument about trade as a zero-sum game (*1688*, 376), but the actual language used by the plaintiff's lawyer Finch was about "rivals," "not only in respect of the Indians themselves, but also in respect of other foreign nations, who are rivals to us in this trade and are ready to take all advantages against us about it."

13. *EIC v. Sandys*, 376 and 421. On free trade, the defendants cited Coke, 3 Inst. 181.

14. *EIC v. Sandys*, 466, 523, 538, 552–53. The notion of "societati mercatorum" relies in part on Puffendorff, *Jure Naturae et Gentium*, lib. 5, f. 655

15. Ogborn, *Indian Ink*, 141. With the coming of the Revolution, the case was printed as a way of trying to stave off attacks against the Company's privileges. See *The Argument of the Lord Chief Justice of the Court of King's Bench concerning the Great Case of Monopolies between the East-India Company, Plaintiff, and Thomas Sandys, Defendant* (London: Randall Taylor, 1689).

16. *The Loyal Protestant* 143 (April 18, 1682), 1; Jenkins to Child, April 24, 1682, TNA SP 44/68/68; newsletter to John Squire, May 2, 1682, TNA ADM 77/2/30; newsletter to Roger Garstell, dated May 2, 1682, TNA ADM 77/2/29.

17. Adam Elliot, *A Modest Vindication of Titus Oats* (London: Joseph Hindmarsh, 1682).

18. September 28, 1687, BL IOR E/3/91, f. 209. The citations go back to Alfred Lyall describing the period as transitional between commerce and "politie." Lyall, *The Rise and Expansion of the British Dominion in India* (London: John Murray, 1894), 49; Philip Stern, "A Politie of Civill & Military Power: Political Thought and the Late Seventeenth-Century Foundations of the East India Company-State," *Journal of British Studies* 47:2 (April 2008), 282.

19. "Charter Granted by the Governor and Company of Merchants Trading into the East-Indies, to the Mayor, Aldermen and Burgesses of Madras," December 30, 1687, in John Shaw, *Charters Relating to the East India Company from 1600 to 1761* (Madras: R. Hill, 1887), 84–96.

20. EIC to Fort St. George, July 2, 1684, E/3/90, f. 329; "Journal of Captain W. Heath

in the ship Defense on a voyage to Fort St. George and Bencoolen," BL IOR L/MAR/B/90. Yale became governor of Madras in 1687.

21. Wills, *1688*, 286–87.

22. See Paul Van Dyke's landmark *The Canton Trade: Life and Enterprise on the China Coast, 1700–1845* (Hong Kong: Hong Kong University Press, 2006).

23. Fu Lo-shu, *A Documentary Chronicle of Sino-Western Relations* (Tuscon: University of Arizona Press, 1966), 1:61, 2:461n.

24. Nikolaas de Graaf, *Reisen* (Hoorn: Ryp; Amsterdam: Wed, 1701), 174–81; see also Wills, *Embassies and Illusions*.

25. Ts'ao Yung-Ho, "The English East India Company and the Cheng Regime on Taiwan," in *The English Factory in Taiwan, 1670–1685*, Chang Hsiu-Jung et al., eds. (Taipei: National Taiwan University Press, 1995), 17; and Leonard Blussé, *Visible Cities: Canton, Nagasaki, and Batavia* (Cambridge: Harvard University Press, 2008).

26. Josiah Child and Benjamin Bathurst to James II, June 29, 1687, BL Add MSS 41822 Middleton f. 107.

27. Hyde had a copy of Verbiest's treatise on the lunar eclipse of 1671, on which he wrote, "This tho small in bulk is a great rarity, it being the only thing of this kinde now in England, 1700." BL Or.74.b.6. It contains manuscript notations by Shen Fuzong. Hyde also had fragments of Verbiest's 1680 Manchu grammar. BL Asian Reg.16.B.3.

28. Justel also wrote to the Royal Society about the publication of Couplet's books (February 2, 1687, Royal Society (hereafter RS) EL/I1/110) and the upcoming translation of Magalhães (June 8, 1687, RS EL/I1/114). Couplet published Verbiest's list of Jesuit missionaries in China first, *Catalogus patrem Societatis jesu* (Paris: R. J. B. de la Caille,1686); along with the controversial dates of Chinese monarchs, Philippe Couplet et al., *Tabula Chronologica Monarchiae Sinicae* (Paris: Bibliotheca regia, 1686); followed by the translation of the "four books," Philippe Couplet et al., *Confucius Sinarum Philosophus, sive scientia sinensis* (Paris: Daniel Horthemels, 1687); Couplet, *Breve raguaglio delle cose piu notabili spettanti al grand'imperio della Cina* (Rome: s.n., 1687); and the Latin life of Candida Hiu translated into French by Père d'Orléans as *Histoire d'une dame chrétienne de la Chine* (Paris: Michallet, 1688). D'Orléans also published a history of the Manchus out of the writings of Martini and Schall, *Historie des deux conquerans Tartares qui on subjugué la Chine* (Paris: Claude Barbin, 1688) and Claude Bernou's translation of Gabriel de Magalhães, *Nouvelle Relation de la Chine* (Paris: Claude Barbin, 1688). The writings of Confucius were first abridged in French from Couplet in Amsterdam as Louis Cousin, ed., *La Morale de Confucius, Philosophe de la Chine* (Amsterdam: Pierre Savouret, 1688), out of which an English translation appeared by way of Randal Taylor, *The Morals of Confucius, a Chinese Philosopher* (London: Randal Taylor, 1691), who also published the first account of the Siamese Revolution and the Sandys case. On the increasing reliance of the French accommodationists on European scholarship rather than dialogues with scholars in Beijing see Jensen, *Manufacturing Confucius*, 111–18. On the rapid dissemination of news about *Confucius Sinarum Philosophus* see Dew, *Orientalism*, 205–34. On Couplet's acquisition of books, some of which came with Shen to Hyde, see Thomas Birch, *The History of the Royal Society*, 4: 426.

29. Thomas Hobbes, *Leviathan* (London: Andrew Crooke, 1651), 714–15.

30. Verbiest, "A Voyage of the Emperor of China, into the Eastern Tartary, in the

year 1682," *Philosophical Transactions* 16 (1686–92), 40, 49, 52, 57. On the atlases see Martini, *Novus atlas Sinensis* (Amsterdam: Joan Blaeu, 1655). Couplet gave a copy of Luo's atlas to Nicholas Witsen in 1684, now Museum Meermanno-Westreenianum ('s-Gravenhage), M.115.B.1. See Marcel Destombes, "A Rare Chinese Atlas," *Quaerendo* 4:4 (1974), 336–37.

31. *Philosophical Transactions* 16 (1686–92), 62.

32. "An Account of a Large and Curious Map of the Great Tartary, lately Publish'd in Holland, by Mr. Nicholas Witsen," *Philosophical Transactions* 16 (1686–92): 492–94. Witsen had particularly close ties to Edward Bernard at Oxford, whom he gave Theodore Petraeus's Coptic dictionary as well as two sets of Coptic and Ethiopian (Ge'ez) types in 1686. He also corresponded with Vossius, but they had fought over whether to let Charles II know about the possibility of a strait between North America and Asia, which led to a failed English expedition in 1676.

33. These events were reported in Europe in A. de Chaumont, *Relation de l'ambassade de M . . . de Chaumont à la cour de roy de Siam* (Amsterdam: Mortier, 1686); Abbé de Choisy, *Journal du voyage de Siam fait en 1685 et 1686* (Paris: S. Mabre-Cramois, 1687); and Guy Tachard, *Voyage de Siam des Pères Jésuites envoyés par le Roi aux Indes et à la Chine* (Paris: Daniel Horthemels and Arnoul Seneuze, 1686). On the Persians, see the account of the embassy by Ibn Mohammed Ibrahim, *Safine-ye Solaymani*, British Library, Asia, Oriental MSS 6942, translated by John O'Kane as *The Ship of Solayman* (London: Routledge & Kegan Paul, 1972); Jean Aubin, "Les Persans au Siam sous le regne de Narai (1656–1688)," *Mare Luso-Indicum* 4 (1980), 95–126; Hiromu Nagashima, "Persian Muslim Merchants in Thailand and Their Activities in the 17th Century: Especially on Their Visits to Japan," *Nagasaki Prefectural University Review* 30:3 (Jan. 30, 1997), 387–99. A second Safavid embassy arrived in January 1688, but Samuel White refused to let them travel on his ships to return to Bandar-Abbas.

34. The formal declaration of war against Golconda was by the Buddhist calendar dated the 21st of the 2nd moon, 2228, or ca. May 1685. See Anthony Farrington and Dhiravat na Pombejra, eds., *The English Factory in Siam, 1612–1685* 2 (London: British Library, 2006), 1071. On Muslim merchant riots in Siam see Michael Smithies, "Accounts of the Makassar Revolt 1686," *Journal of the Siam Society* 90 (2002), 73–100. On the purges of Muslim officials see Elihu Yale at Madras to the East India Company, August 28, 1684, BL IO G/40/3A f. 83v.

35. *Huang yuditu kao* (皇輿地圖考 "Imperial World Map Verified") and *Tonghua jingwei tu kao* (通華經緯圖考 "Verified Complete Chinese 'Warp-Weft' Diagrams"), Bodleian Library, Sinica 123. Along with this came presents of Japanese lacquerware; see John Vaux at Aceh to the East India Company, January 14, 1684, BL IO E/3/43 no. 5007.

36. Ignatio à Costa and Prospero Intorcetta, *Sapientia sinica* (Nanfeng, Jiangxi: 1662). This is possibly now BL C.24.b.2, part of the original Sloane collection.

37. Nathaniel Vincent, *The Right Notion of Honour as it was delivered in a Sermon Before the King at Newmarket, October 1674 published by his majesties special command with Annotations* (London: Richard Chiswell, 1685), 25; Nathaniel Vincent to Samuel Pepys, July 27, 1682; December 11, 1682; May 12, 1688; and Pepys to Vincent, December 23, 1682, in R. Bentley, *The Life, Journals and Correspondence of Samuel Pepys* (London: R. Bentley, 1841), 1: 304, 311, 316; 2: 124. Vincent in the footnotes to

his sermon writes, "There is an ingenious Merchant, a Fellow of the Royal Society, who hath put into the hands of one of his Colleagues, several of Cumfusu's Books brought from Siam, where they were printed, in order to [sic] an English Edition of them, and of a *Lexicon* and *Clavis* to the Language, and to a new World of Learning" (25). On Vincent see Matt Jenkinson, "Nathanael Vincent and Confucius's 'Great Learning' in Restoration England," *Notes and Records of the Royal Society* 60:1 (January 2006), 35–47, who argues that the FRS alluded to by Vincent was Francis Lodwick.

38. Bray, "Introduction," *Graphics and Text*, 39.

39. Lach and Van Kley, *Asia in the Making of Europe* 3:3, 1190.

40. See Ronald Love, "Rituals of Majesty: France, Siam, and Court Spectacle in Royal Image-Building at Versailles in 1685 and 1686," *Canadian Journal of History* 31 (August 1996), 171–98.

41. Guy Tachard, *A Relation of the voyage to Siam performed by six Jesuits sent by the French king to the Indies and China in the year 1685: with their astrological observations and their remarks of natural philosophy, geography, hydrography and history* (London: J. Robinson and A. Churchill, 1688), a translation of the 1686 edition.

42. "A Proclamation for the Recalling of all his Majesty's Subjects from the Service of Foreign Princes in East Indies," July 11, 1686, Windsor and London, TNA CO 77/14/160–61.

43. "King Narai of Siam's declaration of war against the East India Company, Lopburi, August 11, 1687," in *Records of the Relations between Siam and foreign countries in the seventeenth century* 4 (Bangkok: Vajirañana National Library, 1915–21), 183–200. The beginning of the war was reported in the *London Gazette*, 2270, August 18 to August 22, 1687, in a letter from the "Bay of Bengal." The Siamese navy with two ships of about fifty guns largely manned by English mercenaries was not initially perceived as a major threat, "Commission from Sir John Child at Bombay to Captain Joseph Eaton," May 23, 1687, BL IOR E/3/47 no. 5597. An English ship was seized by Siamese ship flying the French flag in May 1688 with "original commissions to her officers in French under the mandareens' seals," Yale to Francois Martin at Pondicherry, May 1, 1688, BL IOR G.19.21 (2) 26–27; Yale to Martin, June 23, 1688, G/19/21 (2) 41–42.

44. For the Company's awareness of this see EIC to William Gyfford at Madras, March 22, 1687, IOR E/3/91, 276–80; EIC in London to John Child at Surat, March 23, 1687, E/3/91, 271–75.

45. Nicholas Gervaise, "Épître dédicatoire," *Histoire Naturelle et Politique du Royaume de Siam* (Paris: Claude Barbin 1688).

46. The *Sapientia Sinica*, f. 1 reads (in parallel text with the Mandarin) "Magnum virorum sciendi institutum constititi in illuminando virtutibus spiritualem potentiam a coleo inditam nempe Animam ut haec redice posit ad originale claritatem quam appetites animales obicubilaverant." Compare the 1687 Parisian version would use the metaphor of the mirror and the improving of the rational self. "Magnum adeoque virorum Principum, sciendi institutum consistit in expoliendo seu excolendo rationale naturam a coelo inditam; ut scilicet haec, ceu limpidissimum speculum, abstersis pravorum appetitum maculis ad pristinam claritem suam redire posit." David Mungello suggests that the *Sapientia* translation was "overspiritualized" and the *Confucius Sinarum* "overrationalized." See Mungello, *Curious Land* (Honolulu: University of Hawaii Press, 1989), 257–58.

47. John Milton, *Paradise Lost* (London: Peter Parker, Robert Boulter, and Matthias Walker, 1667) xi, 115. This is the book of Adam's repentance and submission. See the King James translation of Ephesians, 1: 17: "That the God of our Lord Jesus Christ the Father of glory, may give unto you the Spirit of wisedome and revelation in the knowl-edge of him: The eyes of your understanding being enlightened." On conservative Enlightenment see the Spinozan critique of Boyle outlined in Jonathan Israel, *The Radical Enlightenment: Philosophy and the Making of Modernity* (Oxford: Oxford University Press, 2002), 252–57. On the complex strategies of translation of *dao* in the case of the Jesuits and Kircher, where it becomes the law (*lex*), see Timothy Billings, "Jesuit Fish in Chinese Nets: Athanasius Kircher and the Translation of the Nestorian Tablet," *Representations* 87 (Summer 2004), 1–42.

48. Printed after the Glorious Revolution in Temple, *Miscellanea: The Second Part* (London: R. Simpson, 1690).

49. For the claim of Temple's invention see S. Lang and Nikolaus Pevsner, "Sir William Temple and Sharawadgi," *Architectural Review* 106 (1949), 391–92. For recent attempts see Ciaran Murray, "Sharawadgi Resolved," *Garden History* 26:2 (Winter 1998), 208–13; Michael Sullivan, "Chinese Art and Its Impact on the West," in *Heritage of China*, Paul Ropp, ed. (Berkeley: University of California Press, 1990), 285.

50. Apter (*Translation Zone*, 211) cites Gabriel de Foigny, *La Terre Australe connue* (Geneva: Jacques Vernevil, 1676) as an exemplar of this whose "Australe" grammar was self-translating, making standard language strange to itself and superimposing a private grammatical logic. On "Confucius" as a European representation see Jensen, *Manufacturing Confucianism*, 113–47.

51. Vincent, *Right Notion of Honour*, 15. See my discussion of Temple's "Of Heroic Virtue," in Batchelor, "Concealing the Bounds: Imagining the British Nation through China," in Felicity Nussbaum, *The Global Eighteenth Century*, 87.

52. In particular "Heroic Virtue" and Temple's "Essay upon the Ancient and Modern Learning," seem to be shadowing the work of Isaac Vossius's 1685 essay "De artibus & scientiis Sinarum" in *Variarum Observationum Liber* (London: Robert Scott, 1685), 69–85. Vossius's collection, dedicated to Charles II, also included "De Antiquae Romae et Aliarum Quarumdam Urbium Magnitudine," which described the urban greatness of various ancient societies (Babylonia, Rome, Egypt, etc.), and then devotes all of chapter thirteen to "De magnis Sinarum urbibus," describing the massive population of China in comparison with other cities and countries (56–68).

53. John Locke, *Essay Concerning Human Understanding*, II, ch. 22, sec. 2–7.

54. R[obert] H[ooke], "Some observations made concerning the Chinese Characters," *Philosophical Transactions* 16 (1686), 63–78.

55. See Francis Lodwick, *An Essay Towards a Universal Alphabet* (London: s.n., 1686). The mathematician John Wallis also wrote to Halley in November 1686 about books in Chinese. John Wallis to Edmond Halley, November 8, 1686, read to the Royal Society, November 24, 1686, RS EL/W2/41.

56. See William Poole, "The Divine and the Grammarian: Theological Disputes in the 17th-Century Universal Language Movement," *Historiographia Linguistica* 30 (2003), 273–300, 291; and Hooke's diary entries for June and July 1693 in BL Sloane 4024. Wilkins

printed the Lord's Prayer transliterated and in Chinese characters in his *Essay*, which he had received from Lodwick before the Great Fire (*Essay*, 450–51).

57. See the stress on heterodoxy in Hooke's and Lodwick's circles in William Poole, "The Genesis Narrative in the Circle of Robert Hooke and Francis Lodwick," in Hessayon and Keene, eds., *Scripture and Scholarship in Early Modern England* (Aldershot: Ashgate, 2005); Poole, "Francis Lodwick's Creation: Theology and Natural Philosophy in the Early Royal Society," *Journal of the History of Ideas* 66:2 (2005), 245–63; Steven Shapin, "Who Was Robert Hooke?" in *Robert Hooke: New Studies*, Michael Hunter and Simon Schaffer, eds. (Woodbridge, Suffolk: Boydell Press, 1989), 253–86.

58. Thanks to Haun Saussy for the suggestion that this string of characters looked like a writing exercise. The "China-copy book" (Hyde), which is a calligraphy sampler, Bodleian Sinica 91 is unbound and missing its final pages, possibly Hooke's original manuscript.

59. Such ideas fit in with Hooke's beliefs in catastrophism elsewhere, in which history has radical breaks involving the loss of certain languages and cultures. See Hooke, *Lectures and Discourses of Earthquakes and Subterraneous Eruptions* (London: R. Waller, 1705) and William Poole, "Isaac Vossius, Robert Hooke, and the early Royal Society's use of Sinology" (2008) Oxford Research Archive, http://ora.ouls.ox.ac.uk /objects/uuid:e5acc4b0-968e-45a8-938a-16a8b0f38570.

60. Thomas Hyde, *De Ludis Orientalibus Libri Duo* (Oxford: Sheldon Theater, 1693–94), initially published as *Mandragorias, seu Historia Shahiludii* (Oxford: 1689, 1694) and volume two, *Historia Nerdiludii* (Oxford: 1694).

61. RS, EL/I1/98.

62. George Dalgarno had his bookshop next to Duke Humfrey's library.

63. See the discussion in chapter 4 as well as Hyde's Latin annotations to the tables in the second part of the short encyclopedia, arithmetic primer, and letter book *Zengbu su weng zhizhang zazi quanji* [增補素翁指掌雜字全集, "Essential Supplement for the Miscellaneous Words and Complete Works of the Hand"] Bodleian Sinica 74, 2:12–14. See also Sinica 107 and the copy in the BNF, Département des manuscrits, Chinois 7706, which are also both similar.

64. See BL Asian, Oriental Manuscripts Reg.16.B.3; Reg.16.B.4; Reg.16.B.20. There are also two Chinese dialogues in Oriental Manuscripts, Reg.16.B.21, the second one (f. 20–24) compiled with Shen initially using Chinese characters (the first eleven words) but then simply transliterations. Hyde had the Oxford engraver Michael Burghers prepare some of these as prints, BL Print OR.70.BB.9, for an unrealized book project. Most of the already engraved specimens, including the Chinese compass rose from the Selden Map, show up in the appendix to Sharpe, *Syntagma Dissertationum*.

65. Bodleian Library, MS Hyde 6, "Parvum Vocabularium Sinico-Anglicum, forma oblonga," is an English-Chinese dictionary that does not seem to have been the work of Hyde and Shen; MS Hyde 7, "Aliud parvum Vocabularium Sinense," is a manuscript with numbers in the Fujianese Min dialect, a primer of English and transliterated Chinese words (no characters), and a section on writing numbers. Both probably came from either Banten or Taiwan, although Siam and Tonkin are possibilities as well. Hyde was in correspondence with Christian Menzel about his Chinese dictionary in 1683. Thomas Hyde

to Christian Menzel, February 16, 1683, Glasgow University Library, MS Hunter 299 (U.6.17), "Collectio Sinicorum MSS Opusculorum," f. 195–96 (U.6.21–22). Menzel's correspondence with Couplet is also in this volume (f. 175–95).

66. See British Library, Asian, Sloane 853, f. 23. F. 37 has notes on the northern part of another printed Chinese map from 1679, Bodleian Sinica 92.

67. "Ad Occidentalim hujus muri extremitatem (in Regni Chinensis Mappa quam ab Amplissimo Viro D. Georgio White Mercatore Anglo accepi,) Sinicis characteribus notatum legitur, *Cho chang ching ki cu, chi Liao-tung chi,* id est, *Fabricatio longi muri incipit hic, apud Liao-tung desinit.*" Thomas Hyde, "Sinensium Epistola," in Edward Bernard, *De Mensuris et Ponderibus Antiquis libri tres. Editio altera, purio et duplo locupletior* (Oxford: Sheldon Theater, 1688). See Hyde's and Shen's notes for this in Latin and Chinese in British Library, Sloane 853, f. 37.

68. See the manuscript notations in Hyde's copy of Andreas Müller, *Disquisitio Geographia & Historica, de Chataja* (Berlin: Rungianis, 1671), BL Print 10055.ee.32, esp. f. 40–54. The notation on the forepage by Hyde reads, "In the columnes subscribed Emendatè are the true Characters written by the Chinese which may discover the illness of these in the next columns by Müller and Golius. In hoc exemplari accuratius et emendations seribuntur omnes characteres sinici manu Chinensis nativi Shin Fo-çungh com notisca[illeg.] in Oxino."

69. Shen Fuzong to Thomas Hyde, May 25, 1687, in Sharpe, "Appendix," 519; Thomas Hyde to Robert Boyle, July 26, 1687, in Birch, ed., *Works* 5, 591. The Boyle letter suggests that there may have been another Chinese interpreter at the Bodleian earlier, who knew both Latin and Chinese like Shen but had died by 1687.

70. Couplet and Shen dined in February 1688 with Henry Hyde, Earl of Clarendon, in February 1688, who had fallen out of favor with his brother-in-law James II and would soon be helping William in the Glorious Revolution. Henry Hyde, *The Correspondence of Henry Hyde* 2 (London: Henry Colburn, 1828), 162.

71. Robert Boyle, "Workdiary 36: Entry 69" Royal Society, Boyle Papers, 21, p. 288.

72. Hyde to Boyle, July 26, 1687, *Correspondence of Robert Boyle* 6, 226.

73. Some of the Hebrew work probably occurred in 1689, with the arrival of Isaac Abendana at Oxford. Hyde also worked on a translation of Maimonides's *More Nevochim* ("Guide to the Perplexed").

74. Hyde, "Epistola Dedicatoria," *Mandragorias seu Historia Shahludii* (1694), dated November 9, 1694. At this point Godolphin was a crypto-Jacobite working as first lord of the Treasury, before he had to resign because of the assassination plot against William III in 1696.

75. Hyde had a prose abridgement of the *Shahnama*, BL Asian, Reg.16.B.14, in Farsi, acquired in Surat by the East India Company president there, Captain Aungier. See also Hyde, *Historia Religionis Persarum*, 319.

76. Hyde, *Mandragorias*, 160–78.

77. Lois Schwoerer, *The Revolution of 1688–9* (Cambridge: Cambridge University Press, 2004), 3. Hampden committed suicide on December 12, 1696.

78. Cited in Basil Duke Henning, *The House of Commons, 1660–1690* 1 (London: Secker and Warburg, 1983), 470.

79. "Ade out hic Ludus sorti ac fortunae subjectus sit et obnoxious, uti sunt ubique

omnes Promotiones a Curia Regia dependents, quae non solent dari dignioribus cum Reges plerumque aliorum auribus et oculis audire et videre gestiant" (*Historia Nerdiludii*, 73). On this game that dates to the Tang Dynasty see Andrew Lo, "Official Aspirations: Chinese Promotion Games," in Colin Mackenzie and Irving Finkel, *Asian Games: The Art of Contest* (New York: Asia Society, 2004), 64–75. The only known copy older than the one reengraved by Hyde is in the National Library, Beijing, an early Qing reprint from Nanjing of a Ming edition by Lin Weigong.

80. 1684 was the year in which out of 10,000 total candidates only 73 passed the provincial civil exams in Nanjing, the lowest percentage of any year between 1393 and 1893 for which data survives. See Benjamin Elman, *A Cultural History of Civil Exams* (Berkeley: University of California Press, 2000) 143, 178–79, 681.

81. Lawrence C. H. Kim, *The Poet-Historian, Qian Qianyi* (New York: Routledge, 2009), 92–93, 130–31, 141, 181–82; Qian Zhonglian and Yan Ming, "Qian Qianyi shi zhong de qi yu," *Zhongguo wenzhe yanjiu tongxun* 14.2 (June 2004), 63–91.

82. *Nerdiludi*, 72–73. It is hard not to think of "geworfenheit" here.

83. Fang Qianli, *Touzi xuan ge* (骰子選格 "Rules for Selection through Dice") in *Shuo fu sanzhong* [說郛三種], Tao Zongyi, ed. Cf. Andrew Lo, "The Game of Leaves: An Inquiry into the Origin of Chinese Playing Cards," *Bulletin of SOAS* 63:3 (2000), 392.

84. Hyde spends much less time with Chinese backgammon, pp. 65–68, and Go, pp. 195–201, admittedly easier games to describe.

85. Amartya Sen, *The Idea of Justice* (Cambridge: Harvard University Press, 2009), 315.

86. [Thomas Hyde], *An Account of the Famous Prince Giolo, Son of the King of Giolo, Now in England . . . Written from his own Mouth* (London: R. Taylor, 1692). The reference to the tattoos is p. 27. Hyde's personal copy was originally BL Oriental Manuscripts, Reg.16.B.17, and is now probably BL, Print, 10825 b.47, damaged and rebound in 1930. On Giolo see Geraldine Barnes, "Curiosity, Wonder, and William Dampier's Painted Prince," *Journal of Early Modern Cultural Studies* 6:1 (Spring/Summer 2006), 31–50. Dampier's journal is British Library, Sloane 3236.

87. See the broadsheets as well as the account of the skinning in *Musaeum Pointerianum-Curiosities vol. IV*, St. John's College, Oxford, MS 254, f. 21–23.

88. Birch, *History of the Royal Society* 4: 479–80.

89. Galileo's *Discorsi e dimostrazioni matematiche, intorno à due nuove scienze* (Leiden: Elsevier, 1638), which much more broadly describes his ideas about physics, and the biography by Salusbury that contextualizes these were part of volume two. For the publishing history, see the recent rediscoveries by Nick Wilding, "The Return of Thomas Salusbury's Life of Galileo (1664)," *British Journal for the History of Science* 41 (June, 2008), 241–65. Salusbury's brother was an East India Company factor who turned interloper by the 1670s.

90. Richard Westfall, *Never at Rest: A Biography of Isaac Newton* (Cambridge: Cambridge University Press, 1983), 443–51. For the broader dispute, which occurred first in 1679 and then again in 1686, see H. W, Turnbull, ed., *Correspondence of Isaac Newton* 2 (Cambridge University Press, 1960), 297–314, and 431–448.

91. Cambridge University Library, MS Add. 3990. This was published by John Conduitt as *De Mundi Systemate Liber* (London: J. Tonson, 1728) and translated as *A Treatise of the System of the World* (London: F. Fayram, 1728). See also the Hebraicist history

in Isaac Newton, *The Chronology of Ancient Kingdoms Amended* (London: J. Tonson, 1728), which traces everything back to the Temple of Solomon.

92. See the celebrated discussion of experiments in a shiphold in Galileus Galileus Linceus, *The Systeme of the World: In Four Dialogues. Wherein the Two Grand Systemes of Ptolomy and Copernicus are largely discoursed of*, trans. Thomas Salusbury (London: William Leybourne, 1661), 165–66.

93. See especially D. T. Whiteside, "The Prehistory of the Principia from 1664–1686," *Notes and Records of the Royal Society of London* 45 (1991); as well as the notes to this proposition in Cohen, trans. *Principia* (1999).

94. Hooke to Newton, November 24, 1679, *The Correspondence of Isaac Newton* 2: 297. The outlines of these ideas as a new "System of the World" appeared in Hooke's *Attempt to Prove the Motion of the Earth by Observations* (London: 1674), 27–28. See Alexandre Koyré, "An Unpublished Letter of Robert Hooke to Isaac Newton [9 December 1679]," *Isis* 43 (1952), 312–37. On Hooke's importance here see M. Nauenberg, "Hooke, Orbital Motion, and Newton's *Principia*," *American Journal of Physics* 26 (1994), 331–50; and "On Hooke's 1685 Manuscript on Orbital Mechanics," *Historia Mathematica* 25 (1998), 89–93.

95. A definition used by William Harper, *Isaac Newton's Scientific Method* (Oxford: Oxford University Press, 2011), 50.

96. See especially John Flamsteed, *The Doctrine of the Sphere: Grounded on the motion of the earth, and the antient Pythagorean or Copernican system of the world* (London: A. Gobid and J. Playford, 1680).

97. Edward Bernard to Flamsteed, August 14, 1681, *The Correspondence of John Flamsteed* (1995), 796–807. Bernard noted that he had only consulted half of the works in the Bodleian. See Raymond Mercier, "English Orientalists and Mathematical Astronomy," in G. A. Russel, ed., *The 'Arabick' Interest of the Natural Philosophers in Seventeenth-Century England* (Leiden: Brill, 1994), 158–214. In addition to the constellation map that White donated to the Bodleian in 1684, Hyde also had a larger star map in Chinese, now BL Asian, Oriental Manuscripts, Reg.16.B.26.

98. Nicholas Dew, "*Vers la ligne*: Circulating Measurements Arount the French Atlantic," in *Science and Empire in the Atlantic World*, James Delbourgo and Dew, eds. (London: Taylor and Franics, 2008), 57.

99. The account of the tides in the Gulf of Tonkin is book 3, proposition 24 of the *Principia*. On Davenport's unreliability see George White, *Reflections on a scandalous paper . . . together with the true character of Francis Davenport the said Company's Historyographer* (London: s.n., 1689); and *A Letter to Mr. Nathaniel Tenche in Answer to the Paper Published by him* (London: s.n., 1689). Even the Company's pamphleteer Nathaniel Tenche avoided trying "to prove Mr Davenport an honest man"; see *Animadversions upon George White's Reflection* (London: s.n., 1689). Copies of these and several other pamphlets in this debate can be found in BL IOR Mss Eur D.300. For the data observations at Tonkin see Birch, *History of the Royal Society* 4 (London: 1757), 226–27, for November 21, 1683, and William Dampier, *A new voyage around the world* (London: James Knapton, 1697), 97. See Simon Schaffer, "The Asiatic Enlightenments of British Astronomy," in *The Brokered World*, 72–74; and David Cartwright, "The Tonkin Tides Revisited," *Notes and Records of the Royal Society* 57 (2005), 135–42.

100. *Novum Organon*, 2: xxxvi. See Bernard Cohen, *Isaac Newton: The Principia* (Berkeley: University of California Press, 1999), 242.

101. See Bernard Cohen, "Hypotheses in Newton's Philosophy," *Physis* 8 (1966), 163–84.

102. "Omnis enim Philosophiae difficultas in eo versari videtur, ut a Phaenomenis motum investigemus vires Naturae, deinde ab his viribus demonstremus phaenomena reliqua." The language of phenomena was used in Royal Society circles from the earliest days, as in Robert Boyle, *An Attempt for the explication of the phenomena observable in an experiment* (London: J. H. for Sam. Thomson,1661).

103. Cohen and Whitman, trans. *Principia* (1999), 943.

104. For Bruno Latour's concept of a "center of calculation" as driving scientific development see *Science in Action* (Cambridge: Harvard University Press, 1987), 215–57.

105. I. Bernard Cohen, *The Newtonian Revolution* (Cambridge: Cambridge University Press, 1980), 131.

106. On this and the Newtonian turn towards "singularity" see Peter Dear, *Discipline and Experience*, 48–50, 212–43, as well as the discussion of the problem of newly perceived phenomena for Galileo, 100–107.

107. See Thomas Hearne, "An Extract and particular Account of the rarities in the Anatomy School," Bodleian MS Rawl. C 865, reprinted in R. T. Gunther, *Early Science in Oxford*, vol. 3 (Oxford, Hazell, Watson and Viney, 1925), 264–74; and Thomas Hearne, *Hearne's Remarks and Collections*, vol. 1, ed. C. E. Doble (Oxford, Clarendon Press, 1885), 70.

108. William Petty, "Concerning the proportions of People in the eight eminent Cities of Christendom undernamed," *Five Essays on Political Arithmetic* (London: H. Mortlock, 1687).

CONCLUSION

1. Good examples in different fields are Cohen, *The Newtonian Revolution*; Margaret Jacobs, *The Radical Enlightenment* (London: George Allen & Unwin, 1981); John Brewer, *The Sinews of Power* (New York: Knopf, 1989); Roy Porter, *The Untold Story of the British Enlightenment: The Creation of the Modern World* (New York: Norton, 2001); Pincus, *1688*; P. G. M. Dickson, *The Financial Revolution in England* (London: Macmillan, 1967).

2. For the resources argument see Kenneth Pomeranz, *The Great Divergence: China, Europe and the Making of the Modern World Economy* (Princeton: Princeton University Press, 2000). On the importance of fragmentation and urban concentration against centralization see Jean-Laurent Rosenthal and Roy Bin Wong, *Before and Beyond Divergence* (Cambridge: Harvard University Press, 2011); and for the "industrious revolution," see Jan de Vries and Ad van der Woulde, *The First Modern Economy: Success, Failure, and Perseverance of the Dutch Economy, 1500–1815* (Cambridge: Cambridge University Press, 1997); and most recently Pincus, *1688*, 59 and passim. For classic accounts of the success and diffusion of the European 'world system' against other models see Braudel, *Civilization and Capitalism*; and Immanuel Wallerstein, *The Modern World System* 1 & 2 (New York: Academic Press, 1974 and 1980).

3. See Pincus and James Robinson, "What Really Happened During the Glorious

Revolution," July 2011, http://scholar.harvard.edu/files/jrobinson/files/whatreally happenedfinal.pdf, accessed April 24, 2013. Pincus, *1688*, 367–99, critiques both J. G. A. Pocock, *Virtue, Commerce, and History* (New York: Cambridge University Press, 1985), 108; and *The Machiavellian Moment*, 423–26; and the institutional interpretation of Douglas North and Barry Weingast, "Constitutions and Commitment: The Evolution of Institutions Governing Public Choice in Seventeenth-Century England," *Journal of Economic History* 49 (1989), 815–16; and Weingast, "The Political Foundations of Limited Government: Parliament and Sovereign Debt in Seventeenth- and Eighteenth-Century England," in *The Frontiers of New Institutional Economics*, John Drobak and John Nye, eds. (New York: Academic, 1997), 23. Both Pincus and Robert Markley are sharp critics of Sir Josiah Child's interpretation of events, even though they read him in entirely opposite ways on the issue of exchange.

4. On Asia see Sanjay Subrahmanyam, "Connected Histories: Notes towards a Reconfiguration of Early Modern Eurasia," *Modern Asian Studies* 31:3 (July 1997), 735–62; and more recently John Darwin, *After Tamerlane: The Global History of Empire* (London: Allen Lane, 2007).

5. C. A. Bayly, *The Birth of the Modern World* (Malden: Blackwell, 2004).

6. See for example Peter Taylor, *Modernities: A Geohistorical Interpretation* (Minneapolis: University of Minnesota Press, 1999).

7. Saskia Sassen, *A Sociology of Globalization* (New York: Norton, 2007), 18. Sassen has argued that scholars in an age of globalization have to come to terms with the assemblage—defined perhaps too narrowly and formally by Aihwa Ong and Stephen Collier as "systems that mix technology, politics and actors." Sassen, *Territory, Authority, Rights: From Medieval to Global Assemblages* (Princeton: Princeton University Press, 2006), 5; Aihwa Ong and Stephen Collier, *Global Assemblages: Technology, Politics, and Ethics as Anthropological Problems* (Malden, MA: Blackwell, 2005), 4, 9–14. Certainly the contemporary concept of assemblage owes much to Claude Lévi-Strauss's concept of *bricolage*, set forth in his *The Savage Mind* [*La Pensée Sauvage*, 1962] (Chicago: University of Chicago Press, 1966), 19–21, and perhaps even Francis Bacon's ideas about induction.

8. Burbank and Cooper, *Empires*, 151; Sassen, *Territory, Authority, Rights*, 349; Benedict Anderson, *The Spectre of Comparisons: Nationalism, Southeast Asia and the World* (London: Verso, 1998); and Anderson, *Imagined Communities* (London: Verso, 1991).

9. See especially Amartya Sen, *Development as Freedom* (New York: Anchor, 1999); Carl Schmitt, *The Nomos of the Earth in the International Law of the Jus Publicum Europaeum* [1950], trans. G. L. Ulmen (New York: Telos, 2003).

10. [Sir Josiah Child], *A Treatise wherein is demonstrated . . . That the East-India Trade is more profitable and necessary to the Kingdom of England, than to any other Kingdom or Nation in Europe* (London: J. R. for the East India Company, 1681), 29.